Plenary Addresses, Economics and Applications of Copper

Titles of Related Interest—

Ashby ENGINEERING MATERIALS 1
Ashby ENGINEERING MATERIALS 2
Brook IMPACT OF NON-DESTRUCTIVE TESTING
Koppel AUTOMATION IN MINING, MINERAL AND METAL PROCESSING 1989
Ruhle METAL-CERAMIC INTERFACES
Taya METAL MATRIX COMPOSITES

Other CIM Proceedings Published by Pergamon

Bergman FERROUS AND NON-FERROUS ALLOY PROCESSES
Bickert REDUCTION AND CASTING OF ALUMINUM
Chalkley TAILING AND EFFLUENT MANAGEMENT
Closset PRODUCTION AND ELECTROLYSIS OF LIGHT METALS
Dobby PROCESSING OF COMPLEX ORES
Embury HIGH TEMPERATURE OXIDATION AND SULPHIDATION PROCESSES
Jaeck PRIMARY AND SECONDARY LEAD PROCESSING
Jonas DIRECT ROLLING AND HOT CHARGING OF STRAND CAST BILLETS
Kachaniwsky IMPACT OF OXYGEN ON THE PRODUCTIVITY OF NON-FERROUS
    METALLURGICAL PROCESSES
Lait F. WEINBERG INTERNATIONAL SYMPOSIUM ON SOLIDIFICATION
    PROCESSING
Macmilian QUALITY AND PROCESS CONTROL IN REDUCTION AND CASTING
    OF ALUMINUM AND OTHER LIGHT METALS
Mostaghaci PROCESSING OF CERAMIC AND METAL MATRIX COMPOSITES
Plumpton PRODUCTION AND PROCESSING OF FINE PARTICLES
Purdy FUNDAMENTALS AND APPLICATIONS OF TERNARY DIFFUSION
Rigaud ADVANCES IN REFRACTORIES FOR THE METALLURGICAL INDUSTRIES
Ruddle ACCELERATED COOLING OF ROLLED STEEL
Salter GOLD METALLURGY
Thompson COMPUTER SOFTWARE IN CHEMICAL AND EXTRACTIVE
    METALLURGY
Twigge-Molecey MATERIALS HANDLING IN PYROMETALLURGY
Twigge-Molecey PROCESS GAS HANDLING AND CLEANING
Tyson FRACTURE MECHANICS
Wilkinson ADVANCED STRUCTURAL MATERIALS

Related Journals
(Free sample copies available upon request)

ACTA METALLURGICA
CANADIAN METALLURGICAL QUARTERLY
MATERIALS RESEARCH BULLETIN
MINERALS ENGINEERING
SCRIPTA METALLURGICA

PROCEEDINGS OF THE COPPER 91 - COBRE 91
INTERNATIONAL SYMPOSIUM — VOLUME I
AUGUST 18-21, 1991, OTTAWA, CANADA

Plenary Addresses, Economics and Applications of Copper

Edited by

Plenary Addresses

C. Diaz
Inco Limited
J. Roy Gordon Laboratory
Mississauga, Ontario, Canada

P.J. Mackey
Noranda Technology Centre
Pointe Claire, Quebec, Canada

G.P. Tyroler
Inco Limited
Copper Cliff, Ontario, Canada

Economics

W. McCutcheon
Energy, Mines and Resources
Ottawa, Ontario, Canada

J.L. Mardones
University of Chile
Santiago, Chile

Applications of Copper

E. Gervais
International Zinc Association
Brussels, Belgium

A. Varschavsky
University of Chile
Santiago, Chile

International Symposium Organized by The Metallurgical Society of
Canadian Institute of Mining, Metallurgy and Petroleum
Chilean Institute of Mining Engineers, and
Minerals, Metals and Materials Society of the AIME

30th Annual Conference of Metallurgists of CIM

1991 Fall Meeting of the Minerals, Metals and Materials Society — Extractive and Processing Division

21st Annual Hydrometallurgical Meeting of CIM

PERGAMON PRESS
New York, Oxford, Beijing, Frankfurt, Sao Paulo, Sydney, Tokyo, Toronto

Pergamon Press Offices:

U.S.A. Pergamon Press, Inc., Maxwell House, Fairview Park, Elmsford, New York 10523, U.S.A.

U.K. Pergamon Press plc, Headington Hill Hall, Oxford OX3 0BW, England

PEOPLE'S REPUBLIC OF CHINA Pergamon Press, 0909 China World Tower, No. 1 Jian Guo Men Wai Avenue, Beijing 1000004, People's Republic of China

FEDERAL REPUBLIC OF GERMANY Pergamon Press GmbH, Hammerweg 6, D-6242 Kronberg, Federal Republic of Germany

BRAZIL Pergamon Editora Ltda, Rua Eça de Queiros, 346 CEP 04011, Paraiso, São Paulo, Brazil

AUSTRALIA Pergamon Press Australia Pty Ltd., P.O. Box 544, Potts Point, NSW 2011, Australia

JAPAN Pergamon Press, 8th Floor, Matsuoka Central Building, 1-7-1 Nishishinjuku, Shinjuku-ku, Tokyo 160, Japan

CANADA Pergamon Press Canada Ltd., Suite 271, 253 College Street, Toronto, Ontario M5T 1R5 Canada

Library of Congress Cataloging in Publication Data

ISBN 0-08-041432-X

Printing: 1 2 3 4 5 6 7 8 9 Year: 0 1 2 3 4 5 6 7 8 9

Printed in the United States of America

The paper used in this publication meets the minimum requirements of American National Standard for Information Sciences-Permanence of Paper for Printed Library Materials, ANSI Z 39.48-1984

Editors' Biographies

Carlos Diaz: graduated as a civil mining engineer at the Faculty of Physical and Mathematical Sciences, University of Chile, in 1954. He was then appointed as lecturer in physical chemistry in the Department of Mining Engineering. From there, he went to Columbia University to obtain an M.Sc. in extractive metallurgy in 1958 and then rejoined the Faculty in Santiago to be appointed head of the Chemistry Department. In 1963, he left for London to pursue a Ph.D., also in extractive metallurgy, which he obtained in 1966 from the Imperial College. After his return to Chile he served as head of the Mining Engineering Department, University of Chile. Between 1969 and 1972 Carlos Diaz was director of the School of Engineering of this University. Also during this period and later, he served as a member of the National Science and Technology Council and was a member of the Board of Directors of the Chuquicamata and El Salvador Mines. In 1975, he joined Inco Ltd. in Canada as process engineer and from 1978 onward he has served as head of the Pyrometallurgy Section of the J.Roy Gordon Research Laboratory, Inco Ltd., in Toronto. He has numerous international publications on these subjects. He is the recipient of the 1985 Medal of Merit of the Chilean Institute of Mining Engineers. He was Co-Chairman of the COPPER 87 Conference and held the same position at the COPPER 91 - COBRE 91 Conference.

Phillip J. Mackey received his metallurgical education at the University of New South Wales, Sydney, Australia. After receiving his Ph.D. degree in extractive metallurgy he joined Noranda in 1969 where he helped shepherd Noranda's new continuous smelting process from the pilot stage to full, successful commercialization. He has subsequently worked at a senior level in different divisions of Noranda including:- Smelter Technical Superintendent at the Horne smelter where he was responsible for technical policy, and Program Manager - Noranda Process which carried responsibilities for marketing Noranda Technology worldwide. He is currently responsible for the development of long-term plans to improve operations and sulphur capture at the Noranda smelters. A Past-President of the Metallurgical Society of CIM, he is a co-founder of the international copper conferences, he was Co-Chairman of COPPER 87 and Technical Co-Chairman of COPPER 91 - COBRE 91.

George P. Tyroler graduated in Chemical Engineering with an M.Eng. and a Ph.D. from Technical University, Prague, Czechoslovakia. From 1967 to 1969, he was a post-doctorate fellow at McMaster University, Hamilton, Ontario. He has authored several books, dozens of articles and patents. For a number of years he has been active in TMS/AIME. Since 1969 he has been with Inco Limited in Sudbury, Ontario.

William McCutcheon has a Civil Engineering degree from the University of Toronto and a Masters degree in Engineering (Mining) from McGill University in Montreal in 1974. A Registered Professional Engineer, he worked in engineering and operation positions in underground and surface mines in Canada and Guatamala, the cement industry in Canada and a consulting engineering firm. In 1983 he joined Energy, Mines and Resources Canada as a commodity specialist in the Mineral Policy Sector. He was the copper commodity specialist for the Canadian government from 1984 to 1989; presently he is the Deputy Director of the Specialty Metals and Materials Division, focussing on metals recycling issues.

J.L. Mardones has an M.S. degree in Industrial Engineering at the University of Chile, and holds M.A. and Ph.D. degrees in International Studies from The Fletcher School of Law and Diplomacy, Tufts University, in Massachusetts. He currently is the Director of Finance of ENAMI Mining Co., and an associate professor of economics and finance at the University of Chile. He has researched extensively on mining policies and on application of financial theory to real investments.

Edouard Gervais has a Ph.D. in Metallurgical Engineering. He is the Executive Director of the International Zinc Association, Brussels, Belgium. This newly formed association provides a global forum for zinc industry leaders to understand global issues needing action by the zinc industries. Previously, he worked for Noranda Sales Corporation and the Noranda Technology Centre in marketing, research and development and R&D management. He has about fifty scientific publications including nine patents.

Ari Varschavsky, Electrical Civil Engineer, graduated from the University of Chile in 1965 and completed post graduate studies in materials science at MIT. At present he occupies the position of Professor of Materials Science at the University of Chile and is the editor of a number of international scientific journals related to materials. His numerous publications have been in the field of metal and alloy fatigue and transformations.

Preface

COPPER 91 - COBRE 91 is the second in a series of international symposia devoted to copper. This event, held in Ottawa, Ontario, Canada, followed four years after the highly successful COPPER 87 - COBRE 87 conference in Chile.

First proposed in 1983, the concept to hold a series of international symposia sponsored by the metallurgical and academic institutions of two of the largest copper-producing countries, Canada and Chile, was developed by the Organizing Committee of the 1987 conference. The remarkable success of COPPER 87 - COBRE 87 demonstrated the value of jointly-sponsored symposia. Now, the Minerals, Metals and Materials Society of the AIME joined with the Metallurgical Society of the Canadian Institute of Mining, Metallurgy and Petroleum and the Chilean Institute of Mining Engineers in arranging COPPER 91 - COBRE 91. The Organizing Committee was supported by an International Advisory Committee comprising prominent copper industry representatives from around the world. The organizers believe that the strategic alliance of the symposium's sponsoring institutions holds out to their members increased understanding of the science, technology and practice of the copper industry, truly international in scope. We look forward to continuing and expanding such alliances. Plans are already being made for the continuation of the four-year symposium cycle with COPPER 95 - COBRE 95.

Since the first conference in 1987, many notable developments have taken place in the copper industry and in copper production technology. Among the highlights have been:

• The increasing proportion of world copper produced by the Leach SX-EW route — of the order of 10% in 1990 — and projected to account for more than 15% by the end of the decade.

• Greater use of oxygen in copper smelting — now well above the levels of 1987 when slightly over half of copper smelters worldwide used oxygen enrichment;

• A worldwide increase in the number and scope of modernization and expansion projects at copper smelters and refineries;

• A higher average level of sulphur capture at copper smelters, and improved environmental conditions at many copper plants;

• The application of advanced process control systems in copper concentrators, smelters, and refineries;

• Strong copper prices — averaged somewhat over 100 U.S. cents per pound since 1987 — and new marketing initiatives have helped boost copper's visibility in the international marketplace and its competitiveness against other materials.

COPPER 91 - COBRE 91 has offered a critical review and analysis of these and other developments and an active forum for the exchange of ideas among technical experts and copper industry representatives from all over the world. To provide thematic focus to the technical sessions, plenary addresses given by recognized industry leaders introduced each of the technical sessions. These plenary addresses, which constitute a comprehensive overview of the key issues confronting the copper industry today, are included in Volume I of the symposium proceedings. The Organizing Committee thanks these invited speakers for their contribution.

The four volumes of symposium proceedings, representing some 164 papers from over twenty countries are divided as follows:

Volume I   Plenary Addresses, Economics and Applications of Copper;

Volume II Mineral Processing and Process Control;

Volume III Hydrometallurgy and Electrometallurgy of Copper;

Volume IV Pyrometallurgy of Copper;

It is our sincere hope that these volumes will remain a valuable reference for the copper industry for many years to come. We believe that COPPER 91 - COBRE 91 represents a significant contribution to the literature during a time of ever-increasing rate of change.

The Organizing Committee wishes to sincerely thank all the authors, the International Advisory Committee, the Technical Program Sub-Committee, Session Chairmen, members of sponsoring organizations, the supporting production companies and research institutions, and the staff at CIM Secretariat in Montreal, Canada. We wish to pay special recognition to Perla Gantz and her staff at CIM for co-ordinating the production of the proceedings. She provided outstanding assistance to the editors. We also wish to thank Lucille Green and Nancy Judge at Inco Limited's J.R. Gordon Research Laboratory for their help in maintaining communications with the authors of papers during the various stages of preparation of the Technical Program.

August 1991   The Organizing Committee COPPER 91 - COBRE 91

Foreword

This volume contains the COPPER 91 - COBRE 91 Plenary Addresses and the proceedings of the sessions on Economics and on Markets and Applications.

Copper consumption, coupled with other issues external to the industry itself, drives the demand for mining, smelting, fabrication, and recycling of the red metal. The impact of these complex issues on the copper industry is analyzed in this volume.

In a departure from COPPER 87, Plenary Addresses given by recognized industry leaders introduced each of the technical sessions at COPPER 91 - COBRE 91. These addresses constituted a comprehensive overview of the major issues confronting the copper industry today. The papers included in this volume cover the management of change, operations management, trends in the production of copper from recycled raw materials, new copper marketing strategies for the 90s, and a review of industry competitiveness in the 90s. The editors believe that these plenary papers provide valuable insight into the copper industry and will stimulate serious thought on strategies to be developed by industry management in the future. The editors thank these invited speakers for their contribution.

The economics papers in this volume take the reader on a tour of cost and price analysis, supply and perspectives, and markets and strategies. In these papers, leading authors examine topics such as responses to competitive challenges, price forecasting, future production strategies, the impact of recycling, the concentrate market and custom smelting, and the outlook for specific regions and countries. A wave of modernization programs appears imminent in the industry, and the outlook for the copper business for at least the next five years is generally considered positive.

The final section of this volume contains 17 papers dealing with copper markets and applications, specifically, the fabrication of copper alloys, market development, and product technology. Several interesting case histories of market development initiatives are discussed, including a review of the automotive radiator market, fire sprinklers and a description of the North American Initiative for Copper Architectural Applications, a program originally proposed by Noranda Minerals Inc. Such activities are vital in stimulating and maintaining demand for copper and copper products.

The editors would like to thank all the authors who have contributed to this volume. We are also grateful to Jocelyn Bourassa of Noranda Minerals' Montreal office, Joan Parker at Noranda Technology Centre, Patricia Caine at the Inco Nickel Refinery, Copper Cliff, and Lucille Green at Inco Limited's J. Roy Gordon Research Laboratory for their assistance in the production of this volume.

Montreal
July 1991

The Editors

COPPER 91 - COBRE 91
Organization of International Symposium

Symposium Co-Chairmen
Carlos Diaz, Inco Limited, Mississauga, Ontario, Canada
Ralph H. Nafziger, U.S. Bureau of Mines, Albany, Oregon, U.S.A.
Gustavo E. Lagos, University of Chile, Santiago, Chile

Technical Co-Chairmen
Phillip J. Mackey, Noranda Technology Centre, Pointe Claire, Quebec, Canada
George P. Tyroler, Inco Limited, Copper Cliff, Ontario, Canada
Antonio Luraschi, CIMM, Santiago, Chile

Organizing Committee
Barry Bailey, Noranda Sampling Inc., Rhode Island, U.S.A.
Carlos Barahona, Cia. Minera Disputada de las Condes, Santiago, Chile
W. Charles Cooper, The University of British Columbia, Vancouver, British Columbia, Canada
Glenn Dobby, University of Toronto, Toronto, Ontario, Canada
John E. Dutrizac, CANMET, Ottawa, Ontario, Canada
Anthony Fritz, Inco Limited, Copper Cliff, Ontario, Canada
Edouard Gervais, International Zinc Association, Brussels, Belgium
Guillermo Gonzalez, University of Chile, Santiago, Chile
James E. Hoffmann, Jan Reimers and Associates, Houston, Texas, U.S.A.
Denis J. Kemp, Falconbridge Limited, Timmins, Ontario, Canada
José Luis Mardones, University of Chile, Santiago, Chile
William McCutcheon, Energy, Mines & Resources, Ottawa, Ontario, Canada
Krugger Montalban, Soc. Minera Quebrada Blanca, Santiago, Chile
Chris Newman, Falconbridge Limited, Timmins, Ontario, Canada
Engin Ozberk, Sherritt Gordon Limited, Fort Saskatchewan, Alberta, Canada
Aldo Picozzi, Chilean Copper Commission, Santiago, Chile
Rodolfo Reyes, Sociedad Minera Pudahuel, Santiago, Chile
Nickolas Switucha, East-West Consulting, Ottawa, Ontario, Canada
K. Gie Tan, CANMET, Ottawa, Ontario, Canada
Guillermo Ugarte, CIMM, Santiago, Chile
Ahmed Vahed, Inco Limited, Mississauga, Ontario, Canada
Ari Varschavsky, University of Chile, Santiago, Chile
Pio Vilavella, Codelco-Chile, Santiago, Chile
Andres Zauschquevich, Institute of Mining Engineers, Santiago, Chile

International Advisory Committee
Prof. Eduardo Brocchi, Pontificia Universidade Catolica, Rio de Janeiro, Brazil
Prof. Jacques de Cuyper, Universite Catholique de Louvain, Louvain-la-Neuve, Belgium
Dr. William H. Dresher, International Copper Association, New York, N.Y., U.S.A.
Dr. Douglas S. Flett, Leeds Mineral Services Group, Kirkstall, Leeds, England
Ing. Oscar Gonzalez-Rocha, Mexicana de Cobre, Mexico City, Mexico
Dr. N.W. Johnson, Mount Isa Mines Ltd., Mount Isa, Queensland, Australia
Dr. Michael Koch, Krupp Industrietechnik, Cologne, Germany
Prof. Kalala Ntumba Kumuambi, Gecamines, Kinshasha, Zaire
Ing. Jorge von Loebenstein, Cía. Minera Disputada de las Condes, Santiago, Chile
Dr. Juho Makinen, Outokumpu, Pori, Finland

Dr. Valery V. Mechev, Institute Gintsvetmet, Moscow, U.S.S.R.
Dr. C. George Miller, The Mining Association of Canada, Ottawa, Ontario, Canada
Dr. S.K. Mwenechanya, Zambia Consolidated Copper Mines, Luanschya, Zambia
Dr. Takeshi Nagano, Mitsubishi Materials Corporation, Tokyo, Japan
Ing. Pawel Ofman, Copper Mining and Smelting Group, Lubin, Poland
Prof. Huang Qixing, Central Engineering and Research Institute for Non-Ferrous Metallurgical
 Industries, Beijing, China
Ing. Hipolito Zeballos, Centromin, Lima, Peru

Session Chairmen

VOLUME I. Plenary Addresses, Economics and Applications

Plenary Addresses
C. Diaz
G. Lagos
A. Luraschi
P.J. Mackey
R.H. Nafziger
G.P. Tyroler

Economics
G. Bokovay
W.H. Dresher
I. Marshall
J.L. Mardones
W. McCutcheon

Applications
H.B. Blaber
W.H. Dresher
A. Knapp
H.J. McQueen
A. Varschavsky
S.Y Zhou

VOLUME II. Mineral Processing and Process Control of Copper

S.A. Argyropoulos
G.S. Dobby
G.D. Gonzalez
S. Kelebek
J.S. Laskowski
A. Luraschi
R. McIvor

J.M. Menacho
F. Mucciardi
D.R. Nagaraj
S.R. Rao
J. Skeaff
N.J. Themelis
R. del Villar
W.T. Yen

VOLUME III. Hydrometallurgy and Electrometallurgy of Copper

P.L. Claessens
B.R. Conard
W.C. Cooper
P.A. Distin
E. Domic
D.B. Dreisinger
R. Hackl
J.E. Hoffmann
D.J. Kemp
D. Kerfoot
G.E. Lagos

R. Liemella
H. Leiva
R.G.L. McCready
G.M. Ritcey
J.D. Scott
R.W. Stanley
J. Thiriar
E. Thune
A. E. Torma
I. Wilkomirsky
S. Young

VOLUME IV. Pyrometallurgy of Copper

R.J. Campos
W.G. Davenport
G. Kaiura
G.E. Lagos
C. Landolt
A. Luraschi
P.J. Mackey
K. Murden

C.J. Newman
D.G. Pannell
D. Poggi
N. Santander
J.M. Toguri
G.P. Tyroler
A.E. Wraith

Table of Contents

Table of Contents

Table of Contents

Table of Contents

PROCESS CONTROL AND MODELLING IN PYROMETALLURGY

Plenary addresses

The management of change

D.C. Yearley
Phelps Dodge Corporation, Phoenix, Arizona, U.S.A.

One of the fundamental issues a successful organization must address is the management of change. Without the ability to effectively manage change, business cannot survive.

Copper mining, which dates back thousands of years, has survived innumerable changes through the centuries. "The Management of Change" addresses past, current and future changes in the copper mining industry, and how those changes affect our industry's role in the global marketplace. The lecture also reviews change as it relates to production trends, including the role of technological improvements in the copper industry.

Using specific examples from Phelps Dodge Corporation's transformation from a high-cost copper operation to a worldwide industry leader, "The Management of Change" reviews the philosophy and the dynamics of change in the copper mining industry. The lecture discusses the roles of management, research, exploration, engineering and the individual employee in managing, promoting, implementing and accepting change, and concludes by reviewing strategies for preparing the copper industry for the changes and challenges of the future.

Good morning. I'm delighted to be here and to have the privilege of participating in the COPPER 91 symposium.

How many of you have heard the saying . . . the simpler the subject, the shorter the speech? Considering the fact that I will be addressing "The Management of Change in the Copper Industry," it looks like you'll be here awhile!

Copper is not simple. Neither is the management of change. Copper is the first metal known to man and for several thousand years, was the only metal used extensively by man. Throughout history, civilizations have been shaped by mineral use. Although the copper industry has undergone myriad changes, one thing remains the same: copper is an extremely useful material. Its possible uses are virtually endless.

According to an article published by the University of Pennsylvania, native copper was first used in northern Iraq about 8700 B.C. when it was cold-hammered in the form of a pendant. A copper ornament 10,000 years old! Think of that. Was the pendant offered as a gift? Was it crafted as an offering to a deity? Anything 10,000 years old is shrouded in mystery. My guess is there are a couple of certainties about the pendant. The primitive miner who found the raw copper no doubt spent more than he wanted to in finding it. And he probably sold it for a lot less than he thought it was worth -- starting an unbroken line of thought that endures among copper executives to this day.

By 6000 B.C., the use of native copper had spread over most of the Ancient Near East, enabling man to make weapons, ornaments and utensils for advancing civilizations. Some historians suggest that the next step in the technological process -- the melting of native copper -- led to the smelting of the simple oxide ores of copper such as malachite and azurite.

Today, several millennia later, copper is still advancing civilizations. And the science of metallurgy has undergone countless changes along the way.

As I was preparing my remarks for today's talk, I wondered, who was the first executive confronted with the necessity to manage change? The anthropologists have yet to tell us. I suspect that the chief of a primitive clan of hunters and gatherers found himself bedeviled by change. Perhaps the hunters started returning with less game and the gatherers with fewer berries. What to do? Find a better environment and move the tribe -- thereby directing the first corporate relocation? Banish some members of the tribe so there would be enough food for those who remained -- in a prehistoric downsizing? Try to grow plants close to the village -- thus plunging into the first diversification?

Whatever that first leader decided, philosophers, statesmen, and ordinary people like us have been concerned about change throughout the ages. In the year 500 B.C., the philosopher Heraclitus offered an observation still heard in contemporary conversation, with some variations: He said, "Nothing is permanent but change."

In our own time, President John F. Kennedy cautioned, "Change is the law of life. Those who look only to the past or the present are certain to miss the future."

To those pearls of wisdom, I'd like to add my own adage about change. My message is simple. Manage change or perish. Success is the product of managing change -- ideally before change is forced upon you.

Now, I do not suggest that managing change involves fortune-telling. Some of us remember the forecasts of the 1964 World's Fair in New York -- outposts on the moon, underwater resorts and other naive visions.

How many in this room were not surprised by the demolition of the Berlin Wall, the liberation of Eastern Europe, the collapse of the Soviet economy and political system and the war between the United States and Iraq? All those events held profound significance for the mining industry. Yet few five-year plans -- or even one-year plans -- envisioned them.

Although it is not possible to read the future in tea leaves, it is possible to detect trends and to prepare for the eventualities. For example, it is evident that the environment is high on the list of the world's priorities, and may easily become the foremost issue on political agendas. That being so, companies should pay close attention to environmental considerations in every phase of business.

Similarly, a global economy is taking shape. National economies are being replaced by powerful regional entities in Europe and North America. Once impoverished nations on the Pacific Rim are becoming industrial giants. While the working relationships between these economic spheres may still be vague, the smart manager is preparing to prospect everywhere for business opportunities and make the most of them wherever they are discovered.

The copper industry will have to look harder in more difficult places, and spend a great deal of money, to find those opportunities. The price of copper must be produced at a level that does not encourage substitution of other materials. That means more exploration to meet future demands. We must produce copper at a competitive price, combining the best of the old ways with new technologies.

The utilization of new technologies, however, should be a familiar rerun of our industry's colorful history.

For example, in the early 1900s, several important innovations made underground mining less arduous. The pneumatic drill reduced fatigue and greatly improved output. Electricity provided lighting and ventilating fans.

However, even though underground mining was becoming more "comfortable," so to speak, the concept of open-pit mining opened the doors for extracting large quantities of copper, and virtually revolutionized the industry overnight.

Open-pit copper mining owes its origins to one individual, Daniel C. Jackling of Salt Lake City, Utah. Jackling's experience in the gold mines led him to reason that if block-caving also could be applied on an immense scale, the unit costs of copper mining could be reduced enough to make it practical to mine very low grade copper. From 1907 until World War II, Jackling and the Kennecott Copper Company, which acquired the Bingham Canyon, Utah, mine, were regarded as the world leaders in open-pit copper mining. After the war, open-pit technology shifted to Arizona. In 1931, Phelps Dodge acquired an open pit mine in Ajo, Arizona, and developed its own open pit mine in Bisbee, Arizona in the early 1950s. Our open-pit mine in Morenci, Arizona is widely recognized as one of the world's great mining operations.

During the 1800s, as native copper began diminishing, and ore bodies were being depleted, the concept of concentrating copper was born. But where would our industry be today without the invention of froth flotation? We've all heard the tales of the copper miner whose wife discovered flotation when she was washing his work clothes and the copper stuck to the soap bubbles. Until shortly before World War I we were still concentrating copper through sedimentation, or gravity separation. The process was so crude and inefficient that as much as a third of the copper was flushed off to the tailings along with the waste material.

One of the major breakthroughs in the history of copper mining was the replacement of this inefficient process by froth flotation, which was first developed through experiments on zinc ores in Australia. Minerals Separation Ltd. of London licensed its patented flotation process to Anaconda, and the first successful application in the U.S. occurred in 1915 at Inspiration Consolidated Copper Company of Arizona, at the time an affiliate of Anaconda.

When Daniel Jackling's mines, as a result of trial and error, developed a method of achieving froth flotation with chalcopyrite sulfides, Minerals Separation sued for patent infringement. After years of litigation, the suits were finally settled out of court in 1922 on terms that have never been released.

Some dispute remains about whom deserves the credit for inventing flotation. There is no dispute, however, that the entire industry changed as a result of the evolution of froth flotation.

Since flotation removes most of the copper in sulfide ore, it is extremely efficient. Within six years of its introduction in 1915, every major concentrator in the United States was using this method. The basic concept of flotation has changed very little since 1915. In our age of technology, some outsiders might remark that the process is outdated. While most of my discussion focuses on the need to adopt new technology, at the same time we ought not to make change just for the sake of change. Although the equipment configurations have changed, I think most of us in this room would agree that the flotation process is very inexpensive and effective, and in this case, it's wise to apply the old saying, "If it ain't broke, don't fix it."

Another major industry breakthrough was the emergence of continuous casting of wire rod in 1962, which replaced wire bar. As you know, continuous cast copper rod has superior resistance to breaks when drawn into finer gauge wire. More importantly, the process requires less labor and less energy, contributing to more efficient production. I would be remiss if I did not mention the significant development of the Asarco shaft furnace in the mid-1960s. This efficient cathode melting unit has done much to provide a high quality liquid copper feed to the casting machines.

Still another significant technological advancement was the development of flash smelting in the late 1940s and the early 1950s in both Finland and Canada. Faced with a lack of funds to import fuel oil after World War II, it was crucial that the Finns develop a lower cost, more energy efficient method of smelting copper. Consequently, the Outokumpu Group developed flash smelting, which utilized preheated air and heat generated from smelting to reduce energy costs. Outokumpu's furnace was studied on a pilot scale in 1947 and commercial operations commenced in 1949. Simultaneously, the International Nickel Company of Canada was developing oxygen flash smelting due to the availability of low-cost gaseous oxygen and a market for liquid SO_2. After its pilot program in 1947, the first Inco oxygen flash furnace was brought on-line in 1952.

Phelps Dodge is proud to be the only company in the world to operate flash furnaces of both types. The Hidalgo smelter in New Mexico was the first flash smelting facility to be built in the United States, and it incorporates Outokumpu technology. Our Chino smelter, also in New Mexico, uses Inco flash smelting technology. Both are among the cleanest smelters in the United States in terms of air quality control.

As you know, flash smelting combines the roasting and smelting processes of the conventional roaster-reverberatory series into one continuous process. Oxygen or oxygen-enriched air creates a higher concentration of SO_2 in the process gases and the SO_2 is efficiently captured in sulfuric acid plants. With increasingly stringent environmental regulations, flash smelting offers the greatest degree of air quality control of any smelting technology, and is an important example of how energy costs, labor costs and environmental regulation give us the impetus to run our plants more efficiently.

Other innovations in smelting also are now beginning to emerge. The continuous smelting Mitsubishi process and flash converting are several examples.

Finally, how many of us might not be sitting in this room if solvent extraction/electro-winning hadn't been invented? Since it was first commissioned in 1968 at Rancher's Bluebird mine near Miami, Arizona, SX/EW has proved to be an extremely important, low cost method of recovering copper contained in low grade ore stockpiles. With all the changes I've been discussing, imagine how the mining engineers of the 21st Century will be describing SX/EW technology of the late 1900s!

Tomorrow's history will be made up of today's current events. With changes of cosmic proportions already reality in the copper industry, and more change sure to come, the management of change is essential to the survival of our business. Fortunately, there is nothing mystical about managing change. As we prepare to approach the 21st Century, let's take a look at a few techniques for managing change.

PLANNING

First, let's discuss planning. Planning is a tradition in business. Some companies maintain separate planning staffs who work independent of other areas of the company, an arrangement that probably always was a mistake and is certainly a mistake today. In the global marketplace, change comes with the surprise and force of an avalanche. Remember, it wasn't raining when Noah built the ark. Even slight delay in identifying and swiftly adopting technology and techniques that improve productivity could prove fatal. Today, our industry must strive for planning conducted simultaneously at all levels so that key issues can be identified early and responses initiated swiftly.

A good example of this type of planning is the exercise Phelps Dodge conducted in the mid-1980s. We faced tremendous costs to retro-fit our three aging smelters at Ajo, Morenci and Douglas, Arizona. We were also strapped for cash, coming off three very tough years. How could we solve both problems? Through careful planning, we concluded that we could sell a minority interest in our Morenci operation to Sumitomo Corporation. This transaction generated a large amount of cash. But it also reduced our concentrate supply since Sumitomo takes its share of concentrates in kind for shipment to Japan. No longer needing some of our smelting capacity allowed us to suspend operations at those smelters and invest our cash in SX/EW facilities at Tyrone, New Mexico and Morenci. The success of these facilities allowed us to pursue the next leg of our plan - acquiring additional modern smelting capacity by purchasing a two-thirds interest in Chino Mines Company in New Mexico. This plan was dynamic, it was opportunistic and it was achievable. Most importantly, it was timely. As Peter Drucker has observed: "Long range planning does not deal with future decisions, but with the future of present decisions."

ORGANIZATION

Next, how is the organization structured? Lean and tough. Those words describe the mining company that will meet the challenge of managing change in the 1990s and beyond. Our companies must be guided by a robust organization of generalists, sprinkled with specialists who are familiar with a broad range of company operations. Ideally, they also will be quick learners because the range of company operations may change swiftly.

At Phelps Dodge, we have "flat" management structures designed around our business units so we can quickly recognize and capitalize on opportunities as they present themselves. Old hierarchical traditions and concerns about turf must be discarded to expedite the flow of information to people who need it to formulate and execute strategy.

As speed and agility become more important, more authority should be pushed closer to the operating level. Our employees are successful because they communicate with each other and work together as a team to achieve common goals for the organization. Managers in the branches operate within broad mission assignments, rather than merely administer detailed instructions from higher up.

INCENTIVES

Third, does the company offer incentives to its employees? The mining industry needs to encourage "entrepreneurship" among managers. The company that relies on plodding managers who are satisfied with steady returns generated at low risk might heed the homespun advice offered by baseball great Satchel Page, who said, "Don't look back. Somebody may be gaining on you." To manage change in an industry changing at high velocity, it is vital to encourage innovation.

If imagination and vision are not enthusiastically rewarded, a company will find itself left with the drones while the imaginative doers depart for more promising employment. We believe in Francis Bacon's observation, "A wise man will make more opportunities than he will find."

EVALUATING PERFORMANCE

Another technique for managing change is performance evaluation. Traditional approaches to measuring performance that focus on one-dimensional measurements -- controlling costs, measuring inventory turnover and the like are still valid. But other indicators must be added to meet the challenges of change: Quality for one example. Cost effectiveness, for another. How well do we produce copper at a price low enough to discourage substitution yet high enough to allow a profit? Is management expediting trade-offs necessary to satisfy business requirements? Performance should be measured over a broad spectrum of factors selected to support strategy and goals.

CULTURAL CONSISTENCY

The fifth technique for managing change is the need for management to prepare the corporate culture to accept change. Old practices must be carefully reexamined. More than ever, it is folly to do things because that is the way they always have been done. For example, a soundly devised initiative that emphasizes quality -- doing it right the first time -- could be stopped cold by an incentive system left in place that rewards employees for quantity. If the planning is accomplished by a broad-based task force of operators, obstacles are likely to be identified early. Change will come easier if the organization is kept informed through an aggressive and credible employee communications program.

TRAINING

Another important ingredient is training. The management of change contemplates new enterprises, new techniques, new technology. Obviously, the work force cannot remain unchanged as those transformations occur. Perhaps there is not a better example of how important education and training are than today's Phelps Dodge. We operate mines and factories -- two traditional areas for employees with "strong backs." But our labor gangs laying track in the pit have disappeared -- replaced in part by a computerized mine truck dispatch system operated by highly skilled technicians. Most of the world's businesses have changed, and their employees' educational requirements have changed dramatically. No longer can a worker start mastering skills confident that he will use the same skills on the day before retirement. To remain competitive amid cascading change, we must prepare our employees to perform new tasks. For example, equipment operators also might work with computers, complex communications equipment and sensors. Narrow specialties are as obsolete on the work floor as in the executive offices. Of course I'm not suggesting that we rid our companies of specialists. They serve a valuable need in the organization. We must, however, cross-train much of our work force and expose workers to as many aspects of the business as possible. Because of our size and scope, we must empower our employees and give them as much responsibility as possible.

I believe these techniques for managing change work. Some of them worked at Phelps Dodge years ago, before they became popular items of discussion in the business press. Phelps Dodge is one of the oldest corporations in America, founded in New York in 1834. In fact, copper wire from Phelps Dodge manufacturing plants was used for the first transcontinental telegraph line. Our company has a rich history of corporate courage, displayed in successfully managing change through the years.

For example, Phelps Dodge invested heavily to develop copper resources in Arizona while an Indian war still raged across the unsettled territory. Operating on a frontier, the company could not merely mine and process copper. It had to take on a broad range of challenges to provide housing, mercantile outlets, medical facilities -- almost everything needed by the populations of whole new communities.

Since executive talent was scarce on the frontier, managers of the time were remarkably versatile. For example, one early mining engineer also distinguished himself as a banker, another frontier executive excelled as a publisher. Indeed, Phelps Dodge had an early start on what is regarded today as a new concept: executives who handle diverse tasks simultaneously.

The company prospered and grew as its leadership managed the change brought on by two World Wars, the uncertainties of a tumultuous revolution in nearby Mexico, the Roaring 20s, the Great Depression, floods, droughts, and other calamities. As Phelps Dodge observed its 150th birthday in 1984, however, it looked like the end might be near.

A persistent slide in the price of copper dating back to the 1970s was made worse by a worldwide recession in the mid-1980s. Phelps Dodge found itself in a destructive squeeze. The company was spending 80 cents to produce a pound of copper that sold for 65 cents or so. By the end of 1984, losses over a three-year period totaled $400 million. In December of 1984, Business Week's cover story proclaimed "The Death of Mining" and listed Phelps Dodge among the dying.

It was under those dismal circumstances that the chief executive officer at the time convened a committee of five vice presidents -- all "doers," not planning specialists -- and instructed us to develop a plan that would return the company to profitability. The committee met behind closed doors for one month. Early on, we established three objectives:

1.   Reduce production costs so we could sell copper at the prevalent low prices and still make money.

2.   Strengthen the weak balance sheet by significantly reducing debt.

3.   Initiate a diversification into businesses we understood in order to reduce the company's almost total dependence on copper.

The planning process permeated the company from top to bottom, affecting virtually every asset, subsidiary, investment, department, location and employee. The plan assigned individuals, by name, specific goals to be accomplished in a clearly stated time frame. Every employee -- from management, research, exploration and engineering -- was responsible for managing, promoting, implementing and accepting change.

As a result of the cohesive effort initiated by the plan, Phelps Dodge soon started showing a profit, with no help from the copper market at first. Subsequently, we doubled production, tripled productivity and earned record profits. Today, Phelps Dodge is a thriving company operating mines and plants in 22 countries, producing not only copper, but also other metals, truck wheels, medical devices and carbon black, a vital ingredient in the manufacturing of tires and other rubber products.

That is a lot of change to manage in just seven years. To make the company's copper mining and processing profitable again, Phelps Dodge had to initiate new technologies. Old policies and routines had to be discarded and new ones initiated. Workers had to be retrained to perform new tasks. As diversification proceeded, executives had to learn whole new industries. All of that added up to a significant shock to the corporate culture -- a formidable challenge that was successfully met through effective communication.

To keep in mind the importance of successfully managing change, I sometimes take an imaginary walk through a corporate graveyard in the United States and read the engravings on the tombstones: Baldwin Locomotive. Its epitaph: "Diesel will never replace steam." Curtiss-Wright Aircraft: "Jets will not ground propeller aircraft." Comptometer Company: "Computers will not replace mechanical calculators." There are long rows of those tombstones, mute testaments to failure to manage change. We were determined not to let Phelps Dodge rest in that graveyard.

I hope you notice how closely the turnaround of Phelps Dodge parallels the textbook principles for managing change I mentioned at the beginning of this lecture. Unfortunately, the textbooks were not yet published when Phelps Dodge was struggling for survival. We had to discover them as we went along.

Even so, it could be said that Phelps Dodge is better off because of the crisis it endured in the middle-1980s. If those times had been good, we might have done nothing. For certain we would have done a lot less.

As it stands today, the company is already an experienced participant in the global marketplace. We have refined the lean, tough and versatile organization. The company's executives are thoroughly familiar with the challenges of introducing new technologies and techniques and, thanks largely to partnerships with international firms, they are sensitive to the different ways of different peoples. Phelps Dodge is building a flexible corporate culture demanded by the times. Above all, Phelps Dodge will remain flexible and opportunistic in dealing with the unknowns of copper price, market conditions and acquisition opportunities. Although the company doubtless still has painful lessons to learn, it is solidly positioned in the unfolding global economic structure.

The future promises opportunities. Innovation and hard work will produce others. As the liberated nations of Eastern Europe recover from a half century of state-inflicted economic disaster and start building a modern infrastructure, new markets should open for copper and other items as well.

The Soviet Union remains, as Churchill described it, "a riddle, wrapped in a mystery, inside an enigma." If Russia does not collapse into a multi-sided civil war or become dominated by a new dictatorship, it could become a promising trade partner. China may also offer opportunities, especially if the state economy there is transformed to a marketplace. And, agreements linking Canada, the United States and Mexico into a free trade partnership will enhance the competitiveness of all three economies in world markets.

As Douglas MacArthur said, "There is no security on this earth -- only opportunity." While keeping an eye on developments abroad, business must simultaneously watch for new opportunities and new competition at home. For example, Phelps Dodge and the industry's trade organizations are busily promoting the use of copper as a roofing material in the United States and attempting to guard against expanding use of aluminum in automotive radiators. Phelps Dodge is also actively promoting such experiments as Smart House, a dwelling that makes extensive use of a home computer to operate air conditioning, appliances and lighting through extra wire in the electrical system.

Unlike other industries, where it is crucial to stay several steps ahead of emerging technology, mining must focus its research efforts toward process improvements to existing technology in order to keep costs low and productivity high.

The industry must continuously fine-tune its operations and apply improved technical devices, equipment or controls to improve the way we produce our product. With semi-mobile ore crusher-conveying systems, we can move more rock more efficiently -- millions of dollars more efficiently. New, larger trucks can move more material much faster. Improved furnace operations cut costs and reduce the amount of energy needed. Mine computer programs move more product through the system more cost efficiently.

Although our focus is on improving existing principles, I am not intimating that we become short-sighted. Research and development efforts must maintain a long-range perspective.

We must monitor technical developments, not only in our industry, but in other industries where similar applications of appropriate technology would be beneficial to us. Solvent extraction/electro-winning had its beginnings with the Manhattan project, when upgraded uranium was needed for the atomic bomb. Then again, high-rise elevators evolved from underground mining!

It's important to note that the most recent innovative process changes or upgrades in North American copper mines and metal extraction processing plants were either derived from university and government research, vendor technology or follow-up work on technology conceived before the near collapse of the industry in the mid-1980s.

An example of such a process change is best illustrated by Phelps Dodge's dispatching project with Modular Mining Systems in Tucson, Arizona. In 1980, Phelps Dodge pioneered and licensed Modular Mining's computer truck dispatch system, now used by virtually all major open-pit mining companies in the world. Computers monitor each truck's location and direct its movements to optimize mine haulage. When we initiated the system, we immediately realized a five percent increase in efficiency.

In recent years, Phelps Dodge and others in the mineral mining and extraction business have applied technology developed earlier and held, so to speak, in inventory. A good example is the application of column cells in copper flotation circuits. First invented in the early 1960s by Canadian Pierre Boutain, the copper industry did not take full advantage of column flotation technology until the late 1980s.

In addition to in-house R&D activities, we will continue to use available technology and to transfer technology from other sources -- often from different disciplines. Cooperative R&D programs between industry, government and academia, I predict, will become much more prevalent in the future. In the U.S., the Bureau of Mines and the minerals industry are cooperating now more than ever to develop future technology.

Such future technological innovations being developed by cooperative programs in the U.S. include concepts such as "insitu" leaching, which extracts the mineral underground, leaving the host rock undisturbed and eliminating the need for smelting. Asarco and the U.S. Bureau of Mines are currently determining the technical and environmental feasibility of this new alternative and the industry will be monitoring the program carefully.

The optimization of existing or emerging technology has been an important part of our operations. This strategy should and will continue to be an integral part of our efforts to remain a low cost copper producer.

How can we prepare for the changes and challenges of the future? How will technological development impact copper production?

One thing is certain. Ore head grades will continue to drop dramatically in the next several decades, especially in North America. Other industries do not face the same product declination as mining. Existing technology does not allow us to mine profitably and extract metal values from smaller, very low grade ore bodies, say about 0.2 percent copper or less. In the next century many tailings ponds derived from processing plants of the previous 100 or so years will have copper values at the same level or greater than our current ore bodies. However, there is no developed technology to treat the tailings economically. Since it usually takes 10 to 15 years to take a novel technology from the laboratory stage to the industrial plant, programs looking to fill the anticipated requirements of the 21st Century need to be launched in the next several years.

Efficient use of energy is another area that must concern us as we move toward the 21st Century. Mining, crushing, milling and smelting require a tremendous amount of energy. The cost of power, other resources and the supplies necessary to operate our business will keep increasing. Our innovative ideas must do the same.

As I mentioned earlier, the industry must continue to pursue innovative solutions to environmental issues. We must actively participate in the legislative and regulatory processes and work cooperatively with government agencies to achieve scientific and economically attainable goals. Environmental constraints have dealt serious blows to the industry, therefore we must prepare for the future by continuing to develop effective, rational and equitable programs consistent with sustainable growth. The mining industry has made significant strides over the past several decades to comply with environmental legislation, expending billions of dollars on new and enhanced programs. However, since legislative changes may adversely affect our ability to compete in the global marketplace, we must be committed to finding innovative ways to address these environmental challenges. Efficient mine production cannot and should not exist without sound environmental planning. Complying with the law is not only a requirement of good citizenship, it is also good business. However, since the staggering costs associated with compliance could be detrimental to the stability of the industry and the nation's economy, it is critical that government agencies and industry work together to establish rational solutions to environmental concerns. We must address these concerns thoroughly, aggressively and responsibly.

Finally, we must continue to devote resources to our most important asset -- our people. We must hire and retain the highest quality employees. Intelligent, decision-making, risk-taking, hard working, dedicated, flexible employees. Does such an employee exist? We hope so. Our expectations are high, and so is our productivity. Of course I'm a little biased, but I think, without a doubt, Phelps Dodge has one of the finest work forces in the world. Many of our employees are second and third generation miners. It's in their blood. They understand the ups and downs of the industry and what it takes to succeed.

Our company has been in the copper business since 1834. That's more than 10,000 years after the first discovery of copper. Ten thousand years ago, man was cold-hammering native copper into pendants in Iraq. Today, billions of people throughout the globe rely on more than 10 million tons of newly-mined copper produced worldwide each year. Even more people will be relying on copper in the future.

I don't know who first uttered what is now a slogan: "There are no challenges, only opportunities." Whoever said that certainly was not in the copper industry. Our industry faces a gauntlet of challenges. Change is inevitable. By successfully managing those challenges and changes, we can ensure that copper -- man's first metal -- will continue to shape civilizations for centuries to come.

Operations engineering and management: cradle to grave

W. Curlook
Inco Limited, Toronto, Ontario, Canada

ABSTRACT

A review is presented of a growing number of imperatives, some self imposed, some imposed by society, which the mining and metals industry faces in carrying out its business. These imperatives bear on all activities from the exploration phase, through engineering, design, plant construction and operation, to decommissioning.

INTRODUCTION

The mining and metals industry must learn to cope with, i.e. manage, a growing number of **imperatives.** Some of these, such as the need for responsible stewardship of natural resources, the requirement to respect laws and regulations, the need to provide for safe and healthy workplaces, and the need to provide appropriate returns to investors of capital, have been with the industry for a long time (Figure 1). But the industry now faces additional imperatives that are growing in number and in importance (Figure 2):

LONG-STANDING IMPERATIVES

- RESPONSIBLE STEWARDSHIP OF NATURAL RESOURCES

- RESPECT FOR LAWS AND REGULATIONS

- PROVISION FOR SAFE AND HEALTHY WORKPLACES

- GENERATION OF APPROPRIATE RETURNS ON INVESTMENT

Figure 1: Long-Standing Imperatives

- There is the growing concern about the impact of operations, not only on the **workplace environment** and the **external environment,** but also on the need to monitor not only the chemical concentrations in various effluents but to also monitor the **effects** of such emissions on the environment; i.e., there is a growing concern about the "net" consequences of operations.

- There is the need for on-going **improvement in productivity** and **safety** to compete and survive in a **global economy.** The global business environment has become paramount.

- There is the imperative to **failsafe** all designs, equipment and operational procedures, and having done that to provide for containment of unavoidable spills, and to have ready and in place systems for **emergency response.**

- There is the need, indeed the imperative, to involve all employees in various quality improvement programs.

MORE RECENT (AND GROWING) IMPERATIVES

- IMPACT OF OPERATIONS ON THE EXTERNAL ENVIRONMENT
- IMPACT ON WORKPLACE ENVIRONMENT
- GLOBAL ECONOMIC ENVIRONMENT; COMPETITIVENESS
- PRODUCTIVITY AND SAFETY IMPROVEMENTS; FAILSAFING; EMERGENCY PREPAREDNESS
- TOTAL QUALITY IMPROVEMENT
- **LOWER EMISSIONS;   - VIRTUAL ELIMINATION - ZERO DISCHARGE**
- WASTE MANAGEMENT; REDUCE, REUSE, RECYCLE
- **DECOMMISSIONING**
- SUSTAINABLE DEVELOPMENT

- PRODUCT STEWARDSHIP
- CRADLE TO GRAVE RESPONSIBILITY

Figure 2: More Recent (and Growing) Imperatives

- There is the imperative to face up to "**virtual elimination/zero emissions**"; and

- One must have waste management plans in place to cope with the need to **reduce, reuse, and recycle**; and

- There is the need for **decommissioning plans** for all operations, for the restoration of business habitats to their original states (or better).   Indeed, one must engage in the decommissioning exercise from day one, i.e., right from the time of design and commissioning.

- There is the growing need to reflect on the concept of **sustainable development** in all of our plans, designs and operations. Industry must operate in such a fashion as not to imperil or disadvantage future generations.

- There is the new need to respond adequately to the matter of **product stewardship**. The industry must be concerned about the downstream safe handling and use of the products that it sells to its customers.

- And these all translate into **cradle to grave management responsibility**.

It is not the objective of this paper to discuss all of these imperatives in detail. The focus will be on methods, techniques and actions that the industry must consider in order to respond to the most pressing of these imperatives.

VIRTUAL ELIMINATION/ZERO EMISSIONS

Firstly, there is the matter of "zero emissions". There is a strong philosophy being implanted in the minds of legislators and regulators in countries around the world, of zero emissions from industrial sources, i.e., of gaseous, liquid and solid emissions (Figure 3). There is less emphasis on applying the same rules to the individual citizen. "Virtual elimination" has been identified as a practical interpretation of "zero emissions" in certain circumstances.

**VIRTUAL ELIMINATION
ZERO EMISSIONS**

- GASEOUS EMISSIONS
- LIQUID EMISSIONS
- SOLID EMISSIONS

Figure 3: Virtual Elimination/Zero Emissions

The question is: can the mining industry even hope to be able to comply?

Gaseous emissions

Zero gaseous emission is extremely difficult, if one were to include water vapour (H_2O). However, most regulations exclude water vapour, because it can, and usually does, revert to liquid form whence it came. Similarly, nitrogen emissions can usually be excused, as they originate from air. However, operations such as high temperature combustion of hydrocarbons with air or oxygen can produce some by-product oxides of nitrogen as well as the newly-maligned carbon dioxide. Here, the answer can be "scrubbing" for removal of the nitrogen oxide compounds, but with little possibility of removing the associated carbon dioxide. Another answer could be switching to another process to supplant the hydrocarbons with other sources of energy.

Many in the industry have for ages thought of carbon dioxide as a near-inert gas. Not so any more. A large proportion of the general public has recently become familiar with the so-called "greenhouse" effect (Figure 4). Recently, a high-priced public orator suggested that the mining industry consider shutting down its coal mines and turning to other energy sources.

One need not give up, entirely, on the idea of containing carbon dioxide emissions. In Brazil, for example, there are two ferronickel operations which grow their own trees to provide a source of carbon for both drying and reduction operations (the smelting is carried out in electric furnaces). In effect, these are examples of complete carbon cycles. It can be done.

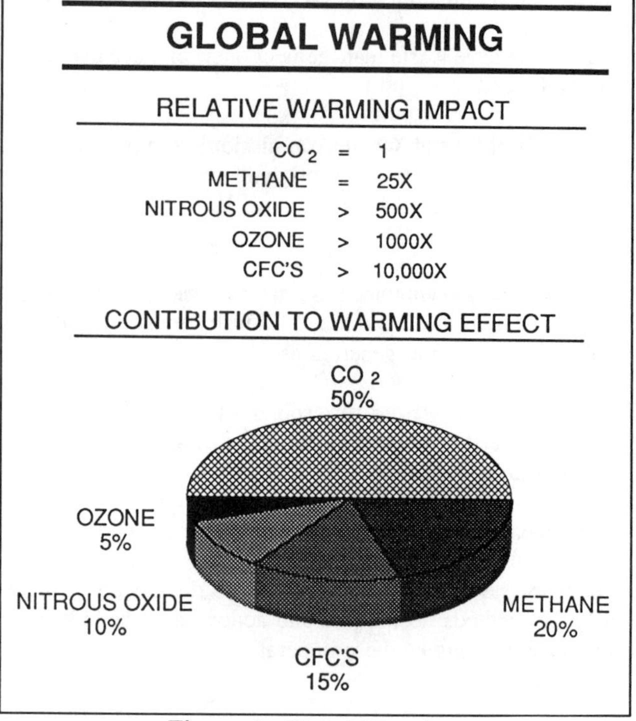

Figure 4: Global Warming

And one could not complete a discussion on gaseous emissions without referring to sulphur dioxide. Sulphur dioxide can be removed from effluents by several known methods such as conversion to sulphuric acid, liquefaction, reduction to elemental sulphur, or by neutralization. "Virtual elimination" can at least be approached.

Liquid emissions

And what about liquid effluents? Achieving zero liquid emissions should not be an insurmountable problem in metallurgical operations. Metallurgical operations should be net consumers of water. Metallurgical operations eliminate water (mostly as vapour) in drying, roasting, cooling, quenching, and dust suppression. Then, where do water effluents come from? Most originate from single pass streams. And, what is the solution? Simply, recycle.

Solid emissions

And what about solids discharges, such as tailings and slags? For starters, one must recognize that non-ferrous feed materials usually contain substantial quantities of sulphides, and that rejects, tailings and slags, contain lesser but significant quantities of sulphides that under certain circumstances of weathering, can give rise to acidic metal-bearing drainage.

With regard to mill tailings one can (and must) consider one or more of the following (Figure 5):

- In the case of open pit mines, one must consider putting the tailings back in the pit, along with overburden rock, and return the contour of the land as closely as possible to its original state. The tailings can provide a suitable base for revegetation.

- In the case of underground mines, one should maximize the use of tailings for backfill.

PERMANENT DISPOSITION OF TAILINGS

- BACKFILL, CONTOURING AND REVEGETATION OF OPEN PITS

- BACKFILL IN UNDERGROUND MINES

- PERMANENT IMPOUNDMENT IN PREPARED AREAS, AND REVEGETATION

- PERMANENT IMPOUNDMENT UNDER WATER

Figure 5: Permanent Disposition of Tailings

- Any excess tailings must be stored in areas that can be properly constructed so as to ensure stable and permanent disposition; and the area must eventually be stabilized and revegetated.

- As an alternative and wherever practical, the tailings should be "put to permanent rest" under water. This is particularly important if the tailings contain a high proportion of reactive sulphides.

Metallurgical slags, on the other hand, should be a lesser problem. Slags have commonly been thought of as being "inert" materials and suitable receptacles for various metals and obnoxious trace elements. Slags have been relatively easy to store, and have proven to be ideally suited for construction purposes and for backfill.

The problem that has reared its head is that many metallurgical slags cannot pass certain regulatory leach tests, and risk being classified as hazardous waste. Thus, one must go back to chemistry and metallurgy books to see what can be done to render the slags non-leachable. Alternatively, the slags would need to be stored in a manner similar to that employed for tailings.

One should consider one or more of the following (Figure 6):

- Modify the chemistry of the slag to make it more inert.

- Remove the sulphide components by slag cleaning processes (thereby recovering additional metal values).

PERMANENT DISPOSITION OF SLAGS

- MODIFY THE CHEMISTRY OF THE SLAG

- SLAG CLEAN

- USE FOR CONSTRUCTION IF CLEANED AND NON-LEACHABLE

- USE AS MINE BACKFILL

- IMPOUND, COVER, REVEGETATE

Figure 6: Permanent Disposition of Slags

- Use slag for construction purposes, if cleaned and non-leachable.

- Use the slag as mine backfill.

- Impound the slag in stable permanent locations, if necessary and cover it with a soil suitable for revegetation.

Gold Operations

Gold milling operations represent some of the simplest metallurgical operations that yield saleable metal products, and as such, they present opportunities for approaching "zero emissions". A portion of the tailings can be placed back in the mine, with the balance impounded and revegetated. By maximizing recycling, water effluents can be kept at zero or near-zero. The most difficult situations are where the mine encounters high ground water flows or where the mine is located in an area of high precipitation (relative to evaporation).

For illustrative purposes reference is made, here, to three new gold mining operations, in Inco's camp. They are good examples of "zero emissions" or "virtual elimination" and are also good examples of tailings being placed in permanent locations "for all time".

Les Mines Casa Berardi: Quebec, Canada (Figures 7, 8)

Because the topography is essentially flat, the tailings are put into constructed impoundment cells and revegetated once filled. Water is recycled to the mill along with mine water for "virtual elimination" of liquid effluents.

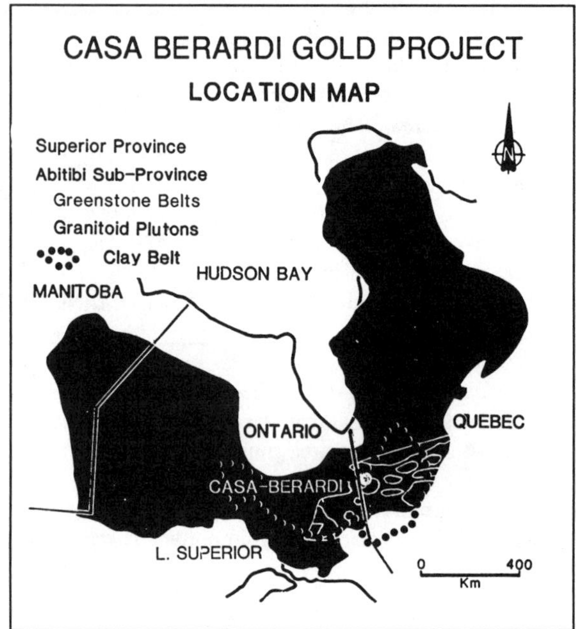

Figure 7: Casa Berardi Location Map

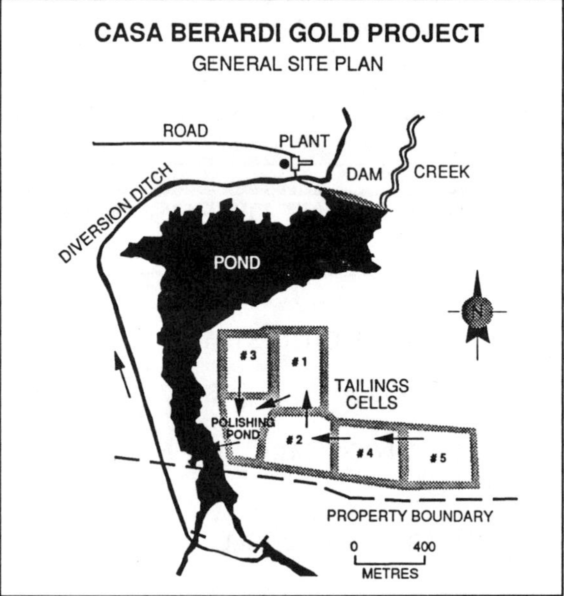

Figure 8: Casa Berardi General Site Plan

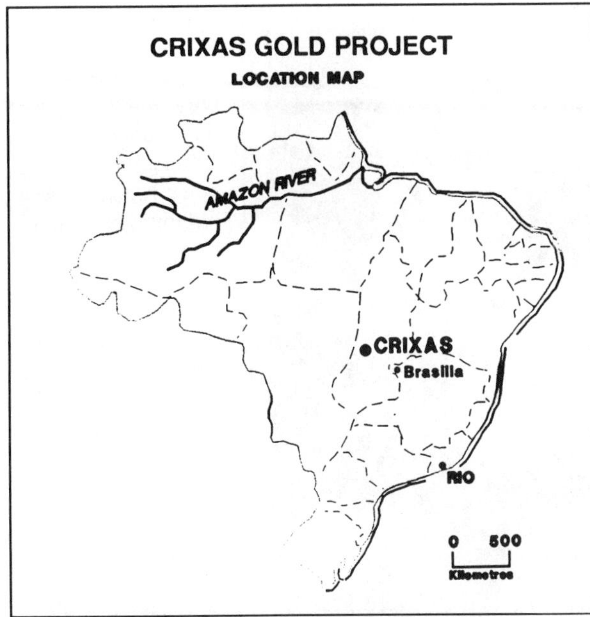

Figure 9: Crixas Gold Location Map

Crixas: Goias, Brazil
(Figures 9, 10)

A stable permanent dam was constructed across a large valley, sufficiently large to hold all of the mine's potential production, to capture the water effluents for recycle, and to provide opportunity for oxidation and destruction of the toxic cyanide used in leaching the gold. Because of the annual hot/dry season, the evaporation from the impoundment area is sufficient to render the project "zero discharge" with regard to water effluent.

Figure 10: View of Crixas Gold Tailings Dam

Mineral Hill: Montana, U.S.A. (Figures 11, 12)

Mineral Hill Mine and Mill, situated on the edge of Yellowstone National Park, receive and deserve the closest public scrutiny. The tailings are filtered and deposited "dry", i.e., trucked out as filter cake, to an area prepared on the mountain slope. An impermeable membrane is laid over a layer of clay after which are placed the dewatered tailings. Progressively, the filled areas will be covered and revegetated. The general contour of the mountain is respected. Water removed at filtration is reused in the milling process. The Mineral Hill Mine has been designed and is operated as a zero discharge facility, and represents the first operation of its kind in the U.S.A.

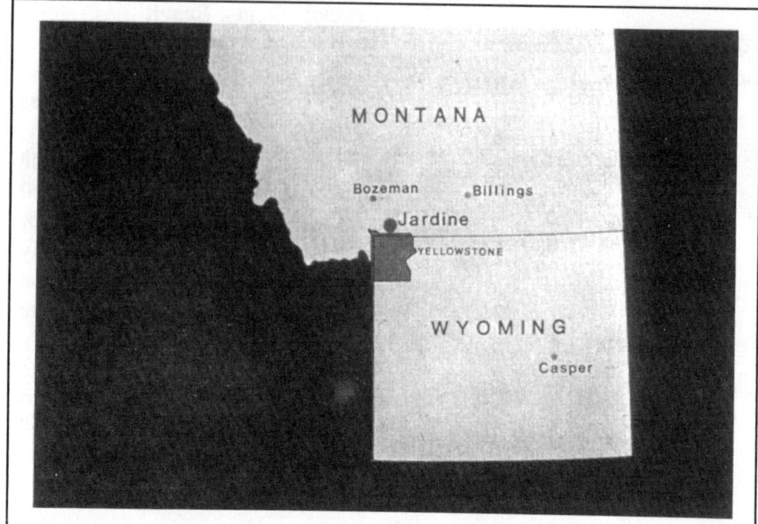

In fact, the area will be in better shape environmentally when the current operation is terminated than when it began. The old tailings containing arsenopyrite, left by previous operators, are being incorporated in the newly-constructed environmentally secure disposal area.

Figure 11: Mineral Hill Location Map

Figure 12: Mineral Hill Placement of Dry Tailings

Oxygen Flash Smelting at Inco

Oxygen flash smelting of sulphide concentrates, development of which dates back to the late 40's and early 50's at Inco, is a modern pyrometallurgical operation which comes close to "virtual elimination" or gaseous and liquid effluents. Oxygen flash smelting is one practical response for replacing fuel-fired furnaces, to better meet many of today's imperatives (Figures 13, 14).

Concentrates are dried in fluid bed units at low temperature, permitting gas cleaning in efficient bag houses. The main effluent is water vapour (along with some CO_2, and nitrogen, products of combustion of the natural gas fuel).

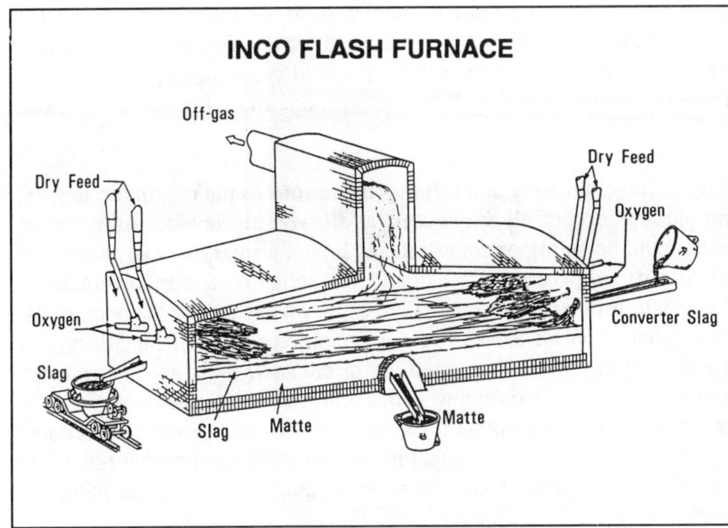

Figure 13: Inco Oxygen Flash Furnace

The dried concentrates are smelted autogenously with essentially pure oxygen (95+%), thus largely eliminating the generation of nitrogen oxides. The gases are scrubbed in several stages to remove all traces of solids, and the SO_2 passes on to either liquefaction or production of commercial grade sulphuric acid and oleum, thus achieving "virtual elimination" of SO_2.

Compared with the old technology based on fuel-fired reverberatory furnaces, the oxygen flash furnace permits quantum reductions of discharges to the external environment, and also provides for significantly lower overall energy consumption, improved productivity and safety, and reduction of overall unit costs of production.

Figure 14: Interior of the Inco Flash Furnace

Oxygen flash smelting goes a long way towards "virtual elimination" of undesirable emissions.

ENVIRONMENT, HEALTH AND SAFETY

Although the mining and metals industry is subject to great pressures from the outside, to reduce emissions to the external environment, it is incumbent on the industry to further improve the **workplace environment.** Wearing breathing apparatus should be the exception to be reserved for abnormal circumstances. The workplace environment must be such that breathing apparatuses are not required under normal stable-state operations.

When talking about health and safety, one must be prepared, at all times, for **emergency responses.** Machines fail and humans fail. One must be prepared for the situation where failure of one or the other could endanger or threaten human life, or could cause damage to the flora or fauna. In order to minimize the possibility of such happenings, one must **failsafe** all designs, equipment and procedures.

In order to failsafe mining, milling, smelting and refining operations to the maximum degree, one must enlist the help and participation of all employees, at all working levels. Adoption of some form of **quality** program could be an important assist. At Inco's primary metals operations the program is called **total quality improvement:** "total" to recognize the involvement of employees at all levels, delving into problems and opportunities, in all aspects of the operations; "quality" to represent positive change in all aspects of the operations be it in metallurgy, in workplace environment, in reducing energy consumption, or in enhancing quality of products; and "improvement" to recognize that it is an on-going endeavour, as there is always room for doing better. By involving employees at the operating levels in reviewing designs and procedures one is enlisting their **ideas,** as well as their physical capabilities and skills; and at that point one is getting all that they've got. One of the most productive types of quality exercises, at Inco, has been the organization of improvement teams focused on reviewing and failsafing specific production procedures. Experience at Inco has shown convincingly that procedures designed to improve productivity do at the same time improve safety. Experience at Inco has also shown that procedures designed to improve safety will usually also improve productivity. One should not be surprised at this. In the process of failsafing, with the help of knowledgeable experienced employees, one usually comes up with simpler more straightforward procedures which in most cases turn out to be the most productive and the safest. Inco's Thompson works is the most advanced in the application of total quality improvement. Their lost-time-accident experience tells an impressive tale (Figure 15).

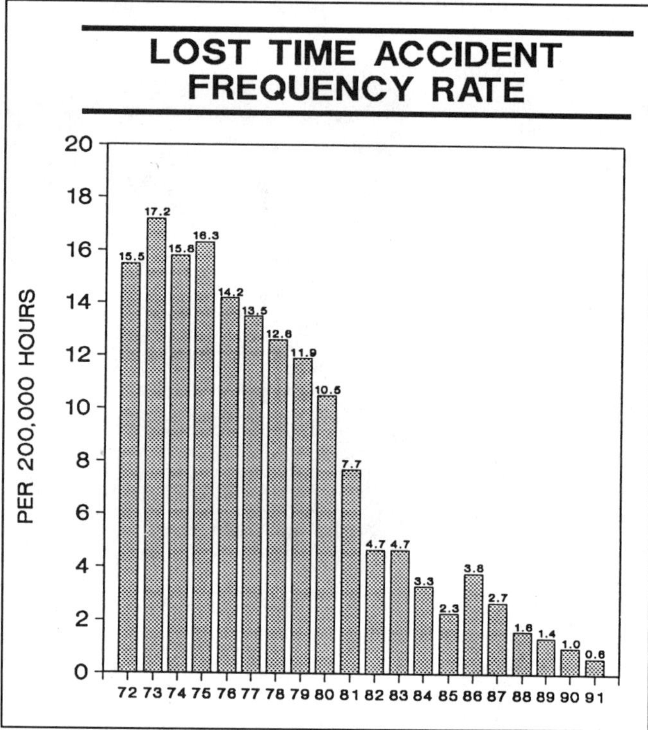

Figure 15: Lost Time Accident Frequency Rate

At Inco's Sudbury operations, during the course of negotiating the latest collective bargaining agreement (May 31, 1991) the company and the union agreed to organize a joint **environmental awareness committee,** with representatives of both management and the Steelworkers union. A mechanism is now in place which will provide opportunity to increase the awareness and involvement of all employees at the operations in reducing the impact of operations on the external environment. Positive results are expected from this new joint initiative.

DECOMMISSIONING

Today, one must draw engineering plans for **decommissioning** at the same time as one is drawing plans for construction of a production facility. Regulations in many provinces in Canada and several states in the U.S.A. require or will require up-front financial assurance to insure an environmentally safe and permanent shutdown. Today, one cannot get by simply by managing the effluents from current operations to ensure that concentrations of various undesirable elements fall below certain regulated levels. Today, one must also be concerned with the **environmental effects,** not only during the period of operation, but also for the period beyond. If one were to undertake a project and could not design for and operate in an environmentally sound manner, and if one could not accommodate appropriate and acceptable means of decommissioning, then the principle of **sustainable development** would be violated. In today's society, one would probably not be permitted to proceed with the project.

And, there is a new, more recent, imperative advancing upon us, and that is **product steward-ship.** The idea is that responsibility doesn't end when the primary product is first sold. There is a measure of lingering responsibility to see that the product is handled safely and in an environmentally sound manner by the customer who in turn must convert it into environmentally sound consumer products (Figure 16).

"LINGERING" IMPERATIVES

- DECOMMISSIONING

- ON-GOING ENVIRONMENTAL EFFECTS

- SUSTAINABLE DEVELOPMENT

- PRODUCT STEWARDSHIP

- CRADLE TO GRAVE RESPONSIBILITY

Figure 16: "Lingering" Imperatives

Finally, it is quite clear that engineering and societal imperatives compel the mining and metals industry to concern itself about its operations from **cradle to grave.**

SUMMARY

The mining and metals industry faces a growing number of imperatives, some self imposed, some imposed by society, regarding the manner in which it carries out its business. These imperatives bear on all activities starting with the exploration phase, through engineering design, plant construction and operation, and finally on decommissioning. Indeed, one must design for decommissioning an operation at the same time that one designs its implementation. The impact of operations both on the workplace and the external environment will, "forever", fall under intense public scrutiny. Initial designs and feasibility studies must be based on the project's full life cycle, incorporating the impacts of all of the foreseeable operational imperatives from commissioning to decommissioning, from "cradle to grave".

Recycling of copper

Klaus Göckmann
Norddeutsche Affinerie Aktiengesellschaft, Hamburg, Germany

1. GENERAL

Copper has been recycled ever since its discovery. Even in Graeco-Roman times recycling of this resistant, and thus valuable, material was carried out as a matter of course. Just take, as an early example, the Colossus of Rhodes: after the figure collapsed during an earthquake, the bronze parts were melted down into ingots and sold.

Copper is a material which can be recycled again and again since copper produced from recycling is not different from copper originating from primary materials. This is mainly due to electrolytical copper refining whereby not only the base metals but also the precious metals can be removed in contrast to aluminium and iron, where only the more electropositive elements can be removed by refining and the resultant scrap must therefore be sorted mechanically into various alloys for subsequent smelting.
In **Fig. 1** the recycling rate for different base metals is given. Despite this ideal prerequisite, the generally quoted recycling rate of copper of about 40% appears rather low. However, it must not be forgotten that the recycling rate is defined as the ratio of the annually utilized quantities of copper scrap and the annual production figures (Fig. 2). Moreover, on account of the economic growth over the last 30 years the availability of copper from scrap materials is lower than its consumption, since copper is principally employed in very long-life consumer durables and only returns to the recycling circuit after many years. The average time until the copper is recycled is detailed for some goods in **Fig.3**.

1.1 The Copper Cycle

Today, under the term "recycling" all the processes are generally included which result in scrap and other materials being treated to a state in which they can be returned to the production circuit. Regarded individually, these processes can be termed reclamation, recovery, recycling and reuse.

Metal recovery from recycling material (secondary products) may be divided into several groups:

a) Fabrication (new) scrap
such as scrap arising in the metal producing and semis manufacturing industries which contains no impurities (e.g. chips, metal cuttings, etc.)

b) Used product (old) scrap
arising from used copper products (scrap cable, computers, cars, fitments, etc.), which have a calculable copper content.

c) Intermediary products or other materials
arising from metals manufacture, processing and treatment. This material is characterized by its varying origins and composition. e.g. filter dust from primary or secondary smelting, slag, skimmings, tankhouse slime or complexly composed material from chemical industrial plants. The material's complexity is often a deciding factor in the cost-effective possibilities of recycling.

In Fig. 4 the life cycle for copper products is represented. The copper products produced in the semis fabrication are sent to the respective industrial sectors: general and consumer goods, electrical sector, construction, industrial transport and equipment. Chips and metal cuttings from machining consist of very pure copper or copper alloys and are offered as new scrap either via the scrap trade or are, more frequently, directly recycled in the resmelting plants of the copper foundry industry = DIRECT USE

The more impure -mainly old copper and copper alloy scrap- from different sources like radiators, electronics, cable or tubes, not directly utilisable, scrap is offered by the scrap trade to smelters and/or refineries, where it is subjected to a refining process before being passed on as pure copper to the semis industry.

The scrap trade functions as a connecting link between the numerous points where this material arises and the recycling plants (smelters, smelting works and refineries in the semis industry). Its task is to compile, sort and process, to purchase and sell, as well as to import and export. The small retailer, who traditionally trades in both metal scrap and other scrap materials, is basically a thing of the past. His function has generally been taken over by private and communal refuse collectors who sort out the utilizable materials and deliver them to the next stage on the scrap market. The scrap trade is thus an important supplier for smelters and smelting works as well as a supplier of better quality scrap for direct utilization in semis and cable works and metal foundries.

Metal scrap and other metal containing materials are, however, principally traded on account of their economic value and not as a means of environmental protection. Market prices are governed by the national and international availability of goods. The close connection between the market value and the availability of scrap is demonstrated in **Fig.5**, in which the average price of copper is compared to the amount of scrap processed at Norddeutsche Affinerie. It can be seen that at times of high copper prices more copper scrap is offered to the refineries. On the other hand, if prices are low, less material is coming on the market and the terms for the smelters get worse. This close connection between price and the amount of recycled material proves that copper scrap is not a waste material but a material with high economic value.

As shown in **Fig. 6** recycling of copper not only saves primary copper resources for the future but also saves energy compared to the production of copper from ore. Depending on the quality of the scrap the saving of energy varies from 50% for alloy scrap to 90% for cable scrap.

2. TECHNICAL PROCESSES
2.1. Conventional Copper Scrap Qualities

Copper scrap can be classified according to purity or chemical composition or to the source of the material, which defines not only the chemical composition but also the physical appearance.

New scrap from the production of copper products has a high purity and can be remelted without any further refining in the semis industry. Also new scrap from brass or other copper alloys with constant composition is used directly by the semis industry at a high value level.

Copper and copper alloy scrap after use normally contain the remains of plastic material or other elements and have to be refined in a smelter. Depending on the chemical composition this material can be processed in an anode furnace, a converter or a shaft furnace.
Besides copper scrap a great deal of copper containing intermediates like slag, flue dust, ash and copper matte are treated in a secondary smelter. These materials contain oxidic compounds and have to be reduced, e.g. with carbon in a shaft furnace or -using the latest available technology- in an electric furnace.

As the conventional types of scrap can be classified according to their quality and chemical composition, nowadays a considerable amount of special copper containing scrap is offered, which has to be treated in a special way. Examples of these very complex materials are cable scrap, electronic scrap, etching solutions or slimes from the galvanic industries.

A flowsheet giving the normal way for the recovery of copper from scrap is shown in **Fig. 7**. According to the minimization of waste material not only the copper is recycled but also most of the impurities or alloy elements can be converted to useful products. Hence an integrated secondary smelter like Norddeutsche Affinerie runs processes both for the refining of the copper and for the recovery and refining of lead, nickel as nickel sulfate, zinc as zinc oxide, tin as tin/lead alloy and minor elements like precious metals, selenium and antimony.

Energy saving was also the main objective in the development of a new process for melting relatively pure copper scrap. The Contimelt process developed at NA (**Fig.8**) consists of a hearth furnace with a shaft and a poling furnace. The main advantage of the Contimelt process is the optimal use of energy. The hot off gas from the hearth heats up the cold copper column in the shaft. This reduces fuel consumption for melting copper by more than 50 % compared to a conventional reverberatory furnace.

Alloy scraps like brass, bronze and German silver contain besides copper, elements like zinc, tin and nickel and are refined in a special scrap converter, where these elements can be removed either by reduction and evaporation or by oxidation and reporting to the slag phase. After the converting stage the blister copper contains about 95-98% Cu. The flue dust of the converter process containing mainly zinc, tin and lead is refined in a special furnace, where tin is recovered as tin/lead alloy used for soldering and zinc is converted to zinc oxide for the pigment industry.

For the converting process of alloy scrap a Peirce Smith converter or a Top Blown Rotary Converter can be used. The TBRC in **Fig. 9** consists of a cylindrical vessel rotating on its mid axis. Tapping occurs by tilting the vessel. Fuel and reducing agents are fed through lances into the reactor.

Whereas most of the scrap contains copper in a compact and elemental form and refining is possible by oxidation in a converter, there exists a lot of mainly complex materials in which copper is either in the oxidic phase or in a powdery form often accompanied by carbon based material.

Examples of such complex materials are:

Drosses

Drosses are materials which form on the surface of molten copper after contact with air, e.g. in a foundry. Drosses not only contain oxides but also drops of metallic particles, which are mechanically incorporated in the viscous oxide slag.

Flue dust

During smelting and refining of primary and secondary copper, material is mechanically withdrawn from the gas phase and collected in the off gas cleaning system.

Catalysts and collector dust

Some organic chemical reactions of industrial importance are catalysed by copper compounds. After a certain time on stream lasting some months or years the catalytical activity decreases and the copper catalysts have to be exchanged. Besides copper the catalysts contain mainly carbon. Production waste from the manufacturing of collector dust also contains carbon and copper powder.

Slimes from electroplating

Waste water from the electroplating industry contains copper and other non ferrous metals. After precipitating with caustic soda or lime the oxidic water containing slimes have to be disposed or recycled.

Metal rich slags

During the refining of copper alloy scrap by oxidation with air not only the impurities are removed, but also some copper is reported to the slag. Mainly the high copper content of converter slags justifies the installation of a reducing process stage in a secondary copper smelter.

Copper Cement

By cementation of copper solutions with iron scrap, copper precipitates as a mixture of an elemental and oxidic powder.

Besides these materials there are many different residues, mainly with a low copper content. Depending on the copper content, the actual copper price and their complexity, these materials can be recycled in an economic way or have to be passed on for waste disposal.

Processing all these residues, slags and other secondary material requires versatile production processes, which are able both to melt and to reduce the oxidic portion to an elemental form, which can be refined in a next step in a converter.

Of the classical processes the blast furnace process is one of the most flexible as regards the different kinds of feed (**Fig.10**). The only feed preparation is pelletizing or briquetting fine divided materials. As a reduction agent coke is used and as a flux silica, lime or iron oxide can be added. For good metallurgical results with a high recovery rate for the valuable metals in the metal phase, it is important to find the right mixture of the various feed materials. As elements like lead, zinc and others are partly dissolved in the copper phase, the product from a blast furnace is relatively impure copper" with about 75-80% copper content. This "black copper" is then refined in a converter. The main disadvantage of the blast furnace process is the high volumes of off gas, requiring a costly and difficult gas cleaning system to meet the stringent national legislation of today.

A totally new technology for melting complex copper containing residues was developed by Norddeutsche Affinerie using an electric furnace. The new 7,5 MW furnace (**Fig. 11**) went on stream on the 8th May 1991. By using electrical energy instead of coal the offgas volume can be reduced by 80% compared to conventional shaft furnace practice and so a very effective off gas cleaning system with an afterburner could be installed. This environmentally gentle technique will gradually replace the different shaft furnace processes at NA.

For an integrated smelter like NA, processing not only secondary but also primary copper from concentrates, the integration of recycling into the normal cycle of processing concentrates is very attractive. **Fig. 12** shows how concentrates proceed to high quality copper and the possibilities of processing scrap in the different stages. Whereas in an Outukumpu flash furnace only fine divided material like copper cement or slime from waste water treatment can be integrated, the Peirce-Smith-converter for converting copper matte to blister copper is a very versatile apparatus for different kinds of scrap like electronic scrap, or packages of scrap copper. Furthermore, the heat generated by the matte converting of the primary material can be used to melt the scrap. Hence, the integration of secondary material into a primary circuit saves considerable energy and is gentle to the environment because of the existing gas cleaning systems for primary smelting. **Fig.13** shows the Peirce Smith Converter at Norddeutsche Affinerie.

2.2. Special Scrap

The world is becoming more and more complicated and so is the scrap of today. Whereas the aforementioned copper scrap can be classified according to copper content and impurity level, there are many different recycling materials, which have to be pretreated either mechanically or thermally before they can be integrated into the flowsheet of a secondary smelter or be directly remelted in a foundry. The range of different materials can be illustrated with three examples:

A very important copper recycling material is cable scrap. Whereas in former times the plastic parts of the cable were removed by burning, the stringent German legislation on emissions has restricted this procedure. Thus, the mechanical dismantling of the cables is predominant (**Fig.14**). The cable is granulated in two or three stages to a size of about 1-30 mm. During granulation the plastics are removed from the copper wire and can be separated by air classification. Iron impurities can be removed by a magnet. According to the kind of cable a very pure copper quality can be achieved. which can be remelted in the semis industry. Cables with impurities of lead or tin have to be refined after dismantling in a secondary smelter.

The most complex copper containing recycling material is electronic scrap. Electronic scrap does not only arise from computers but in an increasing amount from other durables, as today nearly every apparatus contains printed circuit boards. The relatively high gold content makes the recycling of electronic scrap economically attractive. On the other hand electronic scrap not only contains valuable metals, but also plastics, organic flame retardants and ceramics. An approximate composition is given in **Fig.15**. Mechanical dismantling by crushing and separation by means of density or conductivity is one way to process this complicated material. but total separation into metals and other materials is difficult and incomplete and there is no further use for the mixture of the non-metal fraction.

Thermal dismantling can be carried out by burning or pyrolysis. Whereas burning produces a lot of off gas and flue dust, pyrolysis is difficult to handle because the whole system has to be totally sealed from the atmosphere to avoid the exposure of toxic pyrolysis gases.

The most elegant way is to integrate the recycling of these materials into the converter process of a primary smelter, as is the case at Norddeutsche Affinerie. Because of the high temperatures and the excess of oxygen in the off gas system of the converter the plastics are totally destroyed and the ceramics and glassfibre report to the slag. It could be demonstrated in cooperation with the local authorities that the dioxin emissions resulting from this process are extremely low.

Nevertheless. because of the increasing amount of electronic scrap, which is greater than the copper converter capacity for this material, additional processes must be developed to treat electronic scrap in the future.

Each scrap car in Germany contains about 10 kg copper and copper plays an important role in the concept of car recycling, as impurities of copper can decrease the quality of the iron scrap from cars. Additionally, the copper content in cars is increasing. Thus a BMW 750 contains about 35 kg of copper because of the many servo motors. **Fig. 16** shows the methods of copper recovery in automobile recycling and shredder plants. Initially the larger copper bearing parts such as radiators, generators and starters are removed and the cable network manually dismantled. The wreck is then compacted and reduced in a shredder. Plastics and textiles are removed by air separation and form the so called "fluff". Iron can be recovered by magnetic separation. The non-magnetic fraction contains a lot of different alloys and metals. Up to now this mixture had to be sorted by hand, but a new fully automated sorting system has been developed by Metallgesellschaft, which permits each individual part to be analysed quickly and metals and alloys to be specifically sorted.

3. IMPACTS ON RECYCLING IN THE FUTURE

Unfortunately, talking about recycling also means talking about the dangers to recycling, which although probably not intended by the politicians, are the result of:
- stricter regulations for emissions
- higher costs for waste disposal
- the Basle convention

These are factors restricting the recycling of copper.

On the other hand in Germany the so called " Dual System " will force the producers to recycle their goods after usage.

3.1. Emissions Regulations

In Germany the TA Luft regulates the allowed emissions for the different industries. Thus, the sulfur dioxide for most of the plants is restricted to 200 mg/Nm^3; for NO$_x$ it is intended to be lowered also to 200 mg/Nm^3. But the main actual impact on recycling of complex materials is the formation of dioxins during thermal processes which contain -even small- amounts of organic material. The limit for dioxin emission shall be restricted to 0,1 Ng (=Nanogramm)/ m^3 - the so called "Töpfer-Value" named after the German minister for environmental affairs. As the chemical analysis of dioxin takes about 3 months and amounts to about 5000 DM for each sample, nobody in the recycling industry knows the consequences of this regulation, particularly since it is known that even in your chimney at home the values are higher, if you are burning wood.

3.2. Waste Disposal

For toxic wastes, which must be deposited in an underground waste disposal, about 650 DM/t have to be paid including transport and packaging. As more and more materials are declared as toxic, there will a dramatic shortage in safe waste disposal, as the installation of new disposal areas is not acceptable to most of the politicians. On the other hand, the installation of incineration plants will not be accepted by the local population either. The consequence of this dilemma is an increasing price for waste disposal even for non toxic waste material. Hence, the price for depositing fluff -containing textiles and plastics- from automobile recycling is about 250 DM/t and even more in some areas of Germany. Therefore, there is political pressure in Europe to find new applications for recycled plastics, although most of the plastic material cannot be recovered in a relatively pure reusable form, but is separated from the metal only in a complex mixture. As nearly all consumer goods contain certain amounts of plastic electronic material, only new planning permission for incineration or pyrolysis plants can reduce the increasing amounts of waste plastic material.

3.3. The So-Called Dual System

In Germany there are new regulations under discussion, which will force the producer of consumer goods to take back their goods after usage. This so called "Rücknahmeverpflichtung" ("obligation to take back") can be fulfilled either by the producer himself or by a company, authorized by the producer, which is specialized in collecting, transporting and processing the taken back goods.

At present. automobile manufactures. associations and authorities are working on joint projects looking into the possibility of recycling complete scrap automobiles. A good example for this new thinking is the cooperation between Metallgesellschaft and Mercedes Benz in the recycling of automobiles. Extensive preparatory work on the old vehicles is at present necessary and incurs costs - it is also not clear what should happen to the separated scrap parts. Since. presumably. scrap recycling plants cannot be reimbursed for costs incurred in the form of sales revenues, a suitable financing model must be developed. In future, automobile owners can surely expect to participate in the recycling costs of their old cars.

Similar to the automobile industry, the electronic industry and the packaging industry are also looking for solutions to fulfill the "Rücknahmeverpflichtung" of the "Dual System". That means that. besides the traditional established. the consumer good industries enter the "business" of recycling. As the consumer goods industry has bigger public relations department than the commodity based metal industry, from discussions in some media in Germany you can come to the conclusion that recycling has been invented by the consumer goods industry.

3.4 Basle Convention

But besides these national regulations in Germany one of the big problems faced by the recycling industry is the Basle Convention, adopted in 1989 as part of the United Nations Environment Programme.

The regrettable fact that some waste for final disposal was shipped from industrial countries to third world countries has led to the politicians - in my view - overreacting.

53 nations have signed the "Basle Convention on the control of transboundary movement of hazardous wastes and their disposal". which comes into force 90 days after the 20th instrument of ratification has been deposited. Up to now - if my information is correct- 9 nations have ratified. however. if they have also deposited their ratification I do not know.

Today. it is generally expected that the Basle Convention will be enforced in the autumn of 1992. It is clear that a legal and economic chaos is preprogrammed if it is not possible beforehand to agree on joint directives, at least between the 24 OECD members, as to how the Basle Convention is to be implemented .

3.4.1 Definition

The Basle Convention talks about "hazardous waste" and "other wastes". The conceptual limits have been selected so that you can only ask which secondary material does not fall under one or other category of hazardous waste. The so important and natural division into waste destined for reuse. reclamation. recovery, recycling (the famous 4 Rs) versus waste destined for final disposal is for the present of no consequence any more.

The Basle Convention subjects each transboundary movement to a complicated and time-consuming control process between the industry and the authorities in the exporting and importing countries a "prior informed consent régime" with week-long waiting times and, in addition, frequently uncertain results as regards trading in scrap and alloy scrap. In this business fast reaction to commodity exchange developments and exchange rates with the transaction concluded over the telephone is the order of the day. As well prompt actual deliveries are also often expected and needed. If business partners are only permitted to enter contracts promising that many weeks later the approval will be received, recycling will be quickly strangled. Ultimately, the Convention stipulates that after its enforcement trade may only take place between those countries which have ratified the Convention. This can naturally lead to the world being divided into members and non-members and parallel to that a division in recycling.

There are only an few industrial fields in which recycling waste and intermediary products of all types is so important as in the non-ferrous metal producing and fabrication industry. Making recycling so problematic that it almost stops completely was surely not the intention of the UN politicians. But they have presented us with the Convention - with a text which can no longer be changed - and now some practical solutions must be sought. We should not deceive ourselves: there are organisations who want to prevent just that and they put our politicians under tremendous pressure.

3.4.2. Proposals for a solution

For me a solution for scrap should be found by dividing scrap into three lists:
- "Green List" for scrap which should be recycled as before, that means in the framework of existing comprehensive laws and directives for transport, storage and processing of scrap. There should be no additional directives issued by the Convention applicable here.
-"Amber List" for scrap which would be recycled as before but in future the controls would be somewhat more stringent. Perhaps an additional document should be required for each consignment which would be used to notify the authorities but without giving them any time to approve.

All scrap which is not on the "Green" or "Amber" list would automatically fall under the
-"Red List" and be without exception subject to the directives of the Basle Convention. It is therefore not necessary to detail the individual scrap further for this list.

I know the problems really start surfacing when you look at the details, but if all parties concerned show a bit of goodwill it should be possible to reach a practical solution which would safeguard recycling not only for our industry but also for the environment.

I have the impresion as if our industry has been rather late in realising what fateful consequences the Basle Convention can have for us. In any case this is my impression from Europe.

In spite of all the regulations concerning recycling, the political forces and the consumers must know that recycling is practical environmental protection as both resources and energy are saved compared to the primary production of base materials. The only alternative to recycling is waste disposal of materials which could be converted to a material with the same quality as the origin material -as the example copper demonstrates.

What are the conclusions for the secondary, the recycling industry?

Copper scrap is getting more complex than radiators.

I think the pure copper and copper alloy scrap materials will be more and more skimmed off from the market by the primary smelters and by the semis manufacturers who can offer better terms.

That means that the secondary smelters will be left with the more complex materials which are more problematic regarding environmental regulations.

Is this a threat to the secondary copper smelters?
Yes and no!

Yes, if we continue to be the victims of environmental regulations rather than becoming a partner of the legislative and regulatory authorities.

Yes, if we continue to be at the bottom of the industrial production processes rather than cooperating with the manufacturers of metal containing products in developing recycling friendly products and helping them to fulfill their obligations to recycle their old products which will be forced upon them. If we do not do this, they may choose to use other materials instead.

What is needed is **management of change** and I refer to the paper Doug Yearly presented yesterday. There is probably no other industry in this world which has made more changes than the copper industry since its beginnings 10,000 years ago in what is today Northern Iraq. I do believe that the secondary copper industry will be accepting the challenge and will play a bigger role in the industrial cycle of metals.

We should keep in mind that very often those not willing to accept a bigger role end up having no role at all.

	W.EUROPE	USA	JAPAN
Aluminium	30 %	28 %	36 %
Copper	50 %	60 %	47 %
Zinc	29 %	27 %	20 %
Lead	55 %	59 %	38 %

Recycling-Rate
Total Scrap Recovery as Proportion of Consumption

NORDDEUTSCHE AFFINERIE
AKTIENGESELLSCHAFT

Fig. 1

	1980	1990	Change
Mine Production	168	279	+ 66 %
Smelter Production	390	487	+ 25 %
Refined Production	1206	1487	+ 23 %
Copper Scrap Recovery in Refined Production	460	592	+ 29 %
% of Refined Production	38 %	40 %	
Direct Use of Scrap	817	945	+ 16 %
TOTAL SCRAP RECOVERY (A)	1277	1537	+ 20 %
COPPER CONSUMPTION (B)	2724	3052	+ 12 %
(A) : (B)	47 %	50 %	

Copper Production/Consumption in Europe
(1000 t)

NORDDEUTSCHE AFFINERIE
AKTIENGESELLSCHAFT

Fig. 2

Cars
Small Electric Motors
Cables
Buildings
0   10   20   30   40   50   60   70   80
Average Lifetime of Copper for Various Applications (Years)
NORDDEUTSCHE AFFINERIE
AKTIENGESELLSCHAFT

Fig. 3

Old Scrap (such as)
Copper Scrap   Computer Scrap
Fittings   Cu plated Iron Scrap
Radiators   Brass Scrap
Relais Scrap   Gunmetal Scrap
OLD&NEW SCRAP   Scrap Trade   NEW SCRAP
SECONDARY REFINED
Smelters
Refiners
DIRECT USE
Foundries of the
Copper Industry
Copper Products
General&Consumer Goods
Electrical Sector
Construction
Industrial Transport & Equipment
NEW SCRAP
Copper Cycle
NORDDEUTSCHE AFFINERIE
AKTIENGESELLSCHAFT

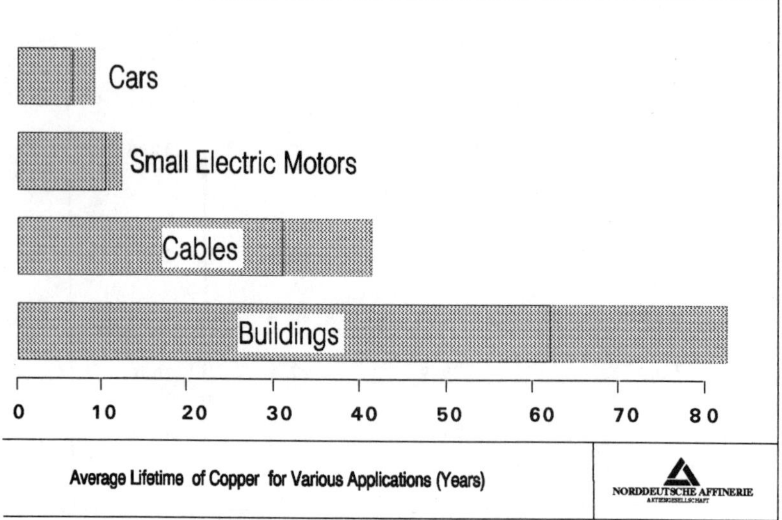

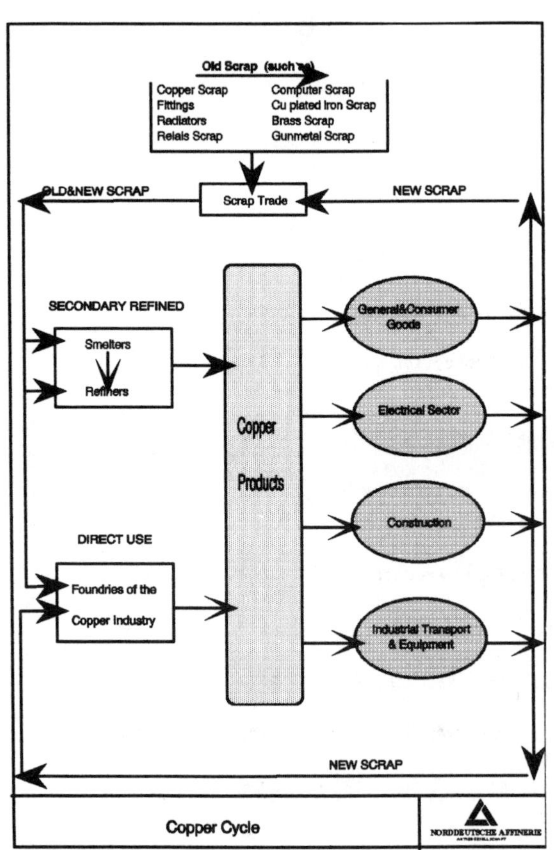

Fig. 4

100,000
80,000
60,000
40,000
20,000
0
Scrap-Input
Copper-Price
500
400
300
100
0
Copper-Price [DM/100 kg]
85/86   86/87   87/88   88/89   89/90
Year
Copper-Price and Scrap-Input at NA
NORDDEUTSCHE AFFINERIE
AKTIENGESELLSCHAFT

Fig. 5

Cables
Granulate
Scrap Copper
Copper (Wirebars)
Copper Alloy Scrap
Copper (Wirebars)
Ore
Concentrate
Anodes
Copper (Wirebars)
10   20   30   40   50   60   70   80   90   100   110
Energy (MJ/kg Copper)
Energy Consumption for Copper Production
NORDDEUTSCHE AFFINERIE
AKTIENGESELLSCHAFT

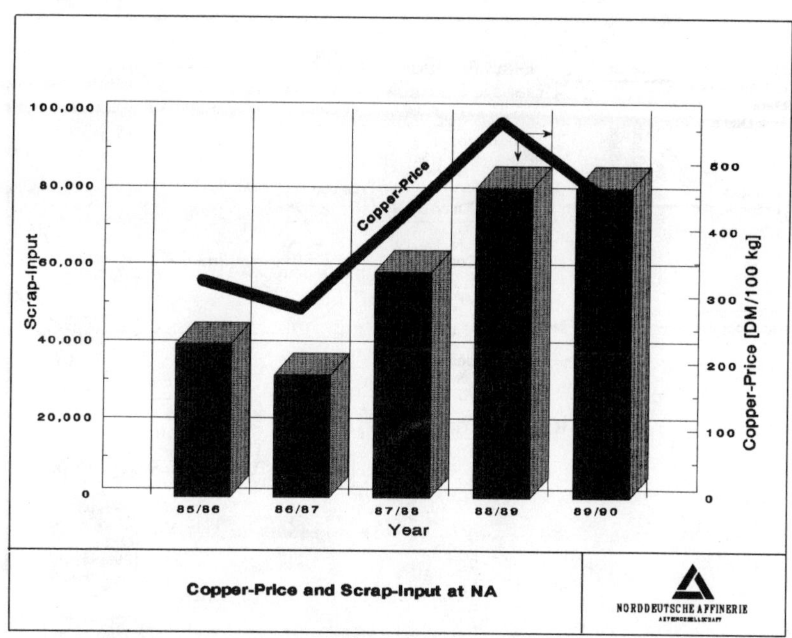

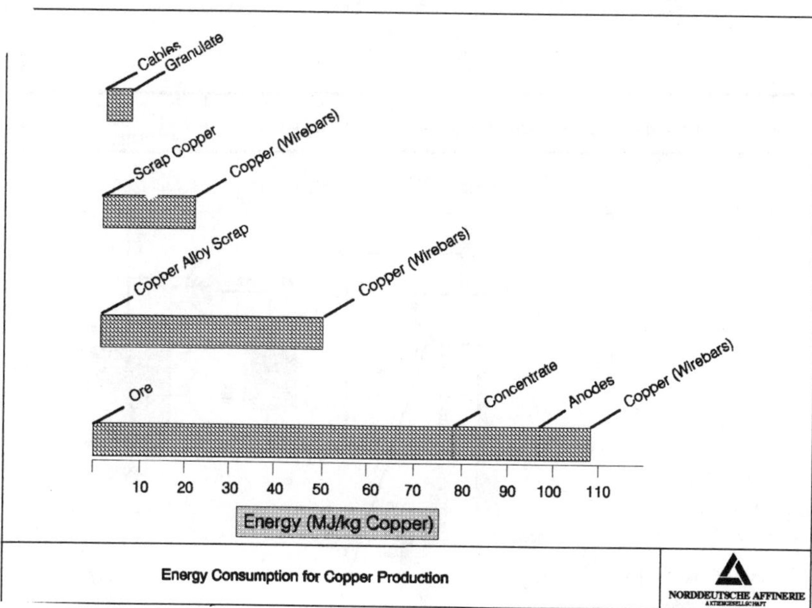

Fig. 6

Residues
Slimes
Electric Scrap
Slags
Flue Dusts
30% Cu
Retour-Slags
Bronces
Brass
40-90 % Cu
Retour-Slags
Scrap Copper
>90% Cu
Blast Furnace
Electric Furnace
Flue Dust
Zincooxide
Slag
Black Copper
70-90% Cu
Scrap Converter
Flue Dust
Blister Copper
96-98% Cu
Tin/Lead-Alloy-Plant
Flue Dust
Zinc oxide
Tin/Lead-Alloy
Anode Furnace
Anodes
98-99,5 % Cu
Sulfuric Acid
Copper Tankhouse
Elektrolyte
Anode Slime
Nickel-Sulfate Plant
Nickelsulfate
Precious Metal Plant
Selenium
Silver
Gold
Platinum
Palladium
Cathodes
Processing of Copper Containing Secondary Material
NORDDEUTSCHE AFFINERIE
AKTIENGESELLSCHAFT

Fig. 7

Cooler
waste gas
baghouse
Poling Fumace
Copper
Contimelt-Process
NORDDEUTSCHE AFFINERIE
AKTIENGESELLSCHAFT

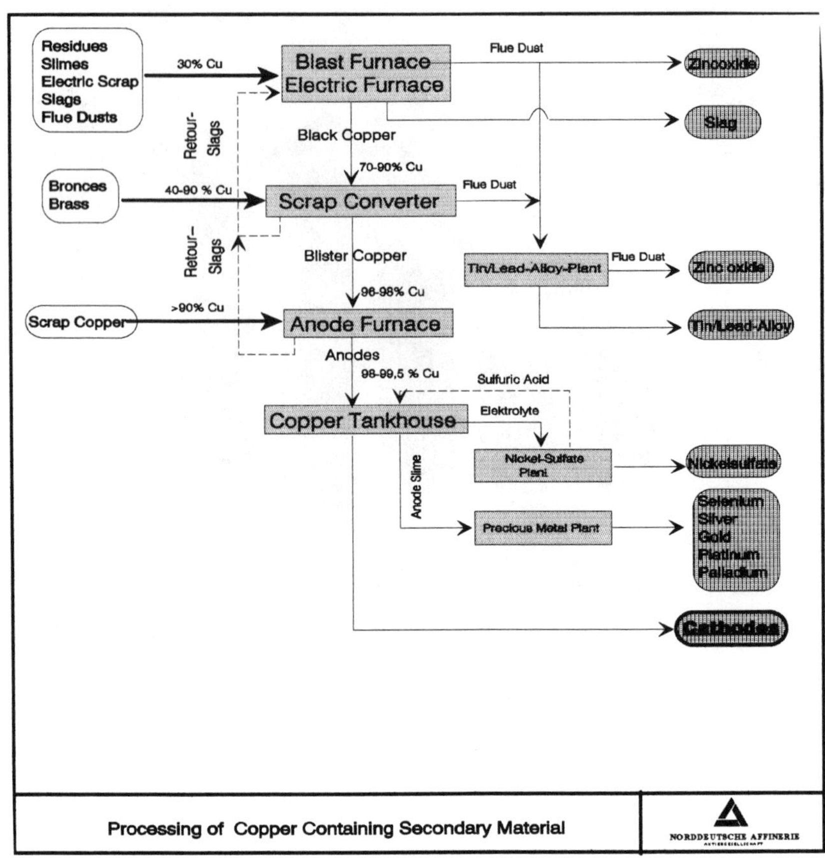

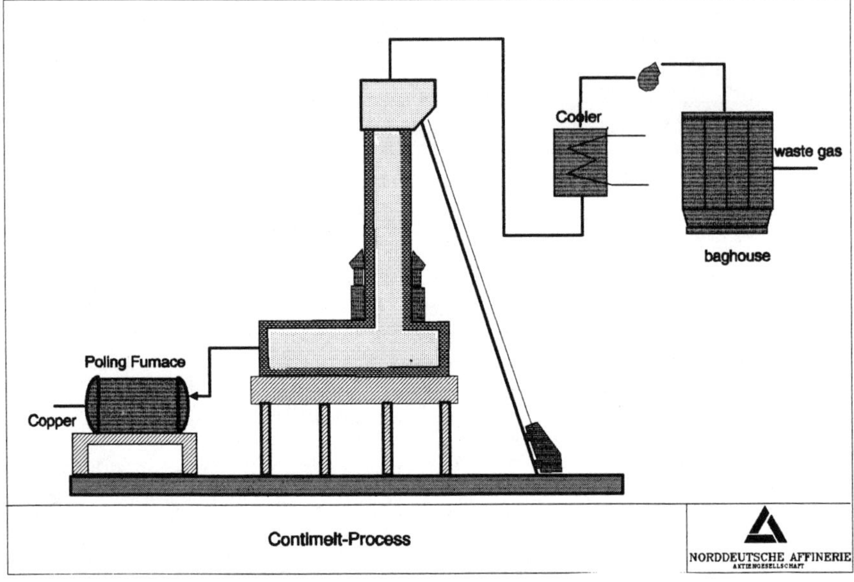

Fig. 8

Off Gas
Reactor
Lances
Oxygen
Reduction Agent
Burner
Oil/Gas
Air/Oxygen
Top Blown Rotary Converter
NORDDEUTSCHE AFFINERIE
AKTIENGESELLSCHAFT

Fig. 9

Off Gas
Feed
Baghouse
Drying Zone
Sintering
and
Reaction Zone
Coolong Water
Air
Smelting Zone
Reaction Products
Blast Furnace
NORDDEUTSCHE AFFINERIE
AKTIENGESELLSCHAFT

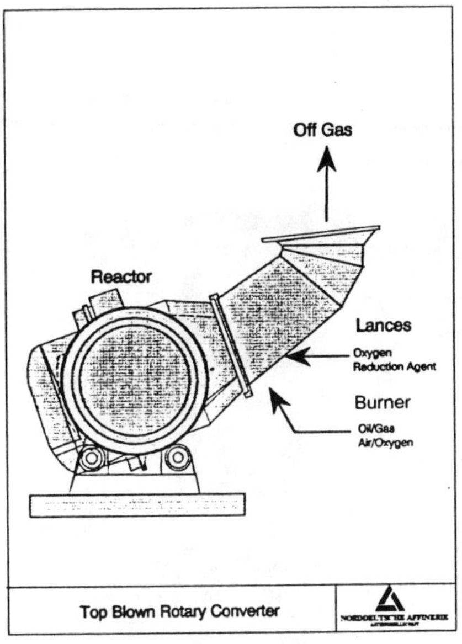

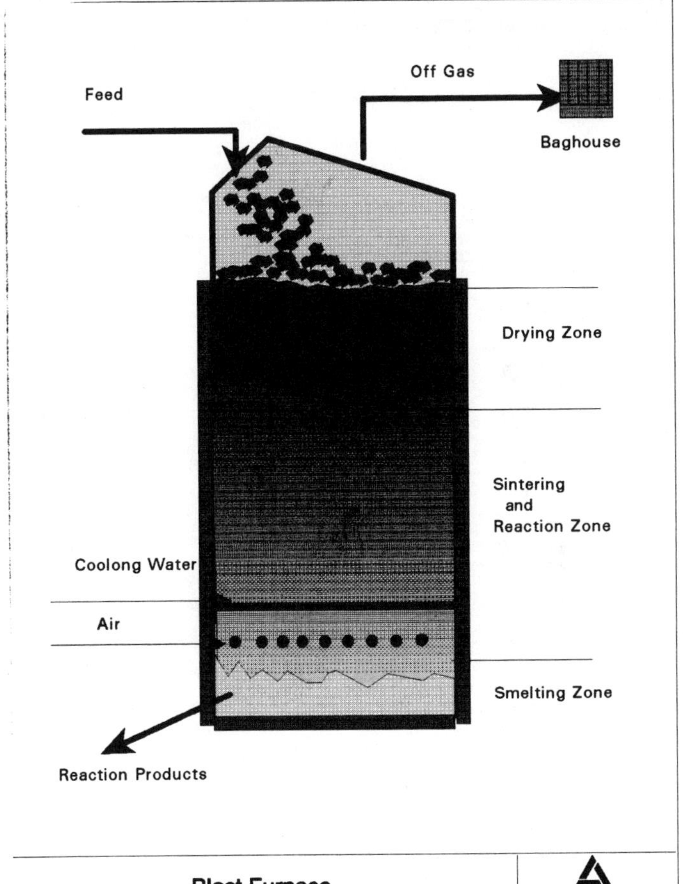

Fig. 10

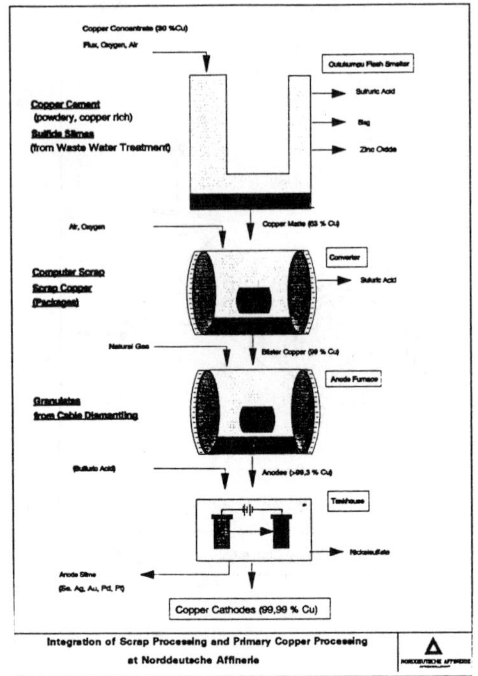

Fig. 12

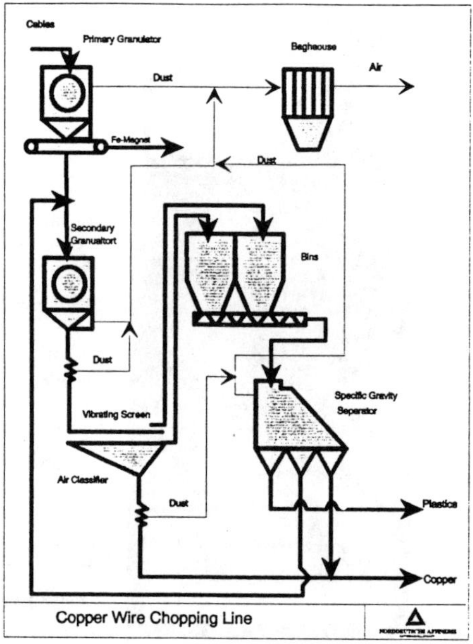

Fig. 14

Organic Material (max. 25 %)
Plastics: Polyethylene, Polypropylene,PVC
              Polyester, Phenol Resins, Polycarbonates
Additives: < 20%
                                              < 5%
Oxidic Material (max. 35%)
Siliciumdioxide   15 %
Aluminiumoxide   6 %
Alkalioxides   6 %
Other Ceramics   3 %
Metals (Max. 50%)
Copper   20 %   Gold   500 g/t
Iron   8 %   Silver   1000 g/t
Nickel   2 %   Palladium   50 g/t
Tin   4 %
Lead   2 %
Aluminium   2 %
Zinc   1 %
Antimony   0,4 %
Composition of Printed Circuit Boards
NORDDEUTSCHE AFFINERIE

Fig. 15

Manual Dismantling
Parts (Cooler, Power Generator, etc.)
Shredder
Waste ("Fluff")
Air Classifier
Fe-Magnet
Non Ferrous Fraction
Iron
Fe-Fraction
Cooled Shredder
Heavy media Separation
Waste
Fe-Magnet
Manual/Automatic Sorting
Al,,Zn,Ni etc
Non Ferrous Fraction
Copper Smelter
Copper/Copper Alloys
Recovery of Copper from Scrap Cars
NORDDEUTSCHE AFFINERIE

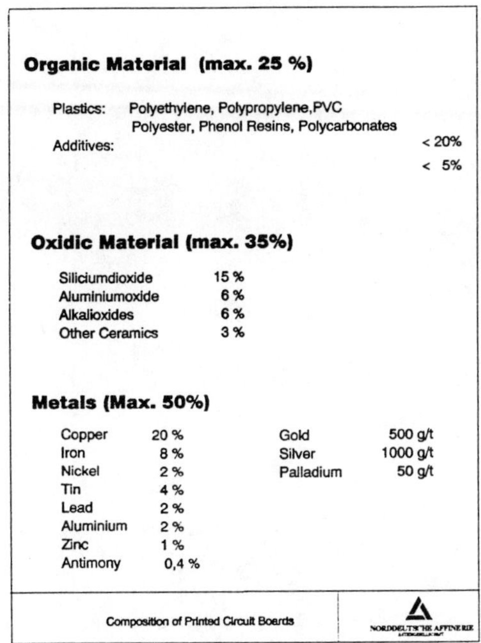

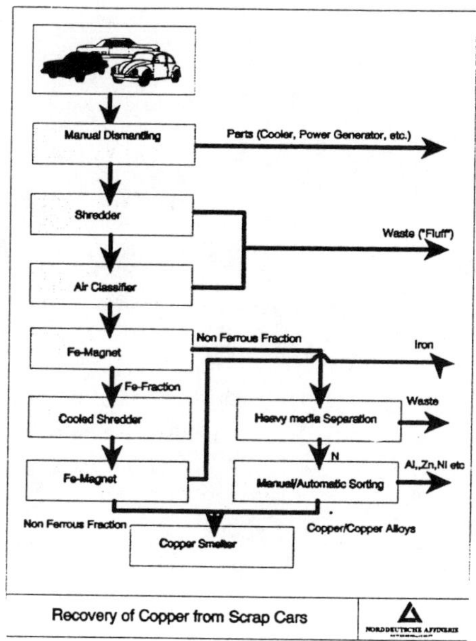

Fig. 16

Global vision 2000 — Decade 90's copper marketing strategies

W.G. Deeks
Noranda Inc., Toronto, Ontario, Canada

ABSTRACT

Copper is in competition in the market place with other metals and materials, a shift in industry and marketing strategies is under way to maintain traditional markets and develop new ones. This address presents a forward-looking vision of copper as a premier global material, building economic value for customers in the supply chain and discusses ideas and strategies the industry can adopt. The vision of copper as a model of product stewardship is presented, with examples given covering all aspects of the processing cycles including production, fabrication and manufacturing, recycling and marketing of copper.

More legroom.
Executive Class.

Classe Affaires.
Plus d'espace
pour vos jambes.

Air Canada

PROLOGUE

As we think about the future, and of copper, we are going to need to:

- **stand tall;**
- **to have flexibility;**
- **to be colourful;**
- **indeed to have lots of leg room as graduates from Executive Class to First Class.**

THE WORLD AHEAD

- ### *A world of real cataclismic change for Metals:*

  - driven by technology;
  - he environment including health, safety and recycling;
  - changing public values;
  - citizen expectations; and perhaps, most important, citizen activism;

- ### *All end use products using Metals are under attack:*

  - issues range from the materials they are made from and how they are sourced;
  - production systems and impact on health, environment and safety;
  - quality;
  - dependability and warranty;
  - health and safety in consumption;
  - recycling and waste management;
  - and more esoteric challenges "do we really need this metal in society?"

THE WORLD AHEAD

- Each Metal is unique in its form and properties but it is usually the <u>co-product </u>of other metals and minerals and the <u>co-component </u>in end use applications with other metals and materials;

- The Metals future is dependent not only on performance criteria and public perception, but on each co-product and co-component; <u>rarely can a metal stand alone, aloof from the issues facing materials and metals,</u> vertically, horizontally and globally;

- <u>Decade 90's is a time of Revolution</u>; the driving force is change, becoming permanent change rather than cyclical change;

- A milestone event will be the <u>United Nations Conference on the Environment and Development,</u>(UNCED), Brazil, June 1992 which will forever change the way business is done.

IN JUNE OF 1992, IN BRAZIL, HERE IS WHAT WILL HAPPEN

The United Nations Conference on Environment and Development (UNCED), June 1-12, 1992, Organized by the United Nations Environment Program (UNEP)

- Heads of Government Conference, 159 States and Observers.
- Secretary General, Maurice Strong, a Canadian.
- Focus is linkage between environment and development.

- ***Objectives:***
  - 4 new Conventions for:
    Climate Change
    Bio-Diversity
    Rain Forests
    Bio-Technology

  - ***An Earth Charter***, a moral global framework on the environment incorporating a developing countries' rights to sustainable development and citizens' rights to clean water, air, and sustainable development.

Expected Output - Agenda 21:

- Environmental action plan for 21st century.
- Coordinated regional and global strategy.
- Forging environment/ development linkage.
- New environmental laws for the planet's ecological balance.

- Plan for drought and decertification.
- Economic and environmental integration.
- Need for new financial resources.
- Environmental technology transfer.
- Human resource development.
- International environmental infrastructure changes.

EACH ONE OF THE AGENDA ITEMS LEADING TO NEW TREATY CONVENTIONS WILL HAVE A DIRECT OR INDIRECT IMPACT ON COPPER

- Climate change.
- Ozone depletion.
- Trans-boundary air pollution.
- Deforestation.
- Decertification, soil loss.
- Biodiversity.
- Biotechnology.
- Marine pollution.
- Living marine resources.
- Fresh water.
- Disposal of toxic and hazardous wastes.
- Poverty and environmental degradation.
- Urban environment.
- Environment and health.
- New and renewable sources of energy.

- Financial resources for environmental protection.
- Technology transfer for environmental protection.
- New international environmental laws.
- New international institutional arrangements.
- Economic incentives for environmental management.
- Environmental education and information.

A REVOLUTIONARY BUSINESS RESPONSE: - THE INTERNATIONAL CHAMBER OF COMMERCE BUSINESS CHARTER FOR SUSTAINABLE DEVELOPMENT

The Business Charter for Sustainable Development of the International Chamber of Commerce as a Committment for the Copper Industry

1. **Corporate Priority**

   To recognize ***environmental management*** as among the highest corporate priorities and as a key determinant to sustainable development; to establish policies, programmes and practices for conducting operations in an environmentally sound manner.

2. **Integrated Management**

   To ***integrate*** these policies, programmes and practices fully into each business as an essential element of management in all its functions.

3. **Process of Improvement**

   To continue to ***improve corporate policies***, programmes and environmental performance, taking into account technical developments, scientific understanding, consumer needs and community expectations, with legal regulations as a starting point; and to apply the same environmental criteria internationally.

4. **Employee Education**

   To ***educate***, train and motivate employees to conduct their activities in an environmentally responsible manner.

5. **Prior Assessment**

   To ***assess environmental impacts*** before starting a new activity or project and before decommissioning a facility or leaving a site.

6. **Products and Services**

   To ***develop and provide products*** or services that have no undue environmental impact and are safe in their intended use, that are efficient in their consumption of energy and natural resources, and ***that can be recycled***, reused, or disposed of safely.

7. **<u>Customer Advice</u>**

To advise, and where relevant **_educate, customers, distributors_** and the public in the safe use, transportation, storage and disposal of products provided; and to apply similar considerations to the provision of services.

8. **<u>Facilities and Operations</u>**

To develop, design and operate facilities and conduct activities taking into consideration the **_efficient use of energy and materials_**, the sustainable use of renewable resources, the minimization of adverse environmental impact and waste generation, and the **_safe and responsible disposal of residual wastes_**.

9. **<u>Research</u>**

To conduct or support **_research on the environmental impacts_** of raw materials, products, processes, emissions and wastes associated with the enterprise and on the means of minimizing such adverse impacts.

10. **<u>Precautionary Approach</u>**

To **_modify the manufacture, marketing_** or use of products or services or the conduct of activities consistent with scientific and technical understanding, to prevent serious or irreversible environmental degradation.

11. **<u>Contractors and Suppliers</u>**

To **_promote_** the adoption of these principles by contractors acting on behalf of the enterprise, encouraging and, where appropriate, requiring improvements in their practices to make them consistent with those of the enterprise, and to encourage the wider adoption of these principles by suppliers.

12. **<u>Emergency Preparedness</u>**

To develop and maintain, where significant hazards exist, **_emergency preparedness plans_** in conjunction with the emergency services, relevant authorities and local community, recognizing potential trans-boundary impacts.

13. **<u>Transfer of Technology</u>**

To contribute to the **_transfer of environmentally sound technology_** and management methods throughout the industrial and public sectors.

14. **<u>Contributing to the Common Effort</u>**

To contribute to the **_development of public policy_** and to business, governmental and intergovernmental programmes and educational initiatives that will enhance environmental awareness and protection.

15. **<u>Openness to Concerns</u>**

To ***<u>foster openness and dialogue with employees</u>*** and the public, anticipating and responding to their concerns about the potential hazards and impacts of operations, products, wastes or services, including those of trans-boundary or global significance.

16. **<u>Compliance and Reporting</u>**

To ***<u>measure environmental performance</u>***; to conduct regular environmental audits and assessments of compliance with company requirements, legal requirements and these principles; and periodically to provide appropriate information to the Board of directors, shareholders, employees, the authorities and the public.

This is the premier voluntary statement of committment for the private sector and is highly relevant to the natural resources industries.

<u>Please</u>:   - *read it;*
           - *consider it;*
           - *sign on!*

It is the path for our economic success for the future.

<u>SO WHAT IS THE ECONOMIC OUTLOOK?</u>

Here is what the OECD Central Planning Bureau believes is the priority model, 1990 - 2015.

Balanced Growth

- Multipolar growth including Latin America and Africa;
- Japan graduates to normal industrial country growth;
- European integration proceeds driven by market forces;
- Slower population growth after 2005;
- Gradual solving of hunger issues;
- Strong technology application;
- Global environmental approach and 50% reduction of energy intensity;
- High rate of globalization of economic and political forces;
- Balance between free market and constructive government economic management;
- Strong economic growth linked to sustainable development and global environmental management.

<u>Comments</u>:

- Note shift in consumption growth to Asia where there is leverage from population mass and economic growth.

- Note expectation of economic progress in Rest of World developing countries.

- Note the lower growth in OECD countries where much of the growth may be driven by exports to developing countries.

- Think about the role of recycling in future copper supply as an industry investment opportunity.

SO WHAT WILL BE THE IMPACT ON THE COPPER PRICE?

Assumptions:
- At the projected percentage of GDP growth, the copper price will remain in *the upper half of the channel*;

- The *copper industry will be proactive in the world of real and permanent change* ahead, finding solutions to its problems and being proactive on new opportunities;

- The copper *price will be at the level needed to sustain citizen's expectations* of copper's value which must and will be aggressively marketed;

- The *copper price will be high enough to sustain and expand supply* as needed including significant commitment to recycling all used and scrap process copper because this will be a value expected by citizens;

- So here is what the historical picture looks like in 1991 dollars:

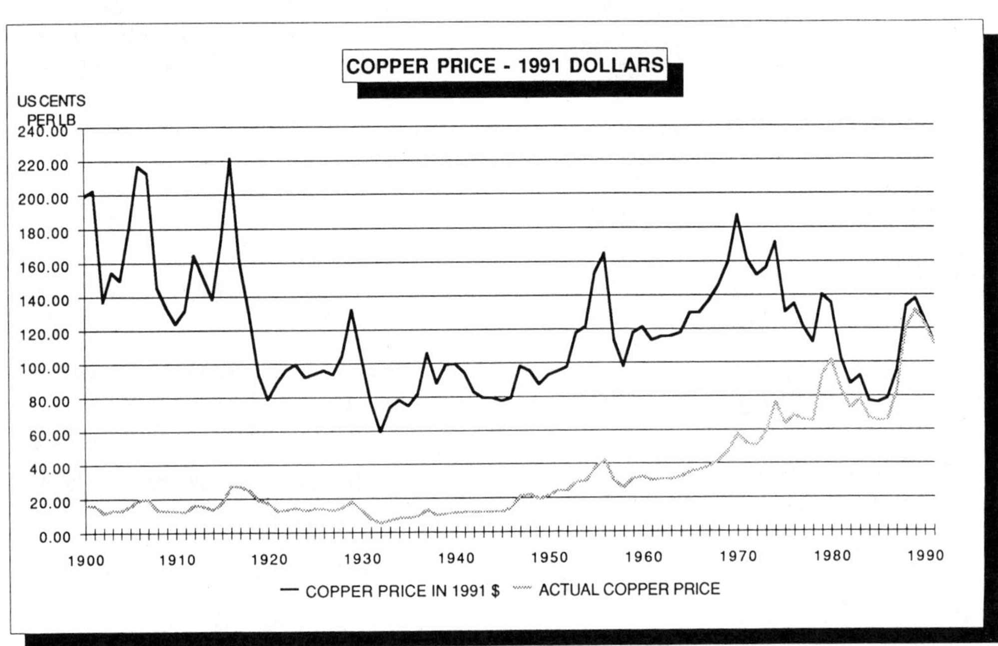

WHAT DOES THIS MEAN FOR FORWARD COPPER PRICES?

- Assume the current channel in 1991 dollars is US$1.40 per lb. to US$0.80 per lb.

- Assume the mid-point of the channel is US$1.10 per lb.

- Then in 1991 dollars the price will average between US$1.10 per lb. and US$1.40 per lb. in the time frame 1990 - 2015.

- This is a reasonable expectation if you buy the assumptions:

    - copper will be proactive on all fronts;
    - copper will be recycled;
    - copper will be environmentally acceptable;
    - copper will engage in aggressive end use marketing;
    - copper will be a leader in product stewardship from exploration to recycling.

HERE ARE THE MARKETS DRIVING COPPERS FUTURE:

- **The transportation industry;**
    - everything that moves;
    - the systems in movement, for cooling, braking, communication, lighting, safety, environment, data management, measuring, signalling;

- **The construction industry:**
    - everything that's built;
    - all the components of construction, materials, communication, heating, cooling, sensing, security, measuring, environment, safety, design, durability, warranty, quality:

- **The communications industry:**
    - everything that transmits;
    - electricity, communication, photography, publication, imaging, recording, electronics, electrical products and components;

- **The agricultural industry:**
    - everything that grows;
    - the materials to nurture and sustain growth and the equipment to facilitate and process agricultural products including fertilizers, fungicides and agricultural equipment.

COMING CHANGES HERE AND NOW!

Now let me walk you through a picture of the changes coming for Copper in its main driving markets. And let me do it by walking you through the February 19, 1991 Society of Automotive Engineers symposium in Detroit, Michigan.

First, let's look at innovation:

- The concept of an experimental racing car as a new materials laboratory innovator;

- The beginnings of a composite materials exhaust free urban transit vehicle;

- The total innovative systems approach to product design and execution;

- A corporate culture, the spirit of innovation;

- The composite materials concept vehicle by Dupont;

- Honda's focus on radical, innovative design with composite materials for motorcycles;

- The innovative flexibility of performance resins and coatings in all products;

- Total free form shapes in innovative designs with thermo plastics;

- Space frame technology, the coming motor vehicle assembly system in engineered glass fibre composite technology;

- A fantastic free-form finished composite materials vehicle;

- The new Lotus Elan on a space frame with a plastic-skinned body;

- Prototyping as a low cost means of seeing the future;

- Even zinc is in the game with an innovative prototype vehicle showing zinc's applications.

All of this is leading to a *technological revolution in materials technology* and competition for end use applications:

- Aluminum is pushing hard especially for cast aluminum engine blocks;

- Zinc is stressing functional applications;

- Zinc is promoting positive public perception;

- Alcan is focusing on a global service network;

- Alcoa is pushing the total aluminum vehicle, the space frame, structural components, and an all aluminum skin;

- Magnesium is a new major player in functional components;

- Thermoplastics have arrived;

- The beauty of a plastic-skinned vehicle is no longer a hypothetical comparison;

- Thermal plastic components are now part of production motor vehicle systems;

- Plastic component systems have arrived as serious engineered products;

- Plastics are everywhere in all of our end use markets;

- Specialty materials engineered for specific applications with optimum properties are the coming rule.

Other characteristics increasingly being sought in products for end use consumption relate to *safety, quality* and the *ability to recycle;*

- A criteria for all new materials in end use products is that they can be recycled;

- Recycling is a measurement of quality and environmental user friendliness;

- Metals are safe and recyclable and we need to publicize this;

- Plastics are concentrating on their ability to be recycled;

- Metal is quality but how often do we say it;

- Quality, dependability, permanence and recyclability will be significant competitive advantages in the future choice of materials;

- Safety is now a measure of quality;

- Safety in motor vehicles, construction, communicating devices, and in agriculture is a high public requirement.

But all of this today is global. Ideas flow without trade restriction and innovations can come from anywhere. Public expectation travels quickly and a process of common international standards of excellence is moving quickly forward. World leadership is a corporate objective in the best companies and world materials leadership is essential for competing materials.

- Success is built on the Toshiba model of safety, upgrading, convenience, comfort, delivered through electronic systems that create satisfaction.

- Materials competition is moving to a world class standard with the optimum technology being the best worldwide. Experimentation, testing, proving, verifying and measuring are all part of the process.

- Motor vehicles are the dynamic application for new technology on mass. Most global enterprises have a component of their innovative focus on the transportation industry. The SAE show is a blue book collection of the world's best in all fields. 3M is such a company.

- Global partnerships are the theme of the future in all industries and a trend you are all observing now is the linking of arms between the Pacific, the Americas and Europe in new synergistic partnerships. The motor vehicle industry is a leader. What about the metals industry?

- Copper was at the SAE show with zinc as brass. Brass used to be the slang for wealth but copper needs to do more than just be the country cousin of zinc at the SAE show or it will not achieve its potential in world economic growth ahead.

CONCLUSION

Copper must launch a global marketing campaign selling its values and building positive public perception as never before. It is individual citizens who will decide whether copper succeeds or fails because of preference for or against it in the products we all buy. A proactive positive marketing initiative by the copper industry collectively and individually is the needed strategy.

Let's march forward together as a global metals industry to compete effectively for our future. And let's support our innovators and protect our messengers!

And let's be expert at what we do. Let's start with our people by being willing to invest in the individual. Let's produce a product to love the world over.

Toyota does. And Toyota is one of our best customers. Toyota uses the world's best copper radiator!

Struggle for competitiveness:
an industry perspective for the nineties

H. Bannach
Codelco-Chile, Chuquicamata Division, Chuquicamata, Chile

ABSTRACT

This presentation focuses on what seem to be the leading indicators of prosperity in our industry during this decade: the geographical distribution of deposits, environmental impact, capital allocation and acid consumption. There is a potential for meeting environmental criteria at a reasonable profit margin, but those of us with limited acid demand within economic range will face hard times to meet growing environmental concerns, resulting in additional mine closures. At the other end, those of us with old waste dumps who are thinking about smelting capacity additions will need capital influx at a much larger rate than ever before. All things considered, we should change the way we compute the world's copper reserves, since there is a growing need to internalize these effects back to the ore blocks before a meaningful estimate of reserves becomes available.

INTRODUCTION

The Copper Industry has undergone several changes during the past decade. Many of those changes were expected to take place since they reflect the economics of aging mines. However, many others took place that shocked the industry because they were not completely under control. The focus of this presentation is to reflect on those changes, and on our market expectations for this decade that somehow depend on their impact.

We will start by reviewing the prospects for new mine developments and expansions from an industry perspective. To do it to a full extent we would need to consider not only the deposits themselves, but also geographical constraints, environmental constraints, labor composition, technology development and use, the availability of capital, ownership issues, and integration within the industry. However, since this agenda would obviously be too demanding, we will concentrate on only some of the key elements that are believed to play a major role in shaping the market in the decade we just begun. Those elements, all interconnected, are:

• the evolving competitiveness of low-grade material amenable to SX-EW practices as opposed to developing new mines,

• the growing environmental awareness of the industry and its impact on investment decisions that affect metallic output and mine closures, and

• the synergistic role of sulfuric acid in bringing those two issues together.

HISTORIC BACKGROUND

To talk about reserves in the industry without talking about costs can be meaningless or at least misleading. Lack of reliable and standardized data sources concerning the evaluation of ore deposits prevent the careful analyst to draw firm conclusions regarding economic deposits. While this feature is common to all mining activities, the case of copper, with reserves in the hands of owners with very different constraints and capabilities since the seventies, is particularly difficult to assess. Recently, an increasingly difficult problem has emerged as an additional confounder, namely, the need to internalize the costs of environmental regulation compliance when calculating the economics of a given deposit.

Our world is increasingly becoming aware of the necessary balance between technological progress and the right of individuals to live in an environment that meets acceptable health criteria. We should therefore expect an increasing pressure stemming from all sources to spend more and more money to keep the environment clean. I am sure we all share that view and would like to make far more progress than has been done to meet higher and higher environmental standards. However, over the last decade many of our mines closed because we were not able to anticipate and properly internalize the costs associated with environmental regulation compliance. As we see this phenomenon take place in other industries as well, it might be that the excuse we had in order to postpone our commitment to a clean environment was that we all felt that since we were all being treated equally, society as a whole would have to end up paying one way or the other. However, where there is potential for diversity - and I argue that is precisely the case - the rule might not apply to all market participants alike. Thus, while some of us might be driven out of business by environmental regulation, some others might even be able to profit from compliance.

We can use some basic figures that have been published elsewhere to make the point. Take, for instance, the outcome of a long standing commitment of the U.S. Bureau of Mines to assess the Western World's copper availability (1). Figure 1 was part of a report they published in 1987. It shows the potential total copper available from market economy countries.

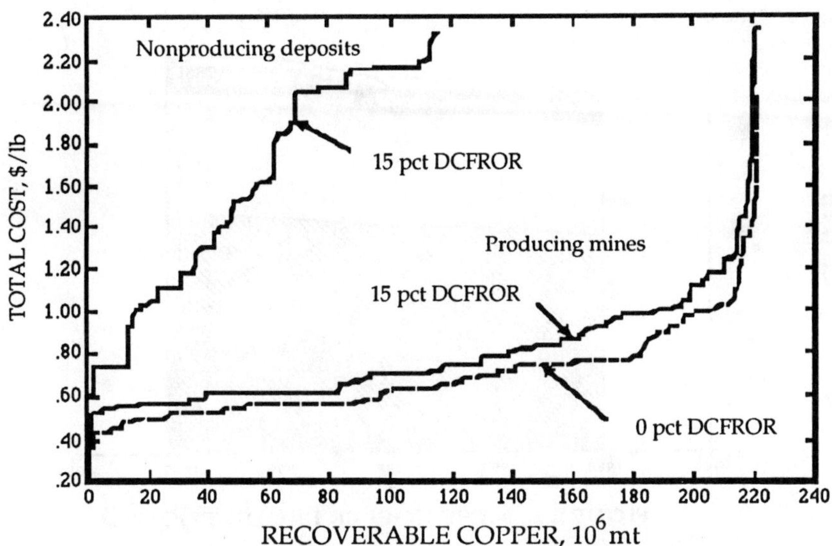

**FIGURE 1: POTENTIAL TOTAL COPPER AVAILABLE
FROM MEC's , IN JAN. 1985 U.S. DOLLARS**

Figure 1 was arrived at after covering with some detail the information about a total of 241 mines (62 in the U.S. and 179 elsewhere). For the moment, overlook the fact that much of that information comes from sources that use non-standard procedures for reserve calculations, since that is a known fact, and one that was acknowledged in that same publication. According to Figure 1, in 1985 dollars the world shouldn't really get much benefit from the longevity of producing mines with cut-off prices at or above the $1.20 level. It also tells us that if the long term price were to remain at that level or higher, new economic deposits at that price level should open (although one can also infer from the graph that in that case capacity additions in excess of capacity retirement would leave the market price at a level below $1.20). We will now advance some thoughts about our key issues that may alter this picture.

TRENDS

To start with, it was precisely during 1985 that 13 of the U.S. mines and 14 other mines closed down. The price of copper that year was around 65 cents a pound, so knowing that the copper market swings, at least a portion of the closed mines were hopeful comebacks after a period of market readjustment. Figure 2 shows the mine copper production schedule for MEC's from 1986 on. When computing the relevant long term costs associated to producing mines, two new elements intervened to complicate our ability to anticipate whether comebacks were really feasible across the industry. One of those elements was that U.S. environmental regulations were being enforced at the very time when the struggle for survival was at its peak, so that the likelihood of making heavy capital commitments in the sector to upgrade smelters was low. The other reason - apparently to the surprise of many - was that capital influx alone would not have been enough: in many cases, by-product sulfuric acid needed a market much closer to smelters than possible.

The net result, after years of adjustment, was that the industry's metallic output ended up depending more heavily on the volatile sulfuric acid market and on transportation costs. If we judge the perception of investors in the sector according to their revealed behavior, we have to admit that there has been a reluctance to invest in smelting capacity beyond reasonable limits. The challenge was and is to invest much more heavily in the industry to stay alive. And indeed, some of us did.

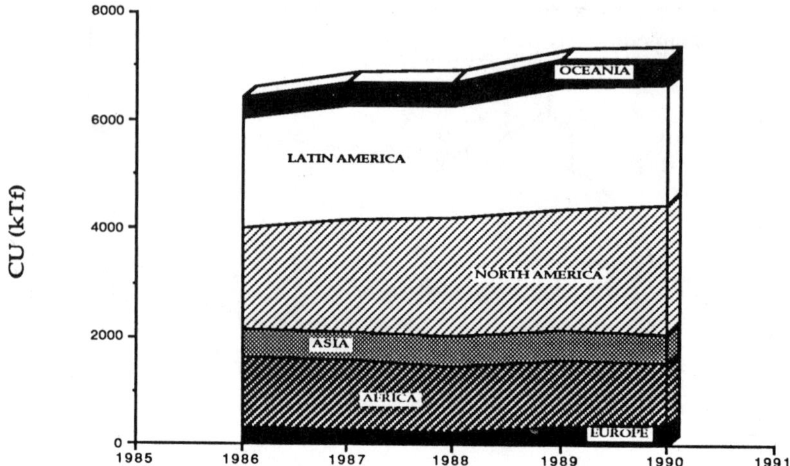

FIGURE 2: MINE COPPER PRODUCTION
MEC's

The shortage of smelting capacity and the surplus of sulfuric acid have driven capital away from environmentally risky smelting capacity investment proposals. This phenomenon has the effect of reducing the chance that a given deposit may become profitable. It also reduces the profitability of current mining operations and produces a net effect that is hard to measure in the industry as a whole, particularly because it is unevenly distributed across producers.

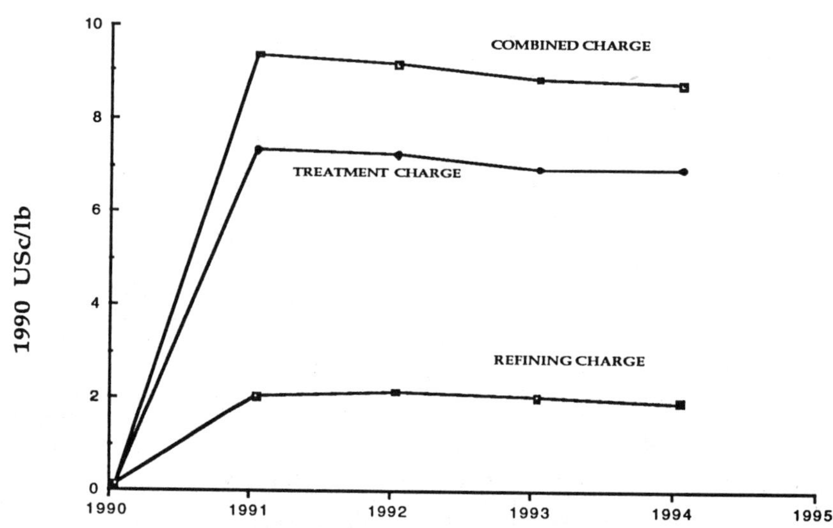

FIGURE 3: PROJECTED DIFFERENTIAL CHARGES

Differentials were calculated with versus without the addition of 200 kt/yr capacity.

Mine production location is at odds with the location of smelting capacity and sometimes with consumer markets. Chile, for instance, has not been blessed with proximity to either one. One might argue that the capital intensiveness of smelting and refining are beyond reach for third world countries in which the economics of producing only concentrates is far better than the marginal contribution of the added value of metallic output. While that has traditionally been the case when the orebody is rich enough and there are plenty of natural untapped resources all available at once, the bonanza for the rest of the industry also appears to be less obvious that one might have thought. In fact, the industry as a whole has dodged installing additional smelting capacity.

The balance looks grim. Treatment charges and refining charges have soared. Clearly, the forecasted addition of smelting capacity proved over-optimistic compared to forecasted concentrate production. The industry should not tolerate concentrate additions that are out-of-phase with adequate smelting capacity: Figure 3 shows the shadow price of the lack of appropriate smelting capacity we were left with after the continuous postponement of a 200 kt/yr smelter (and our estimates show that we need two of those!). In contrast, history tells us that the industry's total smelting capacity and its toll smelting capacity during the past few years remained almost flat at about 8000 ktf/yr and 2500 ktf/yr. respectively.

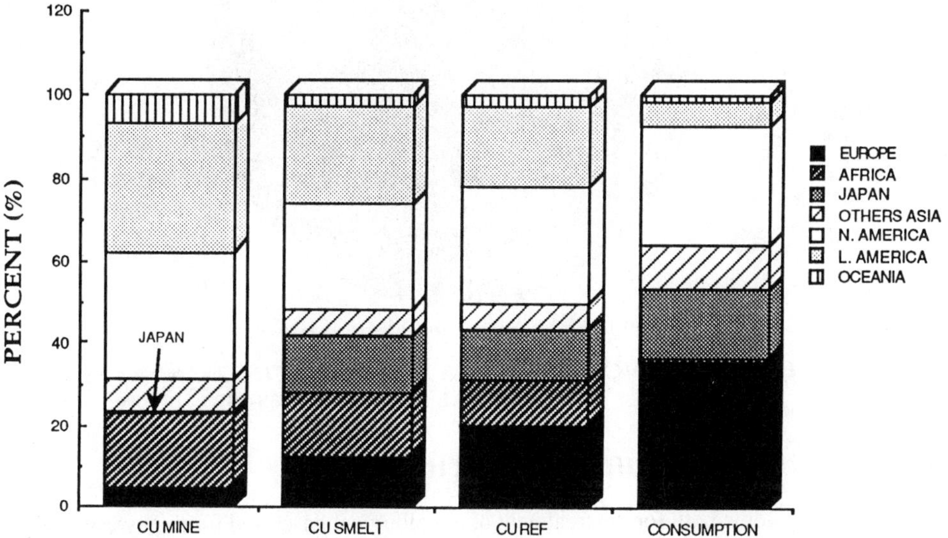

FIGURE 4: COPPER PRODUCTION AND CONSUMPTION
ANNUAL AVERAGE 1985-1990

The location of smelters is another important issue to consider, particularly in the light of a bearish market for sulfuric acid. Figure 4 illustrates the geographic distribution of copper production and consumption. The latter is a valid proxy parameter for smelter-pruduced sulfuric acid consumption in the non-mining sector, since its quality prevents it from being used in the food production chain. It tells us that in order to realize most of the world's copper smelters' sulfuric acid production potential we also need to worry about its logistics. The trouble is that while emerging technologies have enabled the recovery of acid from smelter fumes, distance is still an issue technology has not yet been able to fully help us with. It is common to see smelting capacity additions heavily biased toward countries where acid consumption is more likely or already present, even when stringent environmental regulations might complicate the story as they have, since acid consumers located in regions with high industrial activity also tend to be heavily polluted and do not welcome additional polluters in their neighborhood.

The case of Japan should make us wake-up. Beyond Figure 4, a closer inspection into the available commodity trading information reveals that Japan controls most of the toll smelting capacity as can be seen in Figure 5, a fact that is known to lead to market imperfections. Although reaction from the concentrate producers should not - and probably will not - take long to materialize in the shape of alternative smelting capacity additions of their own or through joint ventures, there is ample room for speculation concerning the costs associated with a concentrate surplus beyond control. All indicates this situation is likely to remain substantially stable at least over the next couple of years, until some of us are no longer able to bear the costs. It is clear that long term contracts subscribed before the TC/RC crisis are paying off.

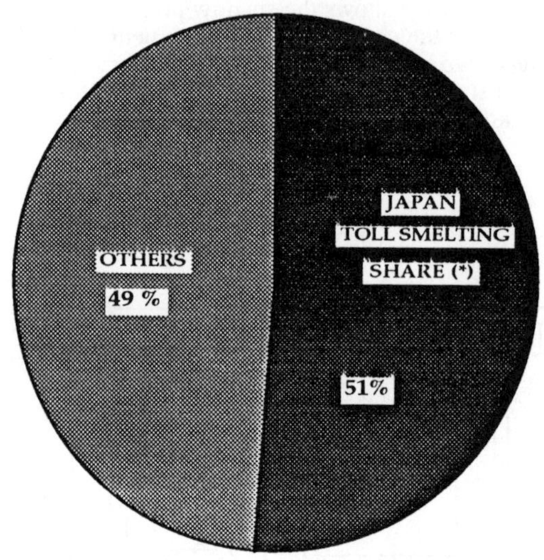

(*): Does not include japanese interests outside Japan

**FIGURE 5: NON INTEGRATED SMELTING CAPACITY
ANNUAL AVERAGE 1985-1990**

THE REBIRTH OF LEACHING OPERATIONS

A solid technology, both for the treatment of metallurgical gases and for the hydrometallurgy of low grade copper-bearing material, has emerged during the last decade as a decisive factor in the market. A 35 year-old technology, SX-EW technology has undergone major revisions that now make it quite attractive. It is now common practice to go out and buy SX-EW plant combinations of all sizes. Chuqui itself has one of the largest ones in operation. Figure 6 illustrates the growth rate of this technology in the last decade compared to the growth rates of refined copper production and mine production. SX-EW plant additions peaked in 1987, but its growth rate remains substantially higher than the average production growth rate in the copper market. Without reservations, this technology is expected to keep giving the copper industry a helping hand in the coming years.

When calculating reserves, then, we should realize that there is a growing amount of leachable material that was once labeled as waste. The term "reserves", as standardized by the US Bureau of Mines (1), applies only to "naturally occurring" material. Thus, old "waste" dumps (i.e., once discarded ore below cut-off grade, crushed leached ore with remaining copper content, and tailings ponds) would not properly qualify as "reserves". However, a good deal of the copper now being produced is coming from those sources and will increasingly do so. One example of that is that the U.S. copper industry is now composed of lower cost producers than a few years ago. Many of those operations are now leaching low grade material using current SX-EW practices.

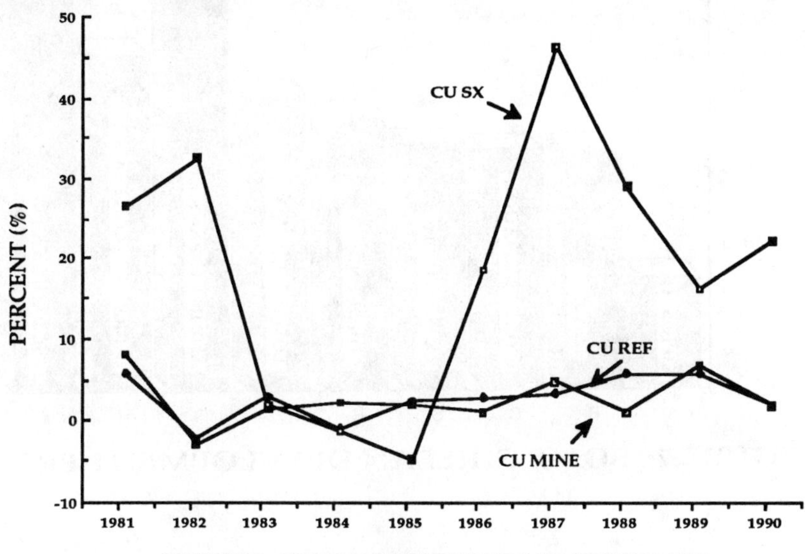

FIGURE 6: COPPER PRODUCTION GROWTH RATE
MEC's

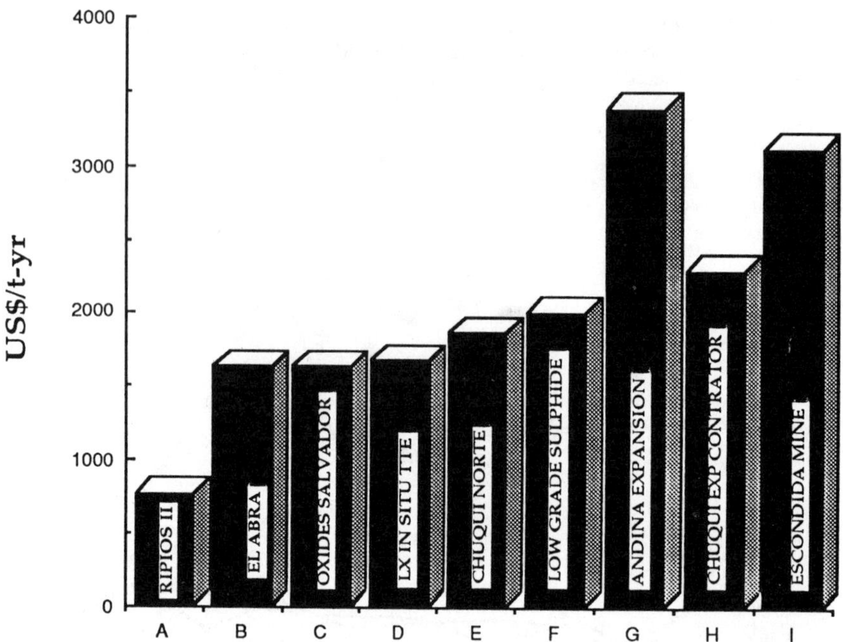

FIGURE 7: SOME CHILEAN DEVELOPMENT PROJECTS

While most of these operations started as side-businesses within their premises, they gradually caught the investor's eye. Figure 7 illustrates that trend as applied to some Chilean projects. In terms of dollars per ton per year, these prospects typically look much better than many other mining alternatives. The first six bars in Figure 7 represent SX-EW combinations. The rest are sulfide projects with investments accounted for only up to concentrate production. If one were to add the necessary capital investment all the way up to metallic output, the advantage of SX-EW projects would become even more obvious.

What are the relevant costs for this new kind of "pseudo-resource"? If we assume acid is available at market prices, then their relevant costs are a function of the market price for sulfuric acid at the plant site. By analogy, the relevant costs for a copper sulfide mining operation should assume credits that reflect the market price for sulfuric acid at the consumer end. We then get an operating cost figure for new orebodies that will consume acid about one third of which could depend on the market price for sulfuric acid. This is yet another reason to compute the world's "reserves" paying closer attention to these issues.

Another issue that is likely to emerge as an important addition to the toolbox of many smelters is by-product recovery from smelter fumes. Gas treatment technology is still under development. While there is a considerable number of appropriate technology alternatives to convert weak gas streams into acid, appropriate technology to enable the economic recovery of valuable condensates is not well spread yet, and less standardized. In retrospect, we probably have to blame it on the lack of time and capital to accommodate the needed research in this area given the short-term demands of environmental regulation compliance. The technology end of the gas handling business is expected to provide us with further credits in the coming years.

With leachable ore in the neighborhood, mining firms have both the potential to still grow profitably and to comply with environmental regulation, but chances are the initial investment will be very substantial.

OWNERSHIP AND THE DECISION TO GO AHEAD

Yet another reason to beware of "reserve" calculations is the fact that we have a diversity of ownership in the industry, making it hard for analysts to forecast the behavior of the different players given demand fluctuations and investment opportunities. Risk perceptions and risk premiums are unevenly distributed among players, with some players able to spread the political risk across various countries and some others constrained to invest in just their own country. The consortia approach advocated by Mikesell (2), blending government's political commitment, technology, capital, and consumption of the commodity seems appropriate, but in the light of this analysis, it might need the involvement of yet another player: the acid consumer or broker.

By way of example and adding things up, it turns out that with appropriate capital additions, the copper production basket from CODELCO might take a shape close to that indicated in Figure 8. From that figure, there is a clear marriage between environmental regulation compliance and the availability of leachable material as suggested here. In many cases, the decision to go ahead will need the commitment of many different agendas.

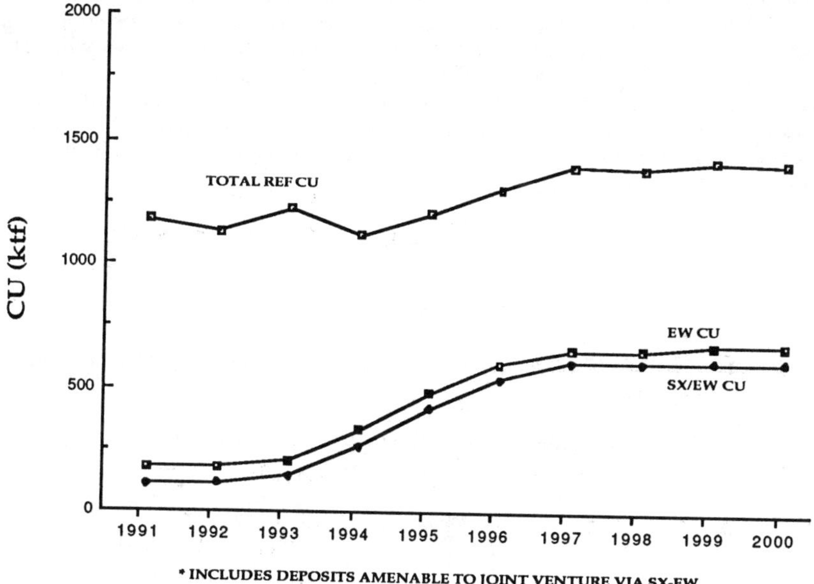

FIGURE 8: CODELCO COPPER PRODUCTION *

1991-2000

The Eastern Block is emerging as a potential open participant in world copper trade. While the economies of the East have been participating in the market indirectly, the heavy commitment to environmental issues in the Eastern Block seem long due and will probably hinder their ability to become competitive producers in the long run, at least not before heavy capital allocation in the sector becomes a reality. Much of its characteristics are still obscure to a western observer, and they will probably remain that way for at least a couple of years until a reliable assessment of their industrial shortcomings and capabilities becomes available. If the trend continues, though, we will probably see an emerging competitor ready to make compromises in terms of ownership. Alas, we should probably expect this last phase to take us close to the end of this decade.

FLEXIBILITY AND REACTION TIME. THE ABILITY TO ANTICIPATE.

The use of technology-intensive exploration methods in recent years is making significant progress in helping us discover and assess new deposits of high grade or extending our knowledge about known deposits. Computer technology has also made significant progress in helping miners develop and use otherwise lengthy calculations. Much of what used to take forever to calculate or program now only takes a few minutes of cheap computer time, which means we no longer have the excuse of calculation lead times to flexibilize our operation and to respond to market conditions. Also, more sophisticated means to calculate the economics of current and potential mining operations are now possible. This is a challenge some of us are currently managing successfully. Under this circumstances, however, a new issue emerges, that of keeping our human resources in pace with technological advances. It calls for a less labor intensive operation, but it also calls for a more sophisticated team.

In addressing all these issues, it is clear that we need to become much more aggressive in the market as an industry that hopes to stay competitive. One good example where we are now spending more than ever is the collective effort of copper producers to promote the uses and advantages of copper in a coordinated manner, and sharing the costs of doing so. Much more of this kind of synergism should take place during this decade.

Finally, let us go back to the basics. We need to internalize all of the above back into the ore block. This approach should result in a more realistic way of appraising our natural resources. One by one, all of us will someday appreciate the need to become actively involved in shaping the copper market, not just reacting to changes that take place or assuming things will not change regardless of our actions. This is an industry that struggles to stay alive in this competitive decade. It may succeed and make us and our constituency better off, or it may fail and see many of us regret the lack of integration of efforts in the past. It takes all of us to make the best out of this decade. We need to improve our flexibility, to anticipate change, and to react much more quickly to it than we have in the past.

REFERENCES

1.   U.S. Bureau of Mines and U.S. Geological Survey, "Resource Reserve Definitions", Geological Survey Circular 831, 1980.

2.   R. Mikesell, <u>Foreign Investment in Mining Projects</u>, Oelgeschlager, Gunn & Hain, Publishers, Inc., Cambridge, Massachusetts.

Economics

Metal recycling and its impact on the demand for metallic ores: the case of copper

R.G. Eggert
Mineral Economics Department, Colorado School of Mines, Golden, Colorado, U.S.A.

ABSTRACT

Secondary copper competes with primary ores and metal to satisfy demand for copper in final products. Thus, future demand for primary copper ores and metal will be shaped by developments in scrap recovery and use. This paper identifies and evaluates five factors that will define scrap's impact on primary producers: (1) the growth rate of demand for copper in final products, (2) the reservoir of copper in old scrap, (3) the recovery rate of old scrap, (4) the generation rate of new scrap, and (5) the recovery rate of new scrap.

INTRODUCTION

Mineral ores and scrap substitute for one another as inputs to the production of metals. To be sure, the degree of substitutability varies from metal to metal. In many cases, for example, impurities or alloys in the scrap limit the amount of scrap that can be combined with other raw materials to produce refined or semifabricated metal. Nevertheless, recycled metal is an important component of the supply of most major metals, and thus future demand for mineral ores will be shaped by trends in scrap metal recovery and use.

The future course of scrap metal recovery and use, however, is uncertain. In the case of copper, growing public sentiment in favor of increased recycling as a partial solution to waste-disposal problems suggests that national governments could be pushed to implement policies aimed at increasing the extent of metal recycling. On the other hand, certain environmental legislation and regulations are raising the costs of, and thus discouraging, recycling. In the international arena, the Basel Convention on the Control of Transboundary Movement of Hazardous and Other Wastes, depending on how it is interpreted and implemented, could restrict international scrap trade and in turn reduce the availability of scrap for some processors. Finally, technological improvements in equipment and design are reducing the amount of scrap generated in the manufacture of semifabricated and fabricated products.

How will these factors affect demand for primary copper ores and metal? A preliminary assessment of this question is the focus of this paper. The purpose is not to generate quantitative forecasts but rather to identify the major factors that will determine scrap's impact on demand for primary ores and metal.

SCRAP'S ROLE IN COPPER SUPPLY

It is important at the outset to distinguish between two types of metal scrap. The first, old scrap, is metal contained in worn-out or obsolete products (e.g., copper in junked automobile radiators and dismantled telephone cables). Over the last three decades, copper from old scrap has supplied 20-25 percent of the apparent consumption of copper in the United States (Figure 1).[1] Copper from old scrap represents a net addition to copper supply because each unit of copper from old scrap reduces the amount of copper needed from primary ores.

The second type of scrap is referred to as new scrap, generated during the manufacture of new metal products; the skeletons remaining after a part has been stamped from strip are a good example. Some new scrap is recycled immediately in the plant that generated it and is known as home or run-around scrap; such scrap is not generally recorded in scrap statistics. Other

---

[1]Apparent consumption is defined as: (primary production of refined copper) + (consumption of old scrap) + (imports of refined copper) - (exports of refined copper), adjusted up or down for changes in stocks of refined copper.

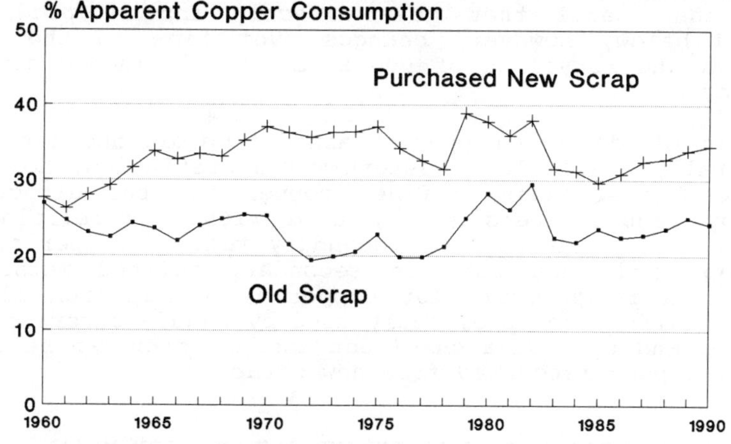

Figure 1 - US Copper Scrap Use as a Share of Apparent Copper Consumption

Sources: References 3 and 11.

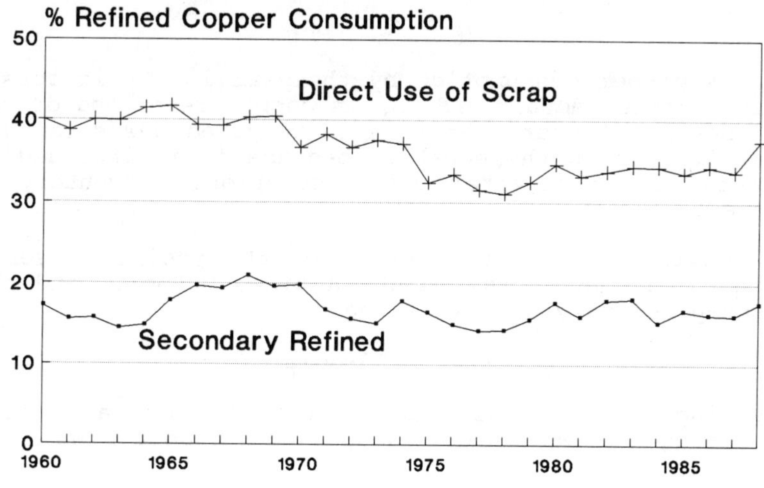

Figure 2 - Western World Use of Copper Scrap as a Share of Refined Consumption

Source: Reference 11.

new scrap is sold before being recycled and is known as purchased new scrap. Statistics indicate that about 30-40 percent more copper is obtained from purchased new scrap than from old scrap (Figure 1).

New scrap, in contrast to old scrap, does not represent a net addition to supply, or alternatively does not lead to reduced demand for primary ores, as long as new scrap is recycled in the same period as it is generated. New scrap is simply metal that takes somewhat longer to be converted from raw ore into a final product than metal that goes directly into a product. As discussed below, however, changes over time in the rates of generation and recycling of new scrap can influence demands for mineral ores.

Worldwide data on copper obtained from old and new scrap are not generally available. Western-world statistics, however, are available for secondary refined copper and the direct use of copper by manufacturers (i.e., used without rerefining)[2]; see Figure 2. More than half of secondary refined copper comes from old scrap, and thus data on secondary refined metal can be considered a rough proxy for copper recovered from old scrap. More than half of scrap directly used by manufacturers comes from new scrap, and thus data on direct use of scrap can be used as a proxy for copper recovered from new scrap.

SCRAP'S IMPACT ON DEMAND FOR PRIMARY METAL

A Simple Model
In many uses, copper recovered from both old and new scrap can be considered a perfect substitute for primary metal. Thus, as a first approximation, the amount of copper used by semifabricating and fabricating mills in period t can be defined as:

$$D_t = P_t + OS_t \tag{1}$$

where D_t is copper consumption by these mills, P_t is consumption of primary refined metal, and OS_t is copper rerefined or remelted from old scrap. Copper contained in recycled new scrap does not appear explicitly in the equation because it is included already in the data for primary refined consumption. Inventory changes are ignored.

Now define OS_t as the product of the pool or reservoir of copper in old scrap at any point in time, $OSPool_t$, and the rate of old scrap recovery, r_t, yielding

$$D_t = P_t + r_t OSPool_t \tag{2}$$

Assuming that old scrap is cheaper than ore as a source of copper-bearing raw material,[3] demand for copper ores is simply

---

[2]Western world is defined as all countries other than the USSR, China, North Korea, Vietnam, Cuba, and the eastern European countries.
[3]Certainly not all copper contained in old scrap has collection and processing costs less than those of primary copper

the difference between total consumption and old-scrap recovery, or

$$P_t = D_t - r_t OSPool_t \tag{3}$$

Add New Scrap

For every pound of copper purchased by semifabricators and fabricators, a certain percentage turns into new scrap; call this the generation rate of new scrap, a_t. A new-scrap recovery rate, b_t, can be similarly defined as the percentage of scrap generated that is actually recycled. If the recovery rate is 100 percent (i.e., all new scrap generated is recycled), then changes in the generation rate have no impact on demand for primary metal and ores. But as long as some new scrap is not recycled, changes in both the new-scrap generation and recovery rates influence the demand for primary metal and ores. In this case, the model needs to be modified. Let

$$P_t = (1-a_t)P_t + b_t a_t P_t + (1-b_t)a_t P_t \tag{4}$$

On the right-hand side of the equation, the first term equals the amount of primary metal making its way directly into copper-containing products. The second term is the amount of new scrap that is generated and then recycled into a copper-containing product, while the third term is the amount of unrecycled new scrap. Substituting equation (4) into equation (2) yields

$$D_t = (1-a_t)P_t + b_t a_t P_t + (1-b_t)a_t P_t + r_t OSPool_t \tag{5}$$

From this equation it is easy to demonstrate that changes in the rates of new-scrap generation, a_t, and recovery, b_t, alter the demand for primary metal, P_t. Let FD_t be the demand for copper in final products. In this case,

$$FD_t = (1-a_t)P_t + b_t a_t P_t + r_t OSPool_t \tag{6}$$

Demand for copper in final products is satisfied by primary metal that goes directly into products (the first term on the right-hand side), metal from recycled new scrap (the second term), and metal from old scrap (the final term). Rearranging again, assuming that scrap metal is cheaper than primary metal,[4] yields the residual demand for primary metals and ores, or

---

production. If this were true, then the recovery rate of copper from old scrap would be 100 percent. A range of recycling costs exist. Low-cost copper from old scrap has few impurities (minimizing processing costs) and is located in large quantities close to processing facilities (reducing transport costs), whereas high-cost old scrap has a larger number of unwanted constituents and often is located far from processing facilities. Nevertheless, large quantities of copper from old scrap are available at costs significantly below those of primary production.

[4]Again, the assumption that recovering copper from scrap is cheaper than primary production in not entirely correct, or else the recovery rate of copper from scrap would be 100 percent. But, as was the case with old scrap, a large amount of copper is recoverable from new scrap at costs lower than those of primary production. An important source of costs savings is in the area

$$P_t = (FD_t - r_t OSPool_t)/(1-a_t+a_t b_t) \tag{7}$$

Assuming that demand for copper in final products is fixed, a decrease in the rate of new-scrap generation results in less demand for primary metal, except in the extreme case when all new scrap generated is recycled. As expected, an increase in the rate of new-scrap recovery, reduces demand for primary metal, and vice versa.

THE CASE OF COPPER

This admittedly simple analysis indicates that five factors will determine scrap's impact on the demand for primary copper metal and ores: (1) demand for copper in final products, (2) the pool or reservoir of copper in old scrap, (3) the recovery rate of copper from old scrap, (4) the generation rate of new scrap, and (5) the recovery rate of copper from new scrap. What can be said about likely future developments of these five factors?

Demand for copper in final products will be governed by a wide range of factors including macroeconomic growth, copper prices, prices of materials that compete with copper such as those of certain plastics, and manufacturing technologies through their impact on the production costs associated with using copper in a particular application. From the perspective of scrap's impact on demand for primary metal, the most important point to note here is that the faster the rate of final-demand growth, the less important scrap supply will be as an input to the manufacture of metal-containing products; scrap's percentage share of copper supply is constrained by the pool of metal in old scrap, which is limited by the level of past metal consumption. In other words, the faster final demand grows, the smaller the pool of metal in old scrap will be relative to the amount of metal demanded for final products. Over the last three decades, world consumption of refined copper grew at an annual average rate of some 5 percent per year between 1960 and 1973, and 2-3 percent per year since then, with significant cyclical fluctuations around these trends (2). For the future, the U.S. Bureau of Mines projects an annual world growth rate for refined copper consumption of 2.7 percent (3).

The second and third factors, the reservoir of copper in old scrap and the associated recovery rate, together define how much old scrap is recycled in a particular time period. The pool of copper in old scrap at any point in time is a function of past production of copper-containing products and the durability (or lifetimes) of these products, minus the amount of old-copper scrap recycled in the past. The worldwide reservoir of copper that is recoverable from abandoned items and products in use is estimated at more than 173 million tons, of which 66 million tons is in the United States (3); the worldwide figure is equivalent to approximately fifteen years of refined copper consumption at 1990 levels. The durability or lifetime of the average copper-

---

of energy. Kaplan and Ness (1) estimate that copper recovered from scrap, both old and new, requires only 3-40 percent of the energy typically needed to produce copper from ore.

containing product has been estimated at about 20 years (3), although the range of lifetimes varies from less than one year for cartridge brass to more the forty years for electrical wiring and copper tubing (4, 5).

There are no good estimates of the annual old-scrap recovery rate, the third factor. But Radetzki and van Duyne (6) estimate a closely related figure, the percentage of copper in old scrap that is ultimately recovered, as 0.65 (i.e., 65 percent of copper in old scrap is eventually recovered). These annual and ultimate recovery rates both reflect the profitability of old-scrap recycling, which in turn is a function of scrap-metal prices and the costs of scrap collection and processing. Higher prices (or lower costs) for old scrap justify collecting and processing lower-quality scrap than would be profitable at lower prices (or higher costs). The price elasticity of old-scrap supply is estimated to range from 0.2 to 0.4 (5, 7, 8), indicating that a one percent increase in old-scrap prices brings forth an increase in old-scrap supplies of 0.2-0.4 percent, holding other determinants of supply constant.

Governments may play a key role in determining future costs of old-scrap supply. At the same time that many people are calling for increased recycling to help solve the problem of bulging landfills, some environmental regulations are making it more difficult and costly to recycle metal. Chopping obsolete cable and wire, for example, generates not only recoverable copper but also tailings of plastic fluff, which typically contain lead in the form of polyvinyl chloride. In the United States, this fluff could be classified as hazardous waste under the recently released Toxicity Characteristic Leaching Procedure of the Environmental Protection Agency. As a hazardous waste, this fluff needs to be disposed of in hazardous waste dumps rather than in the much less expensive ordinary landfills (alternatively, the fluff could be recycled, but the necessary technologies are not yet perfected) (9).

The fourth and fifth factors are the generation and recovery rates of new scrap. New-scrap generation is governed by manufacturing technologies. Actual rates of generation vary significantly from case to case. Up to 65 percent of the brass feed may be generated as scrap turnings from brass screw machines (1). For most other processes -- such as in stamping plants and wire mills -- generation rates are probably much lower, on the order of 20-40 percent. Over time, incentives for cost reduction tend to lead to technical advances that lower rates of new-scrap generation. In recent years, for example, the combination of greater use of continuous casting techniques and improved engineering design have reduced scrap generation rates.

Rates of new-scrap recovery, on the other hand, tend to rise over time, again because of technological improvements prompted at least in part by desires for cost minimization. Recovery rates for new scrap are already high, probably on the order of 90 percent because, compared to old scrap, collection and processing are relatively easy; new scrap is relatively clean, free of impurities, and often located close to processing facilities. As a result, new-scrap supply is less responsive to changes in scrap prices than old scrap. In other words, new-scrap supply is relatively insensitive to changes in scrap prices because most new scrap is recycled already. The quantity of copper recovered from new scrap is determined largely by the level of semifabrication and fabrication that generates new scrap.

As long as new-scrap recovery rates remain high, any changes in generation rates will have relatively little impact on demand for ores. The intuition is that as long as new-scrap recovery rates are high, nearly all of the metal contained in metallic raw materials makes its way into final products. In this case, reducing the rate of scrap generation simply eliminates the need for recycling but has little impact on the amount of copper appearing in final products. To be sure, technological improvements leading to the generation of less scrap may have important implications for the efficiency of the copper industry as a whole (i.e., technical change resulting in lower production costs).

Basel Convention
International trade in copper scrap, although small relative to total scrap use, is not inconsequential. For the United States in 1989, for example, net exports of copper in unalloyed-copper and copper-alloy scrap were equivalent to approximately 15 percent of copper recovered from both old and purchased new scrap (3). The Basel Convention on the Control of Transboundary Movement of Hazardous and Other Wastes could restrict trade in copper scrap and, as a consequence, have a positive impact on demand for primary copper ores and metal by reducing the recycling rates of both old and new scrap.

Negotiated in 1989, the Convention (a) controls international trade in hazardous and other wastes between countries that sign the Convention and (b) prohibits such trade between countries that have ratified the Convention and those that have not. In general, the definition of waste applies not only to materials being shipped for final disposal but also to materials to be recycled, such as much metallic scrap. International trade of hazardous wastes between signers of the Convention would require an agreement between the exporting and importing countries, and that the material be shipped in the most environmentally sound manner. The Convention will not become effective until 20 countries have ratified it. As of early 1991, 50 countries had signed the Convention, while only six had ratified it; but the Convention may become effective as early as 1992 (10).

It is too early to tell what impact the Convention, if it becomes effective, will have on international trade in nonferrous metal scrap. Each ratifying country will develop its own rules for implementing the Convention, and different countries undoubtedly will have different regulatory schemes. An especially important issue is which specific materials will be classified as hazardous wastes under the broad definitions in the Convention.

FINAL THOUGHTS

Several lessons emerge from this preliminary, largely qualitative analysis of scrap's impact on future demands for primary ores and metal. First, attention should be focused on old scrap rather than new scrap. Copper recovered from new scrap would only take away a significant amount of the market for primary copper if new-scrap recovery rates were low and increasing (that is, if the percentage of new scrap generated that was actually recycled was low but rising). But most new scrap is recycled already, making it impossible for new-scrap recovery rates to increase significantly. Consequently, reductions in rates of new-scrap generation as a result of continuous casting and improvements in engineering design simply eliminate recycling of some material that would make it into final products in any case. One exception to this lesson would be if the Basel Convention or other policies led to reduced rates of recycling new scrap.

Second, environmental policies are likely to play an important role in determining scrap recycling (or recovery) rates, particularly for old scrap, which is more likely than new scrap to contain unwanted elements or materials that may be the object of policies. Whether government policies on balance lead to higher or lower recycling rates, however, is unclear. On the one hand, public pressures are growing for governments to do something to encourage greater recycling. Moving in the opposite direction, on the other hand, are some environmental policies that, perhaps unintentionally, increase the costs and thus discourage recycling.

Finally, the growth rate of copper demand in final products will be a quiet yet important determinant of scrap's impact on the demand for primary copper ores and metal. Scrap's contribution to copper supply ultimately is limited by the amount of copper contained in obsolete products. The larger the growth rate of final demand for copper, the smaller the pool of copper in old scrap is relative to copper demand, and vice versa.

REFERENCES

1.   R. S. Kaplan, and H. Ness, "Recycling of Metals," _Conservation and Recycling_, Vol. 10, No. 1, 1987, 1-13.

2.   Metalgesellschaft data, cited in J. E. Tilton, Ed., _World Metal Demand: Trends and Prospects_, Resources for the Future, Washington, D.C., 1990.

3.   J. L. W. Jolly, and D. L. Edelstein, _Copper Minerals Yearbook--1989_, U.S. Department of the Interior, Bureau of Mines, Washington, D.C., 1991.

4.   _Recovery of Secondary Copper and Zinc in the United States_, Bureau of Mines Information Circular 8622, U.S. Department of the Interior, Bureau of Mines, Washington, D.C., 1974.

5.   E. S. Bonczar, and J. E. Tilton, _An Economic Analysis of the Economic Determinants of Metal Recycling in the United_

<u>States: A Case Study of Secondary Copper</u>, Bureau of Mines Open File Report 79-75, U.S. Department of the Interior, Bureau of Mines, Washington, D.C., 1975 (reproduced by the National Technical Information Service). Bonczar and Tilton cite a 1972 report of the Batelle Memorial Institute.

6.   M. Radetzki, and C. van Duyne, "The Demand for Scrap and Primary Metal Ores After a Decline in Secular Growth," <u>Canadian Journal of Economics</u>, Vol. 18, No. 3, 1985, 434-449.

7.   F. M. Fisher, P. H. Cootner, and N. M. Baily, "An Econometric Model of the World Copper Industry," <u>Bell Journal of Economics and Management Science</u>, Vol. 3, No. 2, 1972, 568-609.

8.   M. E. Slade, "An Econometric Model of the U.S. Secondary Copper Industry: Recycling Versus Disposal," <u>Journal of Environmental Economics and Management</u>, Vol. 7, 1980, 123-141.

9.   K. Kiser, "Seeking Wire Chopping Solutions," <u>Scrap Processing and Recycling</u>, Vol. 48, No. 1, 1991, 81-88.

10.   A. McElwaine, "Basel Convention: Threatening International Trade," <u>Scrap Processing and Recycling</u>, Vol. 48, No. 3, 1991, 77-81.

11.   <u>Metallstatistik</u>, Metallgesellschaft AG, Frankfurt am Main, various years.

ACKNOWLEDGMENTS

The research was performed under the sponsorship of the U.S. Department of the Interior, Bureau of Mines Distinguished Young Scholar Award administered by Oak Ridge Associated Universities for the Bureau of Mines.

Beyond 1996, the world copper industry

J.L.W. Jolly and D. Edelstein
U.S. Bureau of Mines, Branch of Non-ferrous Metals, Washington, D.C., U.S.A.

ABSTRACT

While supply surpluses may be a near-term concern for the world's copper producers, attention should be directed toward long-term supply needs. Available world copper supplies from current operations and planned expansions are sufficient to meet the demand forecast through 1996. There is a potential for a large surplus of copper in the years 1992 through 1996, coinciding with the initial production of several large projects. Thereafter, assuming normal economic activity, a sizeable supply deficit could occur to the year 2000. The ability to sustain or expand copper output over the long term will depend upon the industry's willingness to fund long-term mine development, mining research, and metal exploration programs despite near-term economic downturns.

COPPER SUPPLY

PRIMARY SUPPLY SOURCES

Production. - In 1990, 9 million metric tons of copper was mined in 54 countries with the top 11 producing nations accounting for over 80% of the world production. About 7 million tons was mined in the Market Economy Countries (MEC) with the largest contributors, Chile and the United States, comprising 44% of MEC production. Based upon optimistic capacity projections, shown in Table 1, world production, at 85% of this capacity, could exceed 11 million tons by the year 2000, with Chile and the United States contributing 21% and 17%, respectively. As shown in figure 1, most of the anticipated increases in mine capacity will be in South and North America, with Chile being the most important source. In Chile, several new projects are slated for development over this period, as well as anticipated mine expansions at mines such as Escondida, which may be expanded to nearly twice its original size. In the United States, expectations are that only a few new mines will be brought on stream within this period. Several in Montana and Wisconsin are under consideration. In Canada, several projects are projected to come onstream, although some, such as Windy Craggy, could be delayed or even halted by strong environmental objections. Australia, Papua New Guinea and Indonesia also are expected to markedly increase their respective shares of world mine capacity. Smaller production shares are indicated than at present for Canada, Peru, Zaire, and Zambia. Most capacity in 2000 will come from mines now operating or being developed. However, the forecast shown in Table 1 is optimistic in that about 11% of the

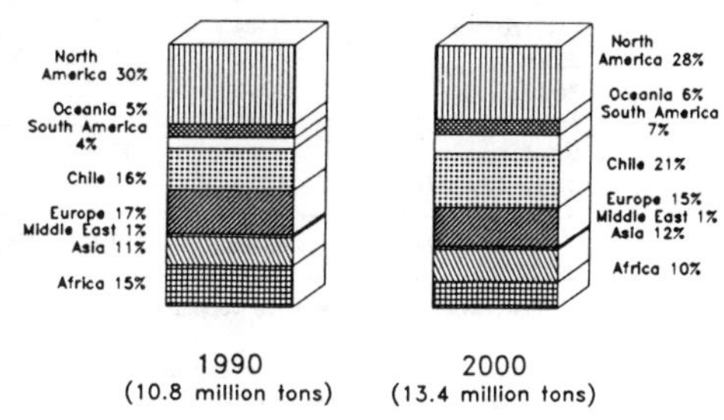

Fig. 1. WORLD COPPER MINE CAPACITY
Percent of World, by Country or Area

Table 1. World Copper Mine Capacity Trends, 1992-2000
(Thousand Metric Tons, Recoverable Copper)

Area and Type of Mine Capacity	Capacity at Operating Mines			Capacity at Developing Mines			Potential Capacity In Exploration			Capacity at Mines Closed in 1991			Total Capacity by Region		
	1992	1996	2000	1992	1996	2000	1992	1996	2000	1992	1996	2000	1992	1996	2000
Africa															
SX-EW Capacity	473	473	453	5	5	5	0	5	5	0	0	0	478	483	463
Other Capacity	1,160	957	731	15	44	44	0	2	2	25	83	93	1,200	1,086	870
Total	1,633	1,430	1,184	20	49	49	0	7	7	25	83	93	1,678	1,569	1,333
Asia/Middle East															
SX-EW Capacity	0	0	0	0	0	0	0	0	0	0	0	0	0	0	0
Other Capacity	1,420	1,493	1,568	2	117	117	0	20	92	0	0	0	1,422	1,630	1,777
Total	1,420	1,493	1,568	2	117	117	0	20	92	0	0	0	1,422	1,630	1,777
Europe															
SX-EW Capacity	1	1	1	0	0	0	0	0	0	0	0	0	1	1	1
Other Capacity	1,869	1,970	1,967	16	19	19	0	20	35	0	1	1	1,885	2,009	2,021
Total	1,870	1,971	1,968	16	19	19	0	20	35	0	1	1	1,886	2,010	2,022
Oceania															
SX-EW Capacity	3	3	3	0	0	0	0	2	2	0	0	0	3	5	5
Other Capacity	548	498	485	20	72	59	0	4	14	100	200	200	668	774	758
Total	551	501	488	20	72	59	0	6	16	100	200	200	671	779	763
North America															
SX-EW Capacity	467	513	473	21	63	63	0	0	51	5	5	5	493	581	592
Other Capacity	2,848	2,511	2,269	73	324	323	0	234	510	34	97	78	2,955	3,166	3,180
Total	3,315	3,024	2,742	94	387	386	0	234	561	39	102	83	3,448	3,747	3,772
South America															
SX-EW Capacity	186	220	186	35	281	421	0	12	67	0	0	0	221	513	674
Other Capacity	2,450	2,277	2,150	54	265	255	0	39	680	0	0	0	2,504	2,581	3,085
Total	2,636	2,497	2,336	89	546	676	0	51	747	0	0	0	2,725	3,094	3,759
World															
SX-EW	1,130	1,210	1,116	61	349	489	0	19	125	5	5	5	1,196	1,583	1,735
Concentrates	10,295	9,705	9,170	180	841	817	0	319	1,333	159	381	372	10,634	11,246	11,691
World Total	11,425	10,916	10,286	241	1,190	1,306	0	338	1,458	164	386	377	11,830	12,829	13,426

Source: U.S. Bureau of Mines, 5/91

Table 2: World Copper Mine Production, Reserves and Reserve-Base.
(Thousand metric tons, contained copper)

Country or Area	Mine Production	Reserves 1/		Reserve-Base 2/			
	1990 p/	1988	1990	1965	1976	1988	1990
United States	1,586	57,000	55,000	30,000	84,000	90,000	90,000
Australia	316	14,000	7,000 3/	1,000	8,000	41,000	21,000 _3/
Canada	780	13,000	12,000 3/	8,000	31,000	23,000	23,000 _3/
Chile	1,603	85,000	85,000	42,000	84,000	120,000	120,000
Peru	334	12,000	8,000	11,000	30,000	32,000	31,000
Philippines	184	12,000	10,000	1,000	17,000	18,000	16,000
Zaire	370	26,000	26,000	18,000	26,000	30,000	30,000
Zambia	445	16,000	16,000	23,000	29,000	34,000	34,000
Poland	380	10,000	10,000	10,000	13,000	15,000	15,000
U.S.S.R.	600	37,000	37,000	32,000	36,000	54,000	54,000
Other Countries	2,236	76,000	68,000	16,000	102,000	109,000	109,000
Totals	8,834	350,000	334,000	192,000	460,000	566,000	543,000

p/ Preliminary, recoverable copper

1/ Copper in demonstrated (measured and indicated) economic ore resources at active and developing mines.
2/ Copper in demonstrated (indicated and measured) resources. Includes economic, marginally-and sub-
economic resources at active mines as well as at undeveloped properties.
3/ Some indicated resources have been reclassified as inferred.

Source: U.S. Bureau of Mines and various country sources.

projected capacity for 2000 is highly speculative, and if successful, will come from deposits currently in exploration stages. These are known deposits that do not have firm commitment for development or financing, and their production status only can be postulated. Under this optimistic scenario, some reopened capacity from currently closed, or uneconomic, mines is also included for this period. World copper capacity from SX-EW production will increase to about 13% of total mine capacity, compared with around 10% in 1992. North and South America will have most of the SX-EW capacity, with South America exhibiting the most significant increase, compared with 1992.

Reserves. - The world copper ore reserve base totals 543 million tons of contained copper. Chile (22%), the United States (17%), the Soviet Union (9.9%), and Zambia (6.2%) have the largest shares of these reserves. Zaire (5.5%) and Peru (5.7%) also possess large copper reserves. Reserve base growth has been slower in recent decades compared with demand growth. The world reserve base, comprised of measured and indicated ore, increased by about 140% from 1965 to 1976, corresponding to a 56% growth in world production and 48% growth in consumption over the same period. Since 1976, however, the ore reserve base has increased by only 18% and production by only 24%, while consumption has grown by 39%. While significant reserve growth has occurred in some countries, such as Chile, Australia and Indonesia, some countries, such as Canada and the United States have exhibited a decrease or stagnation in copper mine reserves in recent years. Current copper reserves (that portion of the reserve base thought to be economically recoverable with existing technology at operating or developing properties) of 334 million tons of copper are sufficient to meet a projected cumulative demand of 118 million tons for primary copper through the next decade. In addition, some of the material already identified in the reserve base, which is currently presumed to be uneconomic to mine, may well become economic with new technology and/or higher copper prices. However, new or expanded production, as currently planned within known reserves, will be insufficient to meet the projected

demand beyond 1996.

SECONDARY SUPPLY SOURCES

Another component of world supply is old or obsolete copper and copper alloy scrap that may be processed into refined copper, or directly remelted in the manufacture of new products. Since World War II, the ever-increasing reservoir of copper products in use, much of which is eventually recycled as "old" copper, has provided annually 19% to 33% of U.S. apparent demand. The recovery of "new" scrap, or scrap such as turnings and cuttings that are generated within the fabrication process, depends upon the level of manufacturing activity; it does not represent a net addition to the copper supply. "Old" scrap, on the other hand, is drawn from a resource of in-service copper, is considered a resource of recoverable copper and, thus, is a source of supply. Historically, old scrap recovery has improved during periods of high refined copper prices, but also increased at any time that primary supplies became scarce, including the deep recessionary years. It is estimated that the world resource of copper in "old" scrap items in use or abandoned in place exceeds 184 million tons. The U.S. scrap resource is estimated to exceed 69 million tons, or 38% of the world's total. The rate of old scrap recovery, however, is limited by copper's long life and use in durable goods. On average, the rate of old scrap recovered in the United States, as a percentage of the total scrap recovery, has declined from 50%-60% in the 1940's to less than 50% since 1960. The decline in the old scrap recovery relative to new scrap consumed correlates not only to an increasing manufacturing base from which to generate new scrap, but also to a changing demand pattern to one in which electrical uses dominate. Copper in electrical uses, which now comprises more than 70% of the market, is less likely to be scrapped. The long service life (greater than 40 years) for utility and building cable, among other reasons, results in a practical limit to the amount and rate at which old scrap from this source can be recovered. For scrappable "old" copper items, the average life has been estimated at about 20 years. Historically, old scrap consumption has averaged 18% of world demand, and it is dubious that this rate will change over the forecast period. Primary copper has contributed about 80% of world copper demand, as shown in figure 2, but is forecast to drop below this mark near the end of the century.

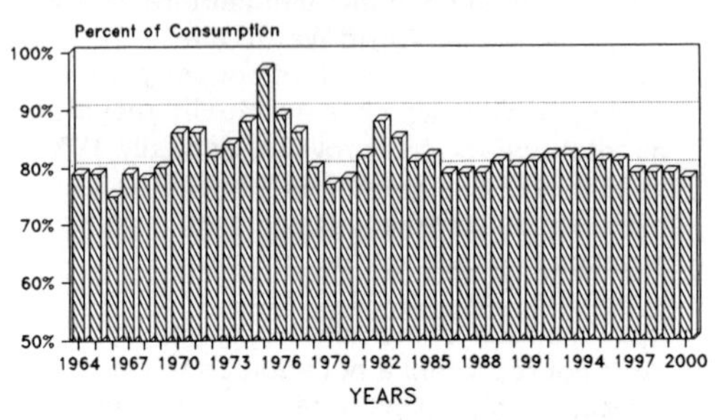

Fig. 2. PRIMARY REFINED PRODUCTION AS A PERCENT OF WORLD CONSUMPTION

U.S. Bureau of Mines, 5/91

COPPER CONSUMPTION

Refined copper is consumed mainly in the industrialized and newly industrialized countries of the world in their semifabrication and fabrication industries. The final products, such as automobiles containing copper, are not necessarily consumed in the country of manufacture. The major copper-consuming countries, or areas, of the world in 1990 were Western Europe (28.5%), United States (19.1%), Japan (14%), Soviet Union (10.2%), and China (5.3%). An estimated 11.3 million tons of primary and secondary copper was consumed in 1990. The forecast trend indicates a higher growth rate for consumption in South America, Asia and the U.S.S.R. compared with other areas, as shown in Figure 3.

Base metals consumption has increased dramatically over the past five or six years in developing and newly industrializing countries. This trend is likely to accelerate. Rebuilding of plants and housing in Eastern Europe will be important, although the ability to raise capital for these improvements may be a limiting factor, particularly considering the current recession.

Historically, world demand has risen steadily with only an occasional setback owing to economic recession, a trend that is expected to continue through 2000. The percentage growth rate, both actual and forecast, for copper consumption is shown in Table 3.

Table 3. COPPER CONSUMPTION TRENDS
Thousand metric tons copper consumed

Period	United States	Western World	World
		Copper Consumed	
1950	1,337	2,502	2,774
1988	2,213	8,299	10,549
1990	2,150	8,920	11,257
2000	2,620	11,100	14,630
		Annual Growth Rates	
1950-1988	1.34%	3.20%	3.60%
1990-2000	1.97%	2.30%	2.66%

Source: U.S. Bureau of Mines, June 1991

Between 1990 and 2000, copper consumption is forecast to grow at nearly 2% per year in the United States and at 2.7% per year in the world. The long-term consumption trend (1950-2000) for the United States is shown in Figure 4.

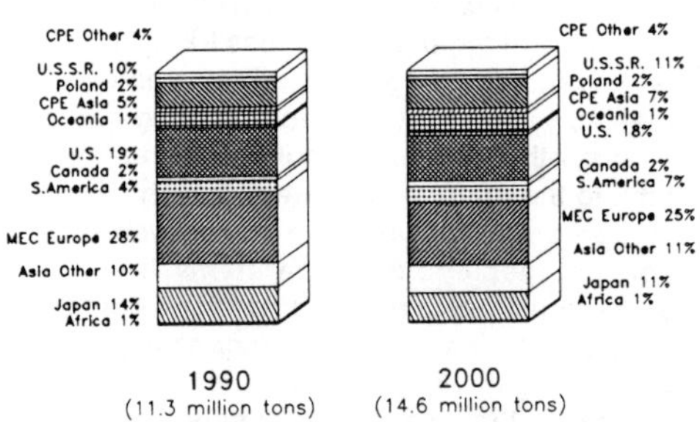

Fig. 3. WORLD REFINED COPPER CONSUMPTION
Percent of World Total, 1990 and 2000

U.S. Bureau of Mines, 5/91

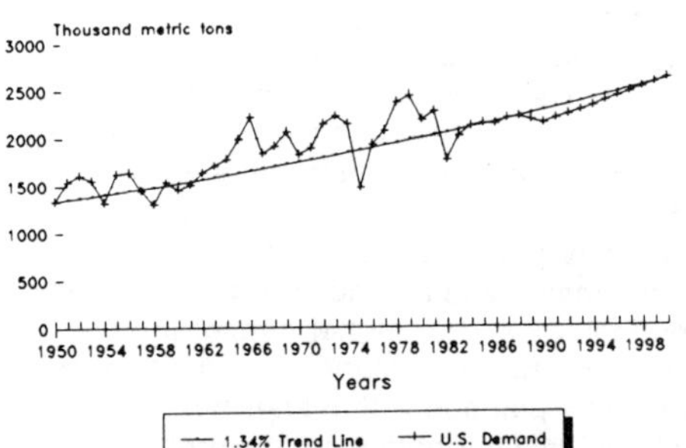

Fig. 4. U.S. APPARENT COPPER DEMAND
Actual and Forecast Trends

Source: U.S. Bureau of Mines, 5/91

In general, the projected growth in copper demand is largely dependent upon continued economic growth throughout the world. High demand for copper products is associated with a high standard of living and increased consumption for electrical uses. The growth of plastics, fiber optic cable, and aluminum as competing materials has affected some traditional copper end uses, as has new technology such as increased microwave transmission of communication signals, and integrated circuit systems. Another factor that has restrained growth of copper demand has been downsizing in machinery and transportation equipment. However, some of these declining trends, such as the decreased use of copper in automobiles, have been reversed as a result of the increased growth in electrical and electronic applications. In the United States, copper in all electrical uses has increased steadily from about 52% of apparent consumption in 1960 to over 70% in 1990, and thus, serves as the backbone to copper demand. In the United States, a high degree of historical correlation between the Federal Reserve Board index for Gross Private Domestic Investment (GPDI) and the demand for copper in electrical uses indicates the strong tie between copper demand and economic growth.

It has been conventional wisdom to view mature societies as less metal dependent. This might be a conclusion reached for U.S. copper demand, if refined consumption by the U.S. manufacturing industry is measured against the Gross National Product. But this comparison is misleading since the service sector of the U.S. economy has grown faster than the industrial sector, thus changing the composition of the Gross National Product. If copper demand by U.S. fabricators is compared with population growth, as shown in figure 5, over the past 40 years a different picture emerges.

U.S. per capita demand for copper, while significantly affected by periodic recessions, has remained remarkably stable, despite substitutions and loss of industry capacity to imports of manufactured goods, such as automobiles and consumer electronics. Global per capita use of copper, which includes copper consumption in all economic sectors, has exhibited a rather steady increase of 1.66% per year between 1950 and 1990. If this long-term growth trend in copper demand per person continues, population pressure alone could result in requirements for about

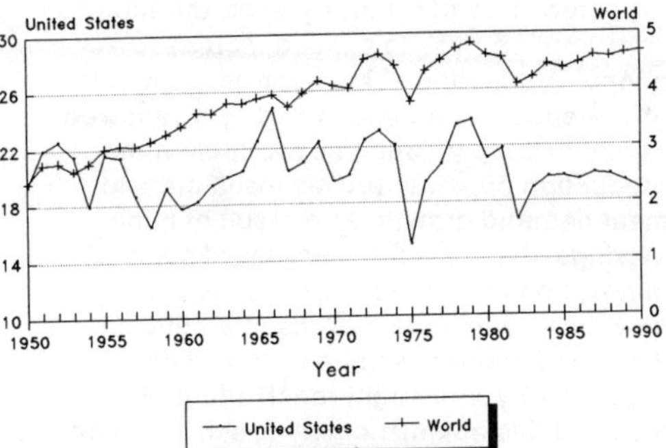

15.7 million tons of copper in 2000, significantly above our current forecast. However, in recent years, per capita copper demand has slowed to less than 1% and, thus, demand in 2000 might be expected to be somewhat lower than the long term trend would indicate, should these lower economic growth rates prevail for the world.

At the same time, however, the noticeable gap between the average 4.7 pounds per person consumed for the world and the 20 pounds per person for the United States illustrates the considerable growth in consumption needed for an equivalent living standard. The potential for increased consumption resulting from a higher standard of living in Eastern European countries and the developing Asian countries could further boost the lower growth rate.

SUPPLY/DEMAND BALANCE

Forecasts for world supply and demand to 2000 indicate a period of inventory surplus over the period 1992-1996 (Fig. 6).

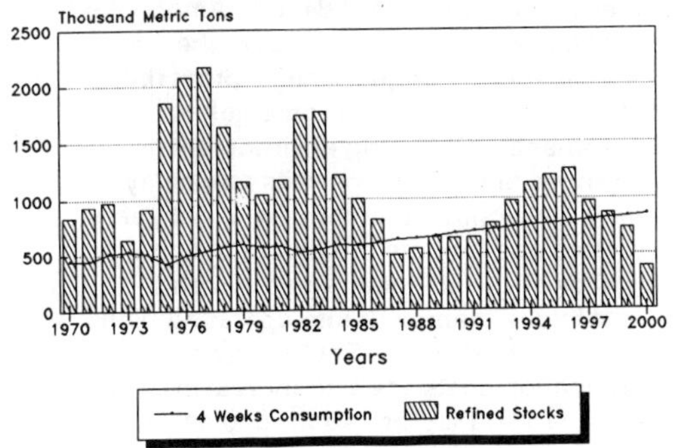

Source: U.S. Bureau of Mines, 5/91

The current economic recession may heighten and/or lengthen the period, but it is expected to be of relatively short duration as consumption rises to exceed increased copper production. This near- term period of oversupply will be followed by a renewed period of supply deficit as available mine production capacity proves insufficient to meet demand growth, as a result of mine closings, decreased ore grades at some mines, and a lack of major expansions or new projects planned for the late 1990's. Projected cumulative mine production over the next 10 years might reach 113 million tons, but this optimistic goal is still 5 million tons short of the demand forecast for the same period. In forecasting primary copper supply, as shown in Figure 2, implied, planned and some purely speculative new primary copper projects were included for the years beyond 1995. Some of the postulated deposits, for example, have not secured firm financial backing or announced certain development, but have been under consideration, and some are still in exploration stages. A projected schedule of development is assumed in the forecast for these properties, but owing to a number of obstacles, discussed below, their development is not assured within this time frame. Even though there is a significant scrap resource, much of it is not available for immediate recovery and, based on a historical recovery rates, copper from scrap will not be sufficient to make up for the sizeable mine supply deficit.

Problems. - In projecting supply, it was assumed that mine development, expansion and production would occur in an orderly fashion with few impediments. However, if the past few years are any example, numerous obstacles could reduce production below forecast and further exacerbate the projected deficit. Among the factors that can negatively impact the development of a project and its profitability are those relating to environmental requirements. Obtaining environmental clearances and mining permits can delay project development for months, if not years. Investors and lenders are aware of spiraling costs for environmental regulation and expect to see them built into operating plans and feasibility studies. Shutdown and site rehabilitation costs are also increasing and upfront capital may be required to assure shutdown compliance. Some financial institutions request environmental audits of operations before approving loans. Securities commissions are pressing for fuller disclosure of environmental risks.

Finding and mining the deposits of the future will take an immense amount of capital. Lack of development capital for some currently planned projects is also of concern. For example, the Saindak, Pakistan project is currently in jeopardy for lack of capital to build the necessary infrastructure. Unfortunately, this is not an unusual circumstance for many parts of the world.

In 1991, higher interest rates, discontinued government incentive programs in Canada, uncertainty about new governmental regulations, and world market conditions have served to slow exploration expenditures in both the United States and Canada. Expenditures for internal and contracted R&D funding was cut substantially as the need to cut costs increased. Increased mining research is important if the industry is to create technologies that minimize the impact of mining on the environment, reduce waste, clean up waste sites and achieve a greater level of safety for workers. Even with the recent rebound in copper prices, there has been little attention devoted to research, exploration and development. Research and development in the mining industry has languished compared with other materials industries. During the 1980's, the focus of mining was on short-term survival.

Uncertainty and disruption of supply has been a hallmark of copper production over history. When a large mine shuts down, such as the Bougainville Mine in Papua New Guinea and Highland Valley Mine in Canada, the market effect is significant. There is no stockpile of emergency supplies, such as existed during the 1960's and early 1970's in the U.S. Government National Defense Stockpile. The large stock accumulations that built up following the last of the stockpile releases in 1974 and the ensuing two (1975 and 1982) recessions are also no longer in existence. These surpluses cushioned production shortfall during the Vietnam War period and during the mid-1980's. Surpluses of this magnitude will not be reached over the forecast period.

Copper production is concentrated in mines that exceed 100,000 tons of copper per year. Twenty-nine mines with greater than 100,000 ton-per-year capacity make up about 58% of MEC mine capacity in only 11 of the world's 54 copper producing countries. About half of these large mines are located in the United States and Chile, where they comprise 74% of U.S. and 80% of Chilean mine capacity. Nine mines of greater than 100,000 tons per year have a Government as majority shareholder. Several of these mines are greater than 300,000 tons per year. These factors become significant when one considers potentially available refined inventories (supply excess to consumption) over the forecast period. Stocks equivalent to around 4 weeks of consumption are usually considered adequate to allow ease of operations without supply constraints and excessive price speculation. Visable MEC refined copper stocks have been below this level at many times within the past 3 years. World inventories, including those in the Centrally Planned Economy (CPE) countries over the 1991 to 1996 period, may accumulate at a modest variable rate that ranges between 60,000 and 200,000 tons per year. At this rate, the peak accumulation will at most amount to about 1.6 million tons, or about 6 weeks of world demand (8 weeks, based on MEC demand only). One mine closed for one year can easily remove a potential excess of 300,000 to 400,000 tons. This compares with 2.2 million tons, or about 4 months of supply in 1977. In addition, because of the immense size of these mines, timing of first entry, or, in the case of Bougainville, reentry into the market is also significant. While some surplus is desirable, an excessive supply at the wrong time can be temporarily disastrous in terms of lower prices, reduced profits and inhibited future development. Increased copper supply for the future will be totally dependent upon new mines. As indicated on table 1, capacity at currently operating mines will decrease despite significant potential increases at some mines, such as Escondida in Chile. Between 5 and 12 years are required for planning, developing, commissioning, and achieving capacity for a major new copper mine. While the major expansions currently planned to begin operations within the next 3 years are more than sufficient to address anticipated near-term demand, there should be serious concern about the lack of planning to meet forecast demands through 2000. Base metal exploration began to renew its vigor with advent of higher prices in the late 1980's, but it has begun to taper off; it was seriously lacking through most of the 1970's and 1980's. The reserve and reserve base estimates for the world increased slowly throughout the last 20 years, indicating a low rate of reserve replenishment compared with high rates of growth since the turn of the century. It is hoped that this slow ore reserve growth rate will be accelerated in the next decade in order to assure future supply. The ability to maintain the momentum necessary for sufficient exploration and mine development will be tested should renewed lower prices and surplus supplies intervene, as might be expected from current forecasts.

REFERENCES

1. U.S. Bureau of Mines. Copper. Mineral Commodity Summaries, 1977-1991 issues.
2. Jolly, Janice L.W. Copper. Ch. in Mineral Facts and Problems, 1985 Edition, Bull. 675. U.S. Bureau of Mines.
3. Jolly, Janice L.W. and Edelstein, Daniel. Copper. Ch. in Minerals Yearbook,   Vol. I, U. S. Bureau of Mines, 1989.

Concentrates—the overlooked market?

G.F. Reynard
Highland Valley Copper, Vancouver, British Columbia, Canada

ABSTRACT

Although a lot of attention is paid to the refined copper market, the market for custom copper concentrates goes almost unnoticed. From its infancy in the early 1960's the trade in concentrates has evolved into a major international business. In recent years the market has been reshaped by a number of major trends, including the increasing importance of a few large producers. This was highlighted by the 1990 startup of Escondida, the world's largest custom copper mine. Despite a current imbalance in custom mining and smelting capacity, the outlook appears bright with both new mines and smelters slated for the 1990's.

INTRODUCTION

While the market for refined copper receives considerable attention from the business press and commentators as an indicator of the health of the world economy, there is another market in the copper business which goes almost unnoticed. This is the trade in custom copper concentrates. Over the last two decades this market has grown to become an important part of the international copper industry. This paper will sketch the history of copper concentrate trade, outline some important recent developments and examine the outlook for this segment of the copper business.

EARLY HISTORY

In the first sixty years of this century the international copper business developed largely on an integrated basis. The European market was supplied with refined copper and blister from the African Copperbelt and Chile, and from producers in Finland and Sweden. The only significant custom smelter in Europe at this time was Norddeutsche Affinerie in Germany.

In North America the market was serviced primarily by domestic integrated producers. Although custom smelting did take place, it was generally a localized business with independent mines in a region providing some of the smelter feed for large integrated producers. The concentrate sales terms (primarily treatment charges) tended to reflect local conditions rather than international market forces. This localized custom smelting activity occurred in the major copper production centres of both the southwest of the USA and eastern Canada. The only North American custom smelter which did a significant amount of international business in this period was Asarco's Tacoma plant which sourced feed from British Columbia, South America and the Philippines as well as from the western USA.

In the early 1960's the Japanese copper industry was fairly self-sufficient with output from the Japanese smelters meeting roughly three quarters of the country's refined copper consumption. Although the Japanese mines only produced about 40 percent of the smelters' feed requirements, the total smelting capacity was low enough that the tonnage of imported concentrate was relatively modest. In the late 1960's this situation began to change and soon Japanese demand for imported copper concentrate was escalating rapidly. This led to the development of the international custom copper concentrate market as we know it today.

DEVELOPMENT OF THE MARKET

With the rapid growth in industrial production in Japan in the late 1960's and early 1970's came greatly increased copper consumption (Table I). The Japanese smelter companies expanded their facilities to meet the rising demand, thus increasing their copper concentrate requirements. However at the same time Japanese copper mine production began to decline due to reserve depletion.

Table I - Japanese Copper Statistics 1965 - 1990

Year	Mine Production (kt Cu)	Primary Smelter Production (kt Cu)	Refined Copper Consumption (kt Cu)	Copper Concentrate Imports (kt - wet)
1965	107	260	428	584
1970	120	501	821	1,565
1975	85	742	827	2,605
1980	53	890	1,158	3,104
1985	43	799	1,231	3,010
1990	13	926	1,577	3,522

In order to ensure an adequate supply of feed for their expanding industry the Japanese smelting companies encouraged the development of new copper mines in such diverse locations as British Columbia, Chile, Papua New Guinea, Indonesia, the Philippines and Zaire. This development of new custom copper concentrate producers was encouraged by offering attractive treatment and refining terms, and by direct involvement in the financing of many of the mines. As shown in Table II, the mining companies responded by developing new mines with an aggregate capacity of about 800,000 tonnes per year of contained copper during the decade of the 1970's.

While it was rapid economic growth in Japan in the late 1960's and 1970's which created the international custom concentrate market, its expansion in the past decade was largely driven by forces elsewhere. Rising copper demand in the newly industrialized countries of the Pacific Rim and Brazil led to the construction of new custom copper smelters. Onsan in Korea began operating in 1979, Caraiba in Brazil in 1982 and PASAR in the Philippines in 1983. In response to the increasing custom smelter capacity, a further 600,000 tonnes per year of custom mine capacity commenced operations in the 1980's.

Closure of Anaconda's Great Falls, Montana smelter in 1980 and the subsequent sale of the company's Butte mine output in Japan heralded the significant involvement of the US copper industry in the international custom concentrate market. This was followed by the closure of Asarco's Tacoma plant in 1985 and four of the Arizona smelters in the 1982-1987 period which released further tonnages of US concentrate onto the custom market.

A third factor leading to the expansion of international trade in copper concentrate in the 1980's has been the gradual conversion of two European smelters from integrated to custom status. Both Rio Tinto Minera's Huelva smelter in Spain and Outokumpu's Harjavalta plant in Finland were built to treat in-house, or at least locally produced mine output. As the local mines have been depleted these smelters have been forced to turn to the international market to secure feed.

The net result of all these events has been the growth of the custom concentrate market from a number of small isolated local businesses in the 1960's to its current status as a major component of international trade. In 1990 this trade totalled some 1.6 million tonnes of contained copper or 5.3 million tonnes of concentrate assuming an average concentrate grade of approximately 30 percent copper. At mid 1991 prices this represents 2.7 billion US dollars per year.

Table II - Development of Custom Copper Mines

Mine Name	Startup Year	Original Capacity	Mine Name	Startup Year	Original Capacity
1960 - 1969			**1970 - 1979**		
Craigmont	1961	25			
Bethlehem	1962	10	Granduc	1970	35
Granisle	1966	10	Andina	1970	60
Marcopper	1969	20	Brenda	1970	15
		---	Island Copper	1972	60
		65	Gibraltar	1972	35
			Lornex	1972	50
			Similkameen	1972	20
1980 - 1989			Bell	1972	20
			Bougainville	1972	180
Dizon	1980	25	Musoshi/		
Teutonic Bore	1981	10	Kinsenda	1972	35
El Indio	1981	15	Ertsberg	1973	60
Highmont	1981	15	Ruttan	1973	25
Amacan	1981	20	Mamut	1975	25
Copper Flat	1982	18	La Caridad	1979	120
Valley	1982	35	Basay	1979	20
Troy	1982	17			-----
Viscaria	1982	30			760
Tintaya	1985	55	**1990**		
Monywa	1985	20			
Continental	1986	40	Escondida	1990	320
Ok Tedi	1987	170	Grasberg	1990	160
Horseshoe	1988	15	Maria	1990	10
Starra	1988	10			-----
Neves Corvo	1989	115			490

		610			

RECENT MARKET TRENDS

In the last few years a number of trends have been shaping the custom copper concentrate market. One of these is the increasing domination of the market by a few very large producers. With the startup late last year of Escondida, the world's largest custom concentrate copper mine, this will be even more the case. In 1991 just five mines; Escondida in Chile, Freeport in Indonesia, Ok Tedi in Papua New Guinea, Neves Corvo in Portugal and Highland Valley in

Canada, will produce nearly 60 percent of the total custom copper concentrate traded. With considerable expansion potential existing at both Escondida and Freeport, this percentage could be even higher in future years.

Another major trend of recent years has been for large integrated copper producers to play a much more active role in the custom market, moving into and out of the market as required to balance their mine and smelter production. Some integrated producers have traditionally been custom market participants. Codelco has long sold concentrate in excess of its smelting capacity while Noranda has participated in both sides of the custom market, purchasing concentrate for its eastern Canadian smelters and marketing output from its British Columbia mines in the Pacific Rim. However, in recent years the market has been entered by many previously solely integrated producers. Perhaps the most dramatic example of this was the emergence of Kennecott as a concentrate seller in 1988. Phelps Dodge and Cyprus have been marketing excess mine production on the international market, while recently Magma has turned to this market to secure feed for its large new flash smelter. Even Palabora, geographically isolated in South Africa, has recently begun to sell excess concentrate in the custom market.

In the last couple of years there has been a shift in the net trade in copper concentrate between the Western World producers and the centrally - planned economies of the USSR, Eastern Europe and China. Traditionally these countries, particularly China, have been net consumers of concentrate produced in the West. However, with reduced copper consumption and a shortage of foreign exchange caused by recent economic turmoil, exports of concentrate from these countries have appeared in the custom market and net trade has become essentially balanced. The closure of smelting capacity for environmental reasons in the eastern part of Germany has also contributed to this change.

THE ROLE OF THE TRADERS

Although most concentrate market transactions are conducted directly between mines and custom smelters, the non-ferrous concentrate traders have played an important role in the copper concentrate business. They provide market liquidity, concentrate swaps, financial hedging, access to difficult markets and general marketing services. While an extensive paper could be developed on the traders' role, only the highlights can be covered here.

First, liquidity; buyers or sellers of last resort. Wherever a miner's regular smelter customers have purchased sufficient feed for their needs, the miner can turn to a trader to take periodic production surpluses. The obverse may occur when a smelter has a sudden spot requirement (perhaps because of a Force Majeure situation) and its other regular concentrate suppliers are sold out for the year. This provides a spot market for concentrates. As well, some mines/smelters deliberately set aside some of their sales/requirements for this spot market. Most is sold/bought via traders because the trader is willing to buy when the mine is ready to sell and sell when the smelter has a supply gap to fill.

Few mines sell 100 percent of their planned production or even fixed percentages of their production to specific smelters. Therefore if production meets or exceeds plan, spot lots appear. Smelters who have been unable to place unwanted metal onto the London Metal Exchange have had to limit their base load concentrate purchases to reflect the amount of metal they can sell year in and year out. This was particularly so in the Far East, prior to the establishment of LME warehouses in Singapore.

Traders can help smelters (or mines) reduce their costs by swapping concentrate for quality, geography or time. For example a smelter with a large base load of clean feed may wish to trade some of this feed for material with higher levels of impurities and a correspondingly higher level of treatment charge and penalty revenue. A mine with a CIF sales contract with a distant smelter (say Ok Tedi to Europe) may prefer to fulfil a trader's sales commitment to Korea. The trader delivers an equivalent quantity to the distant smelter (say from South America) with both sellers achieving ocean freight savings.

A trader can also provide a time swap. For example a smelter with a temporary surplus for whatever reason could divert shipments to a trader who would undertake to deliver an equivalent amount of similar material at a later date. This could be very attractive to the trader who is short nearby but long further out in time.

Traders sometimes provide pre-production financing to new or established mines. They can also provide access to markets/suppliers in/from countries with little or no hard currency or with heavy bureaucratic restrictions on import/export business.

An often overlooked role is the provision of marketing services for copper concentrate producers. While most smelters and the larger mining companies have knowledgeable and skilled commercial organizations, traders can provide sales agency, ship chartering and weighting/sampling/assay representation services for small mine operations or for those mining companies who choose not to employ a commercial group of adequate size.

Finally, traders are a source of market intelligence for those who deal with them regularly.

THE COPPER CONCENTRATE CONTRACT

The copper concentrate sales contract as it has evolved over the years has the following key elements:

Metal Payments;

- Copper - Deduct 1.0 - 1.4 units (%) from Cu assay and pay balance.
           - Typical payment range is 95% for low grade (20% Cu) concentrate to a maximum of 96.75% for high grade (40% Cu) concentrate.

- Silver   - European Smelters   - Deduct 30 g/t from assay and pay balance.
           - Pacific Rim Smelters   - If assay over 30 g/t, pay 90%.

- Gold   - European Smelters   - Deduct 1 g/t from assay and pay balance.
           - Pacific Rim Smelters   - Sliding scale ranging from 90% payment at 1 g/t assay to 97 - 98% payment at assays greater than 10 g/t.

Smelter Charges;

- Treatment Charge   - A charge in $US per dry tonne of concentrate to pay for the smelting of the concentrate.

- Refining Charges   - A charge in US¢ per lb of payable copper and $US per ounce of payable gold and silver to pay for the refining of these metals.

Although in theory the treatment and refining charges should be fixed at levels sufficient to pay the processing costs and provide a margin of profit, in reality these charges fluctuate widely depending upon the supply/demand balance in the custom concentrate market. There are also other terms such as payment dates and quotational periods, the period of time which determines the metal prices paid, which are subject to negotiation and which fluctuate with market conditions.

One contract item which is becoming increasingly common is a price participation clause which provides the smelting company with a stake in the metal price, rather than just a fee for providing a service. With such a clause the copper refining charge is increased by a percentage of the amount by which the copper price exceeds a threshold value. Currently this is 10 percent over the threshold of 90 US ¢/lb. Negative price participation which provides the mine with lower refining charges when the copper price falls below the threshold is often included as well. Tables III, IV and V, illustrate the effects of all these contract terms on the net revenue per tonne obtained from two different types of copper concentrate.

Table III - Sales Return Per Tonne Concentrate - Typical Grade

Assumptions:

Assays		Prices	
Cu (%)	29	Cu	115 ¢/lb
Ag (g/t)	170	Ag	5 $/oz
Au (g/t)	2	Au	375 $/oz

Contract Terms:

Metal Payments		Refining Charges		Other Terms	
Cu	1 Unit Deduction	Cu	9 ¢/lb	Treatment Charge	70 $/t
Ag	If > 30g, 90.00%	Ag	50 ¢/oz	Price Participation	10% > 90 ¢/lb
Au	If > 1g, 90.00%	Au	6 $/oz	Quotational Period	3 Months After Month of Arrival

Invoice Elements:

Payable Metal		Gross Revenue		Deductions		
Cu (lb)	616.963	Cu	$709.51	Treatment Charge		$70.00
				Refining Charge Cu		55.53
Ag (oz)	4.919	Ag	24.60		Ag	2.46
					Au	0.35
Au (oz)	0.058	Au	21.75	Price Participation		15.42
			$755.86			$143.76

Net Return:

Gross Value	$755.86	100%
Total Deduction	143.76	19%
Net Value	$612.10	81%

Table IV - Sales Return Per Tonne Concentrate
Low Grade Concentate With Penalties

Assumptions:

Assays		Prices	
Cu (%)	24	Cu	115 ¢/lb
Ag (g/t)	200	Ag	5 $/oz
Au (g/t)	11	Au	375 $/oz

Contract Terms:

Metal Payments		Refining Charges		Other Terms	
Cu	1 Unit Deduction	Cu	9 ¢/lb	Treatment Charge	70 $/t
Ag	If > 30g, 90.00%	Ag	40 ¢/oz	Price Participation	10% > 90 ¢/lb
Au	If > 10g, 97.00%	Au	8 $/oz	Quotational Period	3 Months After Month of Arrival

Invoice Elements:

Payable Metal		Gross Revenue		Deductions	
Cu (lb)	507.063	Cu	$583.12	Treatment Charge	$70.00
				Refining Charge Cu	45.64
Ag (oz)	5.787	Ag	28.94	Ag	2.31
				Au	2.74
Au (oz)	0.343	Au	128.65	Price Participation	12.68
				Penalties	6.30
			$740.71		$139.67

Net Return:

Gross Value	$740.71	100%
Total Deduction	139.67	19%
Net Value	$601.04	81%

Table V - Low Grade Concentrate Penalty Calculation

Element	Assay (%)	Penalty Level (%)	Penalty Rate ($/%)	Penalty ($/t)
As	0.35	0.20	15.00	$2.25
Sb	0.07	0.05	40.00	0.80
Pb + Zn	5.30	4.00	2.50	3.25
				$6.30

A recent feature of some custom copper concentrate contracts has been one combined charge (in US ¢/lb payable copper) replacing the traditional separate treatment and refining charges. This approach is favoured by some smelters as it tends to maximize their revenues when purchasing high grade concentrate. Of recent years there has been a tendency towards higher concentrate grades due to the use of column flotation cell technology by many mines. In addition some of the larger custom concentrate producers (Escondida, Freeport, Highland Valley) naturally produce high concentrate grades because of the mineralogy of their orebodies.

CURRENT STATE OF THE MARKET

The economic recession of the early 1980's caused by the second oil price shock in 1979 had a severe impact on copper consumption. As a result refined copper inventories ballooned, with reported inventories peaking at 2.1 million tonnes in 1983, and the copper price collapsed. Consequently there were numerous mine closures in the 1981 - 1983 period and a shortage of custom copper concentrate developed. Despite some new mine developments and reactivations in the late 1980's, the custom concentrate shortage persisted to a varying extent throughout the balance of the decade. With the concentrate supply/demand balance favouring the mines, they were able to negotiate relatively low treatment and refining charges from the smelters.

In the past year the market balance has shifted and a surplus of custom copper concentrate has developed. This has resulted from the growth in mine capacity outpacing the growth in custom smelting capacity. Mine capacity has been augmented by such large new mines as Escondida, Neves Corvo and Ok Tedi and by expansions such as that at Freeport. As these mine projects have been coming to fruition, the one new smelter project originally scheduled for this period, Mitsubishi's Texas City plant, has become bogged down in the permit approval process. The Texas City smelter was planned for a mid 1991 startup to coincide with Escondida, but now has been delayed by at least two years. A planned expansion at Caraiba in Brazil has been postponed indefinitely by that country's economic difficulties, and financing has slowed the start of the expansion of PASAR in the Philippines.

The emergence of a surplus of custom copper concentrates has resulted in treatment and refining charges escalating rapidly in the last year. This upward pressure on charges will likely be maintained for the next couple of years until smelting capacity can catch up to mine capacity or until there are reductions, voluntary or otherwise, in mine output. Should the copper price fall much below $1.00/lb the current level of treatment and refining charges is sufficient to cause significant financial pain to the higher cost custom concentrate producers.

FUTURE MARKET OUTLOOK

The future looks bright for the custom concentrate market. Unlike a few years ago, most observers now agree that copper consumption will continue to grow for the foreseeable future, particularly in the rapidly expanding economies of Asia. In order to meet this growing consumption, production of refined copper will have to continue to expand. Currently secondary (recycled) copper accounts for some 17 percent of refined copper production while electrowon copper from leaching operations accounts for a further 9 percent. Although production of secondary refined and electrowon copper will undoubtedly continue to grow and may constitute an increased share of total refined copper production, most copper will still be produced by the smelting or other processing of sulphide concentrates.

In addition to the Texas City plant already referred to, new custom copper smelter projects are being studied in Chile, Indonesia, Portugal, Thailand and Canada as shown in Table VI.

Table VI - Custom Copper Smelter Projects

Project/Location	Parties Involved	Projected Capacity (Cu t/y)
Texas City, Texas	Mitsubishi Materials	182,000
Thailand	Padaeng, Mitsubishi, Marc Rich	120,000
Indonesia	Metallgesellschaft, Local Interests	150,000
Northern Chile	ENAMI, Lac Minerals Caraiba	200,000
Metcob, Portugal	Outokumpu, Local Interests	200,000
Kitimat, B.C.	PRM Resources	150,000

As well, many existing custom smelters are studying the feasibility of incremental expansions which typically have favourable economics. Southern Copper in Australia, just now commissioning its new Noranda reactor, and Lucky Metals Onsan plant in Korea are two such potential candidates for future expansion.

For most of the 1980's the emphasis in mineral exploration was on precious metals. However, in the last few years with declining precious metal prices and relatively high prices for base metals, particularly copper, interest in base metal exploration has revived. This has already resulted in some significant discoveries such as Grasberg in Indonesia, La Candelaria in Chile and Louvicourt in Canada. In all likelihood there will be more such discoveries in the next few years. The partial listing presented in Table VII gives some idea of the diversity of custom copper mine projects currently under evaluation.

With copper demand set to continue to grow in the coming years, and both new mines and new smelters on the horizon, the expansion of the custom copper concentrate market is assured.

Table VII - Custom Copper Mine Projects

Project/Location	Parties Involved	Projected Capacity (Cu t/y)
La Candelaria, Chile	Phelps Dodge, Sumitomo	90,000
Collahuasi, Chile	Falconbridge, Shell, Chevron	40,000
Windy Craggy, Canada	Geddes Resources (Northgate, Cominco)	140,000
Louvicourt, Canada	Aur Resources, Louvem (Teck/Cominco, Noranda)	60,000
Mount Milligan, Canada	Placer Dome	38,000
Montanore, USA	Noranda, Montana Reserves	45,000
Flambeau, USA	Kennecott	30,000
Far Southeast, Philippines	Lepanto, CRA	35,000
Cayeli, Turkey	Metallgesellschaft, Etibank	23,000

The copper industry in the countries of Eastern Europe and in the U.S.S.R.

J. Hennevaux
Fédération des Entreprises des métaux non-ferreux, Brussels, Belgium

INTRODUCTION

Since 1989, the countries of Eastern Europe and the USSR have been marked by profound changes. While change has been rapid on the political front, it has remained difficult on the economic front. Wrongly considered in the past as a homogeneous block, fundamental differences are now appearing between these countries. They face many challenges, and their development will be slow and painful.

While there is a greater openness to communication with the outside world, resulting from a clear political will, it is not easy to form an accurate picture of the situation, options and developments.

In the various sectors of economic life, the data available to us are inadequate. This exercise remains interesting, nevertheless, and this is the goal which we are pursuing today with respect to the copper industry. Here, two countries are particularly significant: they are the Soviet Union and Poland.

GENERAL

a. <u>The Transition to a Market Economy</u>

The overall economic situation, within which our analysis is made, is too important for us to pass over. Certainly, it is difficult to make the transition from a centrally-planned system to a market economy. This clearly has been the experience of the countries which we are considering. Such are the difficulties involved that their economic performance and the standard of living of their people are deteriorating.

This overall phenomenon has many explanations:

- Firstly, after the Second World War, the countries of Eastern Europe based a major part of their development on heavy industries, strictly planned in terms of production standards;

- An active population was concentrated around these activities, much greater than the number employed by equivalent industries in Western countries;

- The manufacturing industry made excessively intensive use of energy and raw materials;

- Anything which acted as a brake on achieving planned production targets was completely abandoned, and there was a large-scale deterioration in the industrial fabric in general, and in the mining and metal processing industries in particular.

- The infrastructure was poorly-maintained, and increasingly frequent breakdowns caused serious problems;

- A high level of outstanding debt was borne by companies, limiting the banks' ability to support healthy sectors of the economy;

- A high level of foreign debt, particularly in Poland, Bulgaria and Hungary, enabled the former regimes to maintain a certain standard of living, despite a decline in economic performance;

- These elements brought about a macro-economic disequilibrium, with rising inflation, shortages and increasingly irregular supplies of goods;

- The dismantling of central planning took place before any of the institutions and structures specific to a viable market were set up, which further accentuated the disequilibria;

- Independence was granted to firms, without imposing any financial discipline or accounting obligations;

- The worsening crisis in the USSR, the main supplier and customer of the other countries of the former Eastern Block, has brought with it a total disorganisation of the supplies of energy, raw materials and other imports essential to these countries. In addition, where these products are available, they now have to be paid for in convertible currency;

- The countries of Eastern Europe are experiencing tremendous difficulties in redirecting their trade towards world markets, in particular in improving the quality of their products. In the past, these countries bought their raw materials from the Soviet Union, and sold it manufactured goods, often of mediocre quality.

This reduction in foreign trade brought with it serious disruption which had multiple effects. Trade between the countries of Eastern Europe fell by 20% in volume last year, while trade between the USSR and its former European allies fell by 15%. In this overall context, in July 1990, at the summit of the world's seven largest industrialized countries in Houston, the Heads of State and Government asked the International Monetary Fund, the World Bank, the OECD and the President Designate of the European Bank for Reconstruction and Development to carry out a detailed study of the Soviet economy, to make recommendations on how it could be reformed, and to determine the conditions necessary for Western economic aid to be able to support this process effectively. Their report was submitted, as planned, at the end of 1990. Some important elements of this report should be mentioned here.

- The index of industrial production, which was between 3 and 4% from 1985 to 1989, changed to a negative index in December 1989. It was –1.2% in 1990, and could reach –10% in 1991;

- Inflation did not exceed 2% between 1986 and 1989. It reached 4.8% in 1990 and could rise to 40% in 1991;

- Unemployment, unspecified before 1990, was recorded at 2% according to official figures in that year and could rise to 10% of the active population in 1991;

- Foreign debt, still at 29 billion dollars in 1985, could rise to 70 billion dollars in 1991;

- The current balance of payments went into the red in 1989, and reached a deficit of nearly 11 billion dollars in 1990. This fall will worsen in 1991;

- Up to 1988, the USSR assessed its growth in terms of net material product (NMP), NMP is equivalent to gross national product (GNP) minus services. According to official estimates, change in NMP was –4% in 1990. A report by Gosplan in February 1991 estimated that change in Soviet GNP would worsen to –15% in 1991.

We do not have the opportunity to go into detail of the economies of the other Eastern European countries, but we believe that these illustrations give an overall idea of the change in the economic situation in this part of the world.

b.   The Situation of Non-Ferrous Metals in the Countries of Eastern Europe

(1) Production

The long-term trend in production of ore and metals in the countries of Eastern Europe has been remarkably stable for more than two decades. Growth in production was strong up to 1975, and strengthened the position of these countries as raw materials suppliers. Since the mid-seventies, the market share of Eastern European countries has been falling. In general, they have been net importers of lead, zinc and tin, and net exporters of aluminium, copper and nickel.

The non-ferrous metals industry contributes to an important extent to earning hard currency. Unfortunately, it also contributes to a great degree to the creation of environmental problems. The high level of pollution of the low value added non-ferrous metals industry means that their closure would be a very cost effective pollution control strategy. Another solution would be to make major financial investments, to modernize an industrial fabric which has been left behind by current technology, and to add equipment which would effectively protect the environment. The latter solution requires large amounts of capital. It is less probable because the current priorities are concentrated on the production of consumer goods, and as a result, the necessary funds will not be available for intensive investment in the non-ferrous metals industry.

(2) Consumption

In all industrialized countries, overall consumption per capita is tending to rise, whereas in the countries of Eastern Europe, it has been falling since the beginning of the 1980s.

As far as non-ferrous metals in particular are concerned, consumption has fallen per capita from 19 kg per capita in 1975 to 17 kg in 1989. Over the same period, consumption rose from 29 kg to 33 kg in Western industrialized countries. For comparison, world wide non-ferrous metals consumption averages 9 kg per capita.

The many problems associated with the present restructuring lead us to believe that, in the short term, the trend for consumption will probably be downward, as Eastern European countries are giving priority to maintaining or even increasing exports in order to obtain foreign exchange, leaving lower quantities available within the country itself.

PRODUCTION AND CONSUMPTION OF COPPER
IN THE COUNTRIES OF EASTERN EUROPE AND IN THE USSR

a. Mining Production

What has been said previously about the falling production of metals in general is certainly true in the case of copper. While the world mining production increased by 770 000 tonnes from 1985 to 1989, the production of the countries under consideration fell by 130 000 tonnes during the same period. This means that their share in world production fell from 18.96% in 1985 to 15.95% in 1989. This reduction is particularly marked in Poland and the USSR.

b. Production of Unrefined Copper

The very similar phenomenon can be observed in the production of unrefined copper. These countries' share of unrefined copper production went from 18.62% of world production in 1985 to 16.43% in 1989, while world production increased by 840 000 tonnes.

c. Production of Refined Copper

Here too, the fall can be observed. With world production rising by 1 200 000 tonnes from 1985 to 1989, the share of East European countries and the USSR fell from 21.10% to 18.23%.

d. Consumption of Refined Copper

World consumption of refined copper increased by 1 300 000 tonnes from 1985 to 1989. The share of East European countries and the USSR fell from 19.62% to 16.13%, with a particularly noticeable fall in the USSR.

USSR

a. Mining Production

A plan was adopted for the whole non-ferrous metals sector, covering the period 1986–1990.

This plan emphasized a substantial increase in mining capacity. We have been above that this did not occur. On the contrary, there was a considerable fall in production from 1 030 000 tonnes in 1985 to 950 000 tonnes in 1989.

The essential objectives of the plan were an acceleration in the creation of new mining capacity, and the optimal use of existing capacity. In order to achieve the latter objective, the plan foresaw progress in the recovery of the metal content, while at the same time eliminating the practice of only exploiting the richest parts of deposits. A considerable improvement in the level of technology was required, as well as the use of better mining equipment, and better ore processing.

In addition, the falling yield from the development of the copper industry is partly attributed to a delay in the construction of new mines and ore concentrators, as well as a lack of modernization of existing plants.

This delay in the development of new mines resulted in intensive production from the richest deposits, and exhaustion of the reserves at existing mines.

In addition, a special effort should be made to protect the environment. A radical improvement was expected at the existing production sites, and new developments will be implemented bearing in mind ecological concerns.

Copper mining production in the Soviet Union is particularly important in the republic of Kazakhstan, in Eastern Siberia, as well as in the Ural Mountains, which separate the European and Asian parts of the Soviet Union.

Kazakhstan is acknowledged to have around half the Soviet Union's copper reserves, and currently provides around one third of its production. There are two very large mining complexes, at *Dzezkazgan* and at *Balkhash*.

Dzezkazgan Complex

Dzezkazgan consists of three underground mining departments, an open cast mining department working two large mines, and two production plants. In the whole complex, considerable efforts have been made to maintain the level of production, i.e. around 200 000 tonnes of copper contained in the concentrates per year, despite a fall in the development of the mine, and a constant fall in the copper content of the ore.

About 80% of the ore extracted in the complex is produced in the underground mines.

In 1988, preliminary exploration started on the new deposit at *Zhamanaybadskoye*. This is located at a depth of 700 metres, and has a copper content two or three times higher than that of the ore currently being produced.

Balkhash Complex

This complex handles all phases of production from ore to electrolytic copper. Some difficulties have been encountered, and it has not achieved its objectives over recent years. Even before the start of the last plan, i.e. in the period 1979–1985, a fall in production of 25% was recorded.

In order to improve performance, ore, blister and anodes were supplied to Balkhash by other mines, while a part of its own ore was sent elsewhere for processing.

Balkhash consists of three mining departments: *Kounrad, Sayansk* and *Vostochno-Kounrad.*

Kounrad

Kounrad is an open-cast mine. The ore has a copper content of 0.4%. The majority of existing reserves have been used. Production at the present time is based on ores which are considered uneconomic. Certain plans have been drawn up to exploit the mine at a greater depth, as considerable reserves have been recorded below the current level of mining. This operation would prolong the life of the mine. Annual production is estimated at around 40 000 tonnes of copper content per year.

Sayansk

This site consists of three open cast mines. Here too the reserves have been largely worked out.

Vostochno-Kounrad

The ore is composed of copper and molybdenum. The greater part of the economic reserves has been worked out, and extraction has had to start in neighbouring areas which are not very promising.

The modernization plans for *Kounrad* and *Sayansk* provide for an extension of their working life until the year 2000. Given the falling content of the ore currently being produced, production could only be maintained at a level close to current output until 1993.

It is planned that the new mining and metal processing plans of *Bosechekul,* which should start operation in 1993, will supply raw materials to *Balkhash.*

At a later stage, more raw materials could be supplied by the *Aktogay* complex, where construction should start this year.

In **Western Siberia,** the mining and metal processing complex of *Norilsk* produces ore with an average content of 0.5% nickel and 0.8% copper. More than half the nickel produced in the USSR originates in the Norilsk region. As for copper, the quantity produced annually is of the order of 100 000 tonnes.

There are many working copper mines in the **Ural Mountains,** which also produce coal, oil, chrome, nickel, etc.

b. <u>Metals production</u>

The production of blister copper fell from 1 140 000 tonnes in 1985 to 1 075 000 tonnes in 1989.

The production of refined copper went from 1 400 000 tonnes in 1985 to 1 355 000 tonnes in 1989, and is estimated at 1 300 000 tonnes for 1990.

The decline in metal production is due to a large extent to the considerable age of the metal processing plants. The fall in mining output has also played a considerable role in this development.

Of the 17 Soviet smelters and refineries, 9 date from before 1940. Of these nine plants, two were modernized before 1950. They are at *Alaverdi*, in Armenia, which was modernized in 1948 but closed at the end of 1989, and *Kirovgrad*, in the Urals, which was modernized in 1945. The *Monchegorsk* smelter, in the Kola Peninsula, was built in 1970, but according to our information, it has been closed. The *Karabash* smelter, in the Urals, which was built in 1920, was also closed in June 1990. The factories closed over the last two years account for a capacity of 200 000 tonnes of smelter and 120 000 tonnes of refinery. It cannot be denied that there will be further closures, largely due to environmental problems.

The table of smelters and refineries in the USSR is as follows:

	Name	Location	Smelter Capacity tonnes p.a.	Refinery Capacity tonnes p.a.	Built in	Closed in
1.	Alaverdi	Armenia	80,000	120,000	1936–1948	1989
2.	Balkhash	Kazakhstan	200,000	300,000	1938	
3.	Dzezkhazgan	Kazakhstan	180,000	300,000	1971	
4.	Irtyshsk	Kazakhstan	70,000	70,000	1977	
5.	Karzakpay	Kazakhstan	100,000	-	1928	
6.	Monchegorsk	Kola	20,000	-	1970	1990
7.	Pechenga	Kola	-	20,000	1975	
8.	Karabash	Urals	100,000	-	1910	1990
9.	Srednouralsk	Urals	100,000	-	-	
10.	Krasnouralsk	Urals	90,000	-	1931	
11.	Kirovgrad	Urals	90,000	-	1930–1945	
12.	Pyshma	Urals	-	280,000	1936	
13.	Kyshtym	Urals	-	100,000	1913	
14.	Almalyk	Ouzbekistan	250,000	300,000	1964	
15.	Nadezdha	Siberia	150,000	-	1980	
16.	Norilsk	Siberia	-	50,000	1950	
17.	Moscow	Moscow	40,000	40,000	1932	

POLAND

Almost the entire Polish output of copper is concentrated in a single mining and metal processing complex at *Lubin*. Under the name KGHM (Kombinat Gorniczo-Hutniczy Miedzi), this complex covers not only mining, smelters and refineries, but also mining equipment, repairs, supplies, transport and research and development. The Poles have a lot of experience in machine-tools, which they produce themselves, and they have more or less total autonomy over mining and metal processing.

The mine complex at *Lubin* contains five underground mines:

		Annual production in tonnes of copper content	Ore grade, % copper
1.	Rudna	170,000	1.87
2.	Lubin	80,000	1.25
3.	Polkowice	115,000	1.56
4.	Sieroszowice	35,000	1.74
5.	Konrad	3,000	1.14

The *Konrad* mine will be closed.

A major programme of conversion was drawn up at Lubin in 1980, concerning the output of the mine right up to the semi-finished product. This plan predicted an output of 500 000 tonnes. It was completely dropped after the strikes in 1982.

During recent years, Poland has exported around 180 000 tonnes of copper per year, of which around 100 000 tonnes are in the form of cathodes.

Poland has four smelters and refineries:

	Plant	Location	Built in
1.	Glogow I	Lubin	1971
2.	Glogow II	Lubin	1976
3.	Legnica	Lubin	1954
4.	Szopienice	Katowice	1946

The mines need to solve serious environmental problems, in particular with respect to water.

An American consultant was asked to carry out a study on the management of the copper industry.

Current production, which has been around the 400 000 tonnes per year level for several years, is falling considerably, as a result of a number of problems linked to the conversion of the economy, and environmental problems.

Refined copper production in 1990 is estimated at 340 000 tonnes, which represents a fall of 15% by comparison with the normal situation, which was still being achieved in 1988.

BULGARIA

The smelter and refinery of *Damianov* is estimated to have an output of 60 000 tonnes per year from home-produced concentrates and certain tonnages of imported concentrates.

The largest Bulgarian mines are the open-cast mines at *Elatzite*, *Medet* and *Assarel*; two less important mines are located at *Chelopech* and *Bratze*.

CZECHOSLOVAKIA

The *Krompachy smelter* and refinery (1965) has copper production estimated at about 25 000 tonnes, based for the most part on a supply of secondary raw materials, and to a lesser extent on production of concentrates within the country. The small mines in the country are working poor deposits and closure seems inevitable. These are essentially the underground mines at *Rudnany* and *Hory*.

The *Krompachy* plant is considered as a high-pollution plant.

HUNGARY

The *Csepel* refinery produces around 15 000 tonnes of copper from secondary raw materials. There is no copper mining.

ROMANIA

The *Baia-Mare* mine in the North of the country, close to the Ukrainian border, supplies a foundry (1966) and a refinery whose production is estimated at around 20 000 tonnes per year. It has major pollution problems.

Another refinery at *Zlatna* (1986) is supplied by small local mines and produces around 10 000 tonnes per year.

ALBANIA

The country has two mines, *Skoder* and *Kukes*, which supply a refinery at Skoder, with a production of around 15 000 tonnes per year.

THE FORMER EAST GERMANY

This is, of course, a special case. As this country has been merged into the Federal Republic of Germany, we only mention it for historic purposes.

The mine and factories of *Mansfeld* were closed in 1990, for both economic and environmental reasons. The *Ilsenburg* plant was closed for similar reasons. The only refinery which is still operating is that of *Hetstedt*, although it also has problems of viability and pollution. It is still operating for political reasons. The region where it is located is suffering from severe unemployment. Its capacity is about 50 000 tonnes per year, and it is normally supplied with imported blister and secondary raw materials.

YUGOSLAVIA

Although Yugoslavia has not been a member of the Eastern Block as such, it is worthwhile having a look at its copper industry.

The entire copper metal processing production is concentrated in the complex at *Bor* (Serbia) which dates from 1936 and has major environmental problems. The production of refined copper is of the order of 150 000 tonnes per year.

Two thirds of this production comes from the three largest mining fields, which are the *Bor*, *Maidanpek* and *Veliki Krivelj* fields, three open-cast mines, of which the copper content of the ore varies between 0.4 and 0.8% Cu.

The *Bor* refinery works on the basis of 80% home-produced concentrates, 13% imported concentrates, and 7% household scrap.

Certain modernization programmes have been drawn up, but there are major problems of pollution, as well as ethnic difficulties between Serbs and Bosnians.

CONCLUSIONS

a. Overall Situation

With its many old factories, and a time-lag of 25 years in the environmental field, the trend of production is definitely downward.

With an imperative requirement for foreign exchange, the producers want to export. Thus there has been a very large increase in exports of refined copper from the Soviet Union. Whereas their exports were around 30 000 tonnes per year in 1985 and 1986, they rose to 170 000 tonnes in 1990, and this trend is continuing in 1991.

In parallel to this development, even if, in the long term, one could anticipate an increase in consumption of consumer goods, and consequently of copper, in the short term only a fall in consumption can be predicted, due to economic disruption and falls in disposable income.

b. <u>The Remedies</u>

The Western community has paid a great deal of attention to the importance of a successful transition from a centrally-planned economy to a market economy. These countries have a large requirement for foreign aid in their efforts to achieve privatization, accompanied at the same time by an organizational infrastructure suited to this process, in the legislative, financial, political, economic, social and environmental fields.

It is with this in mind that, for example, the **European Bank for Reconstruction and Development** (EBRD) was set up. Based on an idea presented to the European Parliament in October 1989, the decision to set up the Bank was taken by 40 signatory states, including Canada.

This was inaugurated on April 15, 1991, with a capital of 10 billion ECU's or around 13 billion US dollars. With such an amount of capital, the Bank can act, in cooperation with the World Bank, with a financial volume of approximately 100 billion ECU's or 130 billion US dollars.

The objectives of the Bank are to promote the transition of the East European countries and the USSR towards a market economy, to promote private initiatives and accelerate demonopolization, decentralization and privatization. A minimum of 60% of the funds will be allocated to private sector companies, while the remaining 40% will be devoted to State infrastructure projects to support the development of the private sector.

It is clear, nevertheless, that such change will take years, and considerable efforts will be needed for such a large-scale operation to be successful.

The greening of copper

S. Hobson
Metals and Minerals Research Services, Bath, Avon, England

My offering to you today is entitled "The Greening of Copper". It doesn't, of course, refer to verdigris, but to environmental protection, the issues it raises in the copper industry and its results and implications for copper producers everywhere.

The history of environmental legislation is a long one in many countries, but there are astonishing divergences in the strength and the stringency of such national laws. At one extreme, some copper producing countries such as Papua New Guinea do not (yet) have tight environmental legislation and even in the United Kingdom our Environmental Protection Act of 1990 appeared more concerned with the problems of litter and dogs fouling pavements - rather than with industrially-generated hazardous toxic wastes.

At the other extreme you have the US copper industry which appears to function only with the aid of vast armies of lawyers. They have to contend with (amongst others) the National Environmental Policy Act, two Clean Air Acts, the Clean Water Act, the Safe Drinking Water Act, the Resource Conservation and Recovery Act, the Comprehensive Environmental Response, Compensation and Liability Act (no wonder that one's called "Superfund" for short), the Worker Health and Safety Act, a whole range of public land acts, the Wilderness Act, the Antiquities Act of 1906 and, of course, the Bald Eagle Protection Act.

The EPA's recently published semi-annual regulatory agenda contained a listing of environmental regulations in the making which required (Slide 1) 89 pages in the Federal Register, and most of the hundreds of regulations required under the USA's 1990 Clean Air Act haven't yet even been listed. And the onslaught continues.

But if the USA is making the running, the rest of us are catching up, at various speeds. While the European Community has yet to completely formulate its environmental legislative proposals (these will be pan-European from January 1992 onwards, and will be extremely tough if Dutch proposals win the day), environmental concerns are now being aired in Mexico, Peru, the Philippines, India, South Korea, Thailand and Brazil. In Australia, the environmental lobby (financed by federal government funds, as well as by private donations) is becoming exceptionally vocal and Greenpeace, the international environmental pressure group, has been monitoring one of the country's major lead/zinc complexes - Pasminco's Risdon.

In Chile, meanwhile, some of the tightest environmental laws in the world exist for copper producers. Indeed, so stringent are these laws that, if they were applied to the country's copper companies, many would be forced to close. Thus we have the bizarre situation that Chilean environmental laws relevant for copper producers are simply not applied: however, these laws are now being re-written.

The central issue that has to be faced by both legislators and metals producers was neatly summed up by a Chairman of Amax when he said (Slide 2) "One of our most demanding challenges is to balance the needed development of minerals resources with demands for environmental protection". This

Chairman was Sir Ian MacGregor and he was saying it 20 years ago in 1971. Progress has been long and hard; much has been achieved - but at enormous costs. These costs have not, by and large, been shouldered by metal consumers (via the copper pricing mechanism on the LME and Comex) neither, by and large, have copper producers received much help from their governments.

It has been for the polluter to pay and he has achieved this in two ways: firstly he has diverted profits and made investments into retrofitting, plant upgrading and new smelter construction programmes. Illustrative of this is the fact that of the non-Socialist World's (Slide 3) 64 primary copper smelters now in operation, 22 were built after 1970 using environmentally clean technologies (usually some form of flash smelting).

Secondly, he has done this by improving productivity, thereby compensating for the added investment costs and, sometimes, greater operating costs involved in environmental clean-ups. This accounts for the fact that in the last copper market recession LME copper quotations in 1985 plunged to a low of 54 cents per lb, their lowest real level since the Great Inter-war Depression, while *at the same time* the copper industry in many countries - and particularly in the USA - was in the middle of major environmental upgrading.

The costs involved in this environmental upgrading have been truly staggering - and they are continuing. Here in Canada, Inco will construct 2 flash furnaces, both exceptionally clean environmentally, for commissioning in 1991 and 1993. The cost of this and the Clarabelle mill expansion will be above the earlier projected (Slide 4) 494 million, Canadian Dollars.

Since the early 1970's Falconbridge has spent around (Slide 5) $300 million Canadian Dollars to reduce sulphur dioxide (SO_2) emissions from 6.5 tonnes per tonne of nickel produced in 1975 to 1.9 tonnes per tonne of nickel last year. A further $25 million Canadian Dollars investment is earmarked.

In the past 10 years Noranda has spent (Slide 6) 276 million Canadian Dollars on environmental expenditures ($34 million Canadian Dollars of this in 1990, alone): the company's metal discharges have fallen by 51 per cent since 1985 and in 1990 the 200 tonnes of metals which were discharged from Noranda Minerals' operations were 65 per cent lower than maximum levels allowed by government regulations.

On the other side of the border, improving air quality at copper smelters has also been fearsomely expensive. In the 13 years to 1985, Phelps Dodge (P.D.) spent (Slide 7) 385 million Dollars on air quality improvement facilities at its 4 copper smelters and from that year until the end of next year P.D. will have spend around a further 100 million Dollars.

These massive capital expenditure programmes are far from the only examples I could use of the enormous costs involved in environmental upgrading. Kennecott, for example, has offered to spend 200 million dollars in reducing metallic particulate emissions and Cyprus Minerals, to the south, is replacing its existing smelter with an ISASMELT unit (as is the process's licensee, MIM Holdings at Mount Isa in Australia). RTB Bor in Yugoslavia is raising its SO_2 capture rate from its current 75 per cent and is involved in large expenditures in new sulphuric acid-making facilities. Others have yet to feel the environmental spotlight on them, however.

Southern Peru Copper Corp. emits large tonnages of SO_2 and has no sulphuric acid-making facilities at all, neither does the Cananea smelter in Sonora, Mexico, and President Salinas is now flexing his not inconsiderable environmental muscles. There are other smelters in Latin America and elsewhere which are old and environmentally questionable and it is only a matter of time and money before they are forced to upgrade and retrofit.

But environmental attention is now turning away from SO_2 emissions to water quality: (Slide 8) this is the next major issue and encompasses the acrimonious debate over the EPA's storm water ruling (which applies to active *and* inactive mines), acid mine waste waters, contamination of aquifers, and discharges into rivers and seas. US laws are already tough on this: Asarco's mill expansion at Ray in Arizona is being held up by talks with State and Federal EPA's and the US Army Corps of Engineers to obtain a permit for a new tailings area, while Texas Copper (a Mitsubishi Materials subsidiary) is - as I am sure you are aware - only slowly emerging from a costly mass of permitting for its new 180,000 tpy smelter.

To the north, Geddes Resources ran into large scale problems - particularly from the EPA - over acid mine drainage at its Windy Craggy project in north-west British Columbia. So far 45 million Dollars has been spent on this venture: fully 11 per cent of this has been environmentally-related.

It is not only the highly visible and huge capital costs of plant upgrading programmes and delays caused by environmental permitting which deserve your very close attention, for environmental issues are now making and breaking companies in competitive situations. For a bidder for licenses to develop new properties to be successful, he must - in virtually all countries - demonstrate successful environmental management, while "due diligence" in mergers and acquisitions now includes careful inspection of the environmental liabilities of each of the prospective partners.

It is becoming increasingly onerous to both open *and* close copper producing facilities. Regarding new mine developments consider, for example, the case of Kennecott at Flambeau. In order to mine the Flambeau copper property in Rusk Country, Wisconsin, Kennecott had to struggle through 3 years of public hearings and environmental reviews. The company has had to post a 9 million Dollar bond to cover reclamation costs. After Flambeau's 6 year mine life, 19 months of reclamation will see the mine site restored to its original contours and all soil, glacial overburden and waste rock will be returned to its original sequence and the entire site planted with grass and trees.

This may be an extreme example of a State's stringent environmental regulations. But could it not also be a portent of things to come? And could you ever imagine Kennecott having to undertake Flambeau-style mine reclamation at it huge Bingham Canyon mine in Utah? Or Highland Valley in British Columbia or Chuquicamata in Chile?

There is one final thought about Flambeau which will not have gone unnoticed by government authorities. Once the reclamation process has been completed here, Kennecott will be permanently responsible for monitoring and maintaining the Flambeau site. And permanently means *for all time.*

As far as copper mine closures are concerned, these are getting to be extremely expensive, and companies in all industrially developed countries now make provisions against earnings to cover final mine closure and reclamation costs. Asarco's current provision with respect to its Superfund sites brings total reserves for future environmental costs to 127 million

Dollars, while Noranda has, to date, made provisions against earnings of 108 million Canadian Dollars to cover final mine closures. Even little Equity Silver - Placer Dome's subsidiary in British Columbia - which produces around 6,000 tpy of copper in concentrates, has had to provide 31 million Canadian Dollars for post-closure treatment of effluents, and even this amount may be insufficient.

It would be a very brave (as well as an entirely foolhardy) person who did not factor-in the (Slide 9) cost of mine development *and* mine closure into his or her discounted cash flow analysis for new projects.

Not only are environmental laws becoming more complex and stringent, but also they are springing up everywhere. Take, for example, the Great Lakes Critical Programs Act, which is to do with water quality and waste discharges. This was quietly passed by the US Congress as part of an appropriation bill near the end of the last session. Only recently has this act come to the attention of the metals industry. Meanwhile, in the south, the Mexican version of the EPA - SEDUE - and the EPA itself have joined forces to jointly inspect sites on their border and are jointly and quietly monitoring air quality at El Paso and Juarez.

And what has the response of the copper industry - the producer of an easily workable, environmentally-friendly metal - been to all this? It has been a near perfect example of Professor John Kenneth Galbraith's Theory of Countervailing Power, where one power-bloc (in this case governments and their agencies) indirectly encourages the formation of a strong opposing power-bloc. This opposing bloc is the new International Council on Metals and the Environment (ICME): it sprang from a North American initiative but it now includes virtually all major copper producers. The ICME will monitor, amongst other things, the environmental laws in all countries relevant to them.

The ICME is a long-overdue organization; J.K. Galbraith propounded his theory in the late 1960's and it has taken the copper industry over 20 years to discover that it applies to them.

The ICME will have to take note of the fact that laws which regulate copper production are now, in the words of the EPA's Scientific Advisory Board "Reflective of public perceptions of risk (rather) than of scientific understanding of risk". In this context, I don't want to enter the acid rain debate, but I merely note that it is ironic when so many North American smelters are being forced to reduce their SO_2 emissions, at gigantic costs, that the federally-funded 10-year National Acid Precipitation Assessment Program concluded "There is no evidence of a general or unusual decline in the forests in the USA and Canada due to acid rain".

Finally, while in the future the quality of custom copper concentrates will have to be high, (Slide 10) with increasingly low noxious or toxic metallic contents, and while we will see a drift of new mine developments into the financially-powerful hands of those who can match the environmental provisions required of them, we will also see a drift of new smelters away from urban areas and closer to the mines that supply them. We shall also see continual changes in environmental laws on copper producers everywhere.

In this context it would be helpful if the ICME could point out to federal and provincial governments and to large international organisations such as the European Community and the OECD that environmental laws and proposals should be laid out for as far ahead as possible. This would assist copper producers to plan, finance and implement their environmental upgrading schemes: the reverse of this - constantly changing environmental laws - makes life for the producer of this vitally important metal most difficult. For it is very hard to play the game when the goalposts move.

While I am here, and as I see I have 30 seconds left before the plug is pulled out from my microphone, I would like to congratulate Noranda Minerals on the production of its Annual Environmental Report. It publishes the good and the bad about the environmental aspects of its mining and metals producing operations. It is an extraordinarily brave document to publish at a time when metals producing industries are becoming increasingly seen as environmentally guilty until proved innocent: but such openness is also one way to win!

U.S. copper trends: exports will be sustained

J.F. Champagne
Magma Copper Company, San Manuel, Arizona, U.S.A.

ABSTRACT

Since the mid-1980's world demand for refined copper has been at record levels, led by growth in the Far East. Mine production gains have outstripped smelting and refining capacity, resulting in bottlenecks and imbalances. High capital costs, low profitability and lengthy environmental permitting of new smelter projects continue as major barriers to entry to the business of copper smelting. Treatment and refining charges (TC/RC) must be maintained well above recent levels -- to the historic real (inflation adjusted) highs of the 1970's of 40 cents per pound -- to encourage new smelter/refinery projects. These increased charges will add about 20 to 25 cents per pound of copper to the cost of production from the mid-1980's cost levels. The increases in the TC/RC may force some high-cost mines to close or curtail production and may revive interest in processes for leaching sulfide concentrates. To assure the necessary investment in mining, the additional TC/RC should eventually be reflected in the price of copper.

To reduce costs, US producers have rationalized their copper flows resulting in increased exports of refined copper to copper-deficit areas, particularly the Far East. The export gains are based on real production and transportation efficiencies and are expected to be sustained in the future. Continuing liberalization of trade policy will cement these gains and yield benefits for producers and consumers.

INTRODUCTION

The unequal distribution of the world's resources encourages and requires trade. Resources are not only mineral reserves, but include capital, technology, a skilled workforce and an established relationship between producers and consumers. In the copper industry this trade involves shipment of large volumes of copper concentrates to smelters and a similarly large trade in refined copper. Smelter products -- anodes and blister copper -- are sold in international trade, but involve a lower volume than concentrates or refined copper.

Copper is a growth metal. There was considerable pessimism about the outlook for copper during the mid-1980's, leading to forecasts of flat and even negative growth for the metal. These viewpoints were conditioned by an extended economic downturn, combined with industry-wide changes in inventory-holding practice. The fact is that the demand for copper has continued to grow. The demand for copper is based on its electrical and physical properties which are particularly appreciated in a world that is becoming increasingly quality-conscious. The expanding use for copper indicates an increase in copper trade, fostered by a rationalization of past and current copper flows, tariffs and trading practices.

Recent trends in world trade present a paradox of movement toward increased liberalization of trade and the continuing development of regional trade blocks. The upcoming removal of barriers to European Community trade at end-1992 has received great attention. The Canada-US free trade pact has been in effect now for nearly three years, with noticeable impact on the copper industry. The expansion of the trade pact to include Mexico will further rationalize the North American copper trade, with overall favorable effect on copper miners, smelters, refiners and consumers.

REFINED COPPER BALANCE

Over time the demand for refined copper is satisfied by production. Cyclical production deficits are made up by reductions of inventories until they are exhausted; any surplus will increase inventories. The refined copper balance for the western world for 1978-1992 is shown in Figure 1; Figure 2 shows the year-end inventories. The balances shown include net imports of refined copper from the centrally-planned economies (CPE's, which include members of the former COMECON plus China).

Figures 1 and 2 show the large surpluses that were generated in the early 1980's, which resulted in an accumulation of large inventories. These inventories depressed the copper market throughout much of the 1980's. This extended period of very low copper prices forced rationalization of production; mines closed in North America and elsewhere. Most producers deferred planned investments and implemented cost reductions and productivity improvements.

Perhaps the most dramatic changes in production took place in the US (Figure 3). During the ten year period from 1981 to 1991, the number of people employed in the US copper industry declined by more than fifty percent. This drastic restructuring caused a sharp decrease in production from 1981 through 1985. By mid-decade, however, the trend in US refined production became positive and increased from a low of 1.44 million tons in 1985 to nearly 2.1 million tons (estimate) in 1991[1]. Increased capital investment in more efficient technologies and

---

[1] All figures are in metric tons of 2,204.62 pounds

substantial changes in labor practices have significantly improved productivity. This trend of increasing production and improved competitiveness is likely to continue for several years in the US.

Over the last five years US copper production has grown faster than US consumption. The gap between consumption and production that has traditionally been filled by imports has been narrowing. In 1986, US consumption of 2.10 million tons exceeded US refined production of 1.48 million tons by 0.62 million tons. By 1991 (estimate), US consumption had changed little, while production grew to 2.04 million tons, reducing the gap to only 0.06 million tons (Figure 4).

Furthermore, geographic forces and the US-Canada free trade agreement have encouraged Canadian copper producers to focus on the US market. In contrast to US copper producers which are located in the western United States, Canadian producers in Quebec and Ontario are relatively close to key US copper-consuming industries concentrated east of the Mississippi River. Canadian copper has been increasing its penetration of this part of the US market.

With expanding domestic production and increased Canadian competition in the US market, the western US producers have a natural incentive to export copper to the dramatically growing markets of Asia, particularly the Japanese and Taiwanese markets. For some US producers, transportation factors even make it easier to export to Asia than to the eastern United States. US producers have both the capacity and the commercial incentive to be long-term suppliers to Asian markets (Figures 4, 5 and 6).

From the Asian customer's perspective, several factors make US refined copper commercially attractive. First, ocean shipping between US and Japanese or Taiwanese ports offer shorter routes and transit times, more frequent service and greater vessel availability than from alternative supplier countries in Latin America and Africa. The faster delivery from the US fits the "just-in-time" inventory management style of the Asian customers. A US producer can ship through Long Beach, California, to Osaka in 2 or 3 weeks. Shipments from Chile to Osaka require 6 to 8 weeks, while shipment from land-locked Zambia to Osaka takes 8 to 13 weeks.

Asian consumers have found an additional advantage in the shorter US shipping times given the persistent "backwardation" of copper prices in which the futures market price has been lower than the spot price. In such a situation, a buyer wants to avoid contracting now at the relatively high spot price for copper which will not be delivered until several weeks or months in the future when futures market trends suggest that the price may be lower. The shorter US shipping time reduces this problem.

Rapid and efficient shipment from the United States also minimizes the financial cost of having material tied up in long periods of ocean transit. The cost of working capital for material in transit can exceed the actual freight cost for long voyages.

Although the United States exported significant volumes of refined copper to the Far East in the pre-World War II period, such exports dwindled post-World War II. No significant exports to this region took place in the post-war period until about five years ago. Since 1986, refined copper imports from the US by Japan increased dramatically. By the first-half of 1991, the US had displaced Zambia as the second-largest supplier of refined copper to Japan accounting for 23% of all imports. Remarkably, the surge in US exports to Japan has occurred in the face of a tariff of Yen 15,000 per ton, equivalent to five cents per pound at the current exchange rate of Yen 136 per US dollar. Japan's "Generalized System of Preferences" (GSP) permits some refined copper to enter Japan duty free, but US copper is not eligible for GSP benefits.

Figure 1. REFINED COPPER BALANCE

Thousand tonnes refined copper

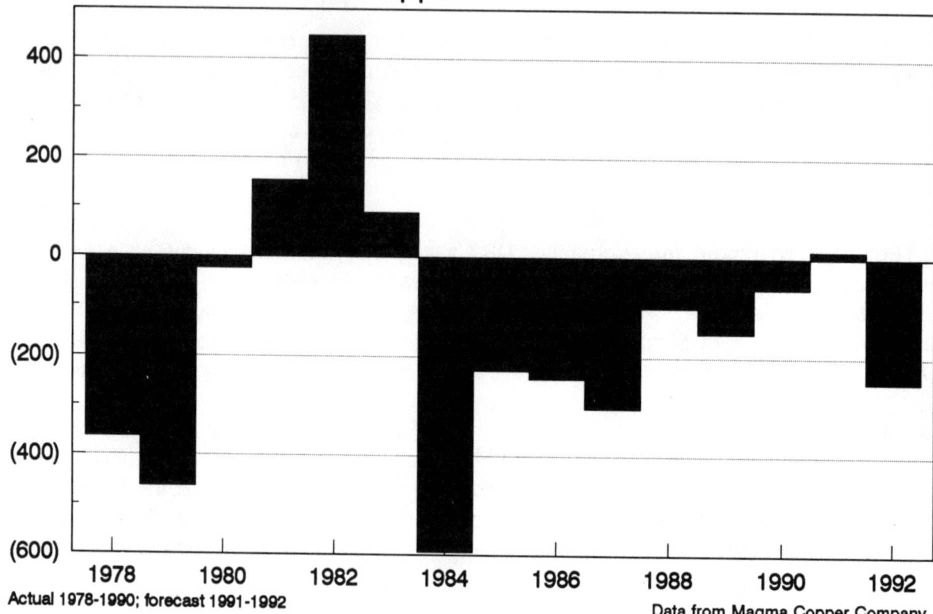

Figure 2. REFINED COPPER INVENTORIES
Year-end Reported Stocks

Thousand tonnes refined copper

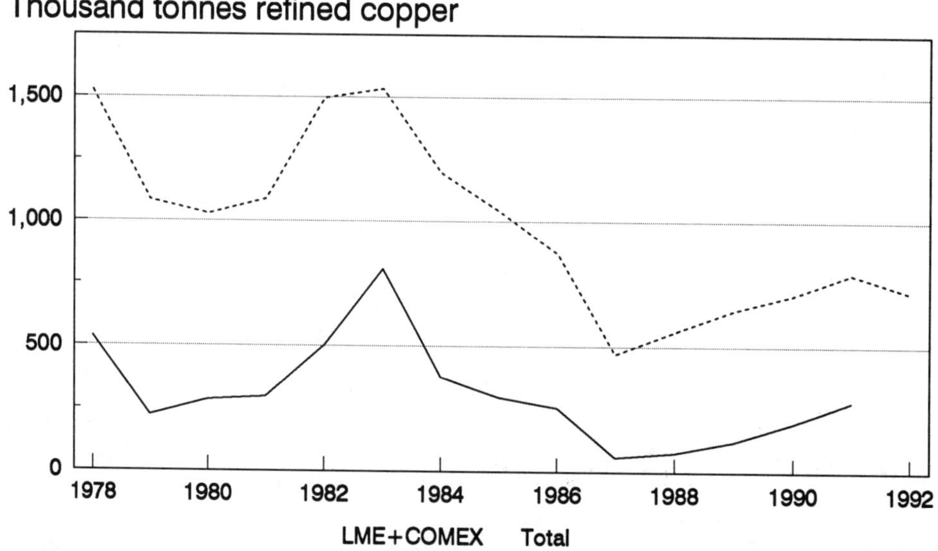

Figure 3. USA REFINED COPPER PRODUCTION

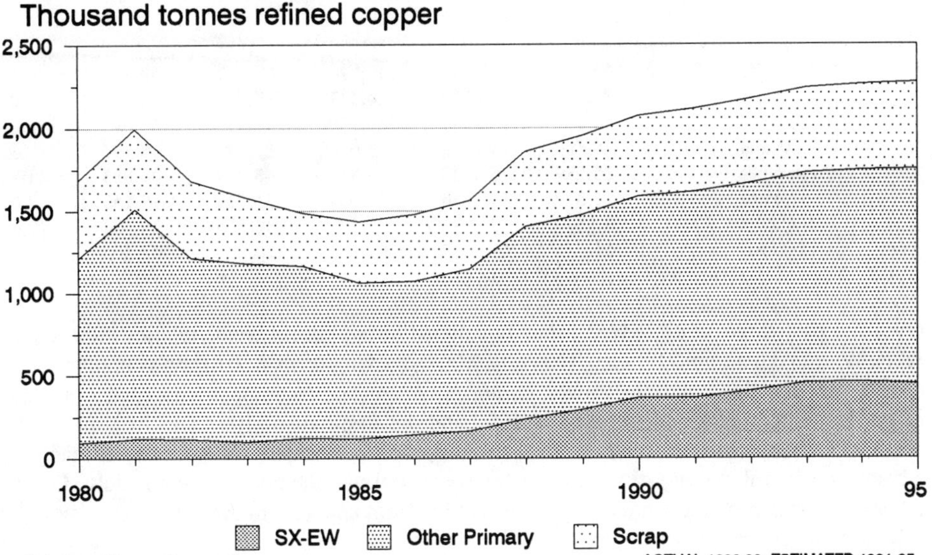

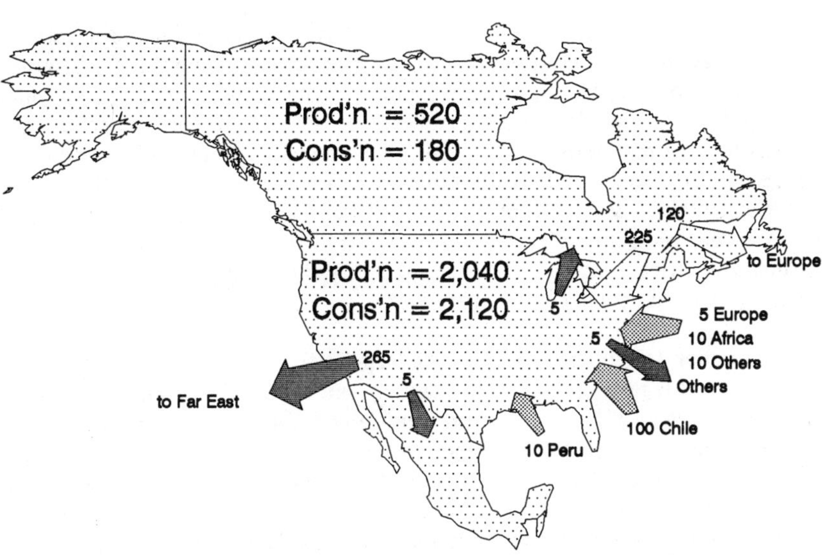

Figure 4. NORTH AMERICA REFINED COPPER FLOWS In 1991

Thousand tonnes refined copper

Data from Magma Copper Company

JAPANESE IMPORTS OF REFINED COPPER
BY COUNTRY, 1986 - JUNE, 1991

	Jan-June 1991	1990	1989	1988	1987	1986
Chile	109,004	193,953	142,074	97,687	68,890	46,333
USA	**80,343**	**103,019**	**48,377**	**34,515**	**5,941**	**285**
Zambia	70,451	155,749	145,550	152,031	167,693	138,611
Australia	26,563	40,719	26,056	27,647	16,795	11,172
Philippines	25,889	57,630	50,372	48,729	43,883	41,343
Peru	17,372	31,786	25,009	21,710	18,159	17,083
Others	22,349	34,981	45,205	38,515	26,304	17,610
TOTAL	351,971	617,837	482,643	420,834	347,665	272,437

Sources: WBMS; Japan Customs Office reports

Over the past three years, Magma has developed strong commercial relationships with major fabricators in Japan and Taiwan, stressing Magma's long-term reliability as a supplier of high quality electrorefined and electrowon cathodes. According to company estimates, at least 70% of the US copper exported to Japan in 1990 and the first-half of 1991 was Magma production. As Magma's mines and smelter/refinery complex are located in Arizona, it has a freight advantage when shipping refined copper for export through the port of Long Beach. Other major Arizona copper producers must ship their anodes from smelters in Arizona and New Mexico to refineries in Texas, incurring additional freight and financial costs.

CONCENTRATE BALANCE

The focus of copper investment during the past decade has been on new mine development. Smelter development has lagged, in large part due to high capital costs, environmental constraints and expected low returns on investment. Modernization of plants in the United States, Chile and Japan has added little to the world smelting capacity because the expansion gained by the modernizations has in large part been offset by reductions forced by pollution abatement and by the closing of obsolescent smelters. Such closings have been centered in the United States, dictated by environmental as well as economic considerations. (Smelters at Anaconda, Ajo, Morenci, Douglas, Ray, McGill and Tacoma have closed in recent years). Modernizations in Chile have yielded a significant increase in smelter capacity. However, this gain has been well below the increase in Chilean copper mine production.

Significant, recent smelter developments include the installation of Outokumpu flash furnaces at La Caridad (Nacozari), Magma Copper (San Manuel) and at Chuquicamata, as well as the current construction underway at Naoshima, Port Kembla in Australia, El Paso, TX and Miami, AZ in the United States, and PASAR in the Philippines. Exxon is proceeding with the modernization of its Chagres smelter. A new smelter is under consideration for northern Chile for completion in 1996, involving a five-company consortium (Fundicion y Refineria del Pacific). A major modernization/expansion program is being evaluated by ENAMI for their Las Ventanas and Paipote smelters and at least one of these plants is expected to undergo a major

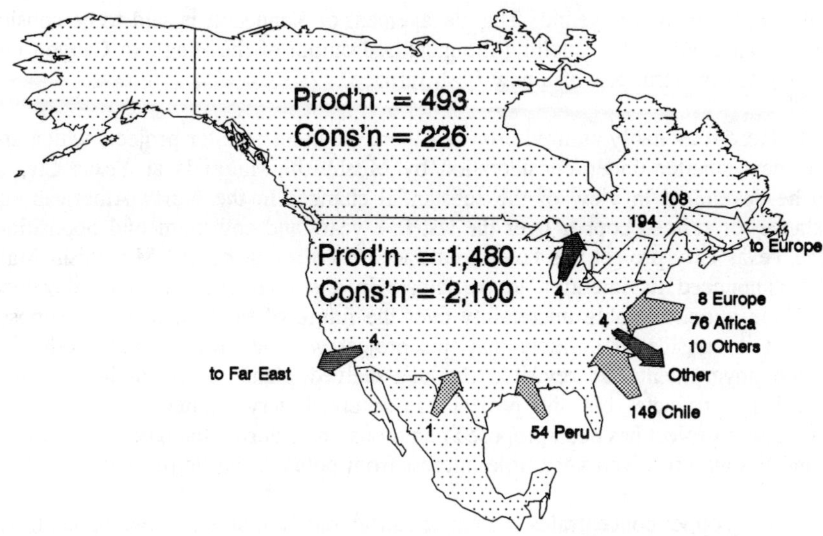

Figure 5. NORTH AMERICA REFINED COPPER FLOWS In 1986
Thousand tonnes refined copper

Data from USBuMines, WBMS

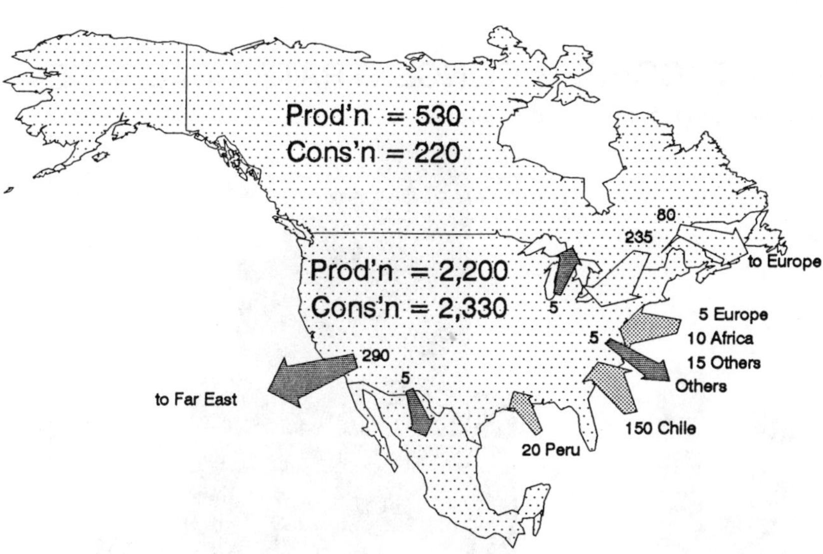

Figure 6. NORTH AMERICA REFINED COPPER FLOWS In 1995
Thousand tonnes refined copper

Data from Magma Copper Company

upgrading during the next five years. The management of Kennecott is said to be considering several possible alternatives for smelting the increased output of the Bingham Canyon mine in Utah, including the construction of a new flash furnace.

In addition to these reasonably assured projects, there are other smelter projects which are less certain. The new smelter under consideration by Mitsubishi Materials at Texas City seems unlikely to be developed in view of the substantial changes in the North American supply/ demand balance for refined copper over the last five years and environmental opposition. In addition, the Texas City project has failed to attract US co-investors and Mitsubishi Materials has recently announced its intention to participate in a similar smelter to be developed in Thailand. There is considerable uncertainty over the future of the new smelter proposed for Portugal and it is considered unlikely that this project will be brought into production. In Indonesia, the government has revisited its long-shelved plans for a smelter to treat the Ertsberg/Grasberg products, but the project has received very limited support. A similar government-backed project has been proposed numerous times during the past decade for British Columbia and has also received very little interest from potential participants or investors.

The availability of copper concentrates is a major consideration in smelter development. The net World Outside Communist Area (WOCA) copper concentrate balance for 1985-2000 is graphically summarized in Figure 7. This net concentrate balance is determined by summing individual smelter concentrate requirements, then subtracting scrap use and the available mine production of concentrates and precipitates (SX-EW production is not included). Supply deficits require drawdowns of inventories or imports, while a surplus will increase inventories. These balances are reduced by copper concentrates exported to the CPE's, notably China.

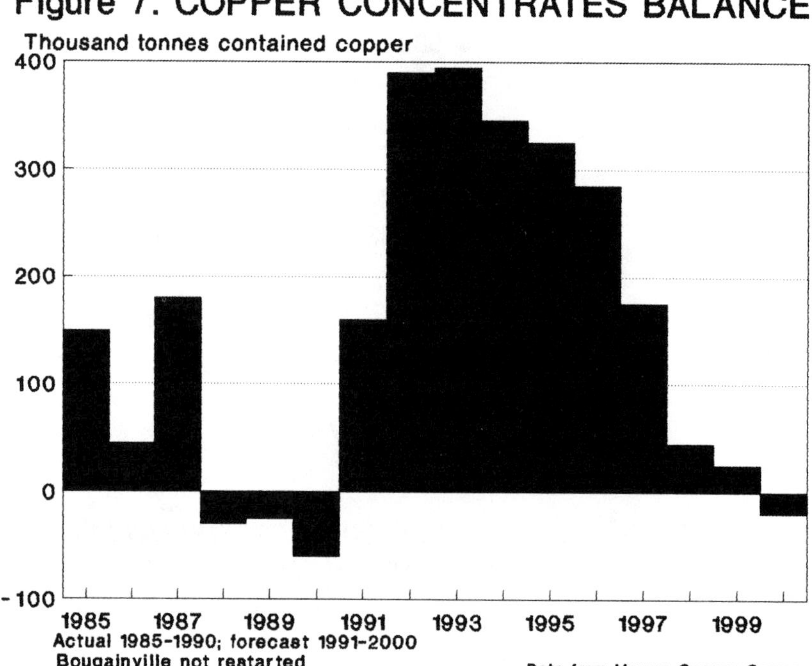

Figure 7. COPPER CONCENTRATES BALANCE

The net WOCA copper concentrate balance has been in deficit for the past three years. The negative balance would have been more modest in 1989 with a probable surplus in 1990, except for the shutdown of the Panguna mine on Bougainville in mid-1989. However, even in the absence of production from Bougainville it can be seen that the concentrate balance has developed a marked surplus in 1991 and is expected to remain in surplus for much of this decade. The immediate cause of the shift in the balance is the development of the Escondida mine in Chile. The first Escondida product was shipped on December 31, 1990, six months ahead of the original schedule; the mine reached full production in February 1991. During the first seven months of 1991, Escondida produced over 420,000 tons of copper concentrates with an average copper content of 40.69%. Expansion of production at other mines in Chile, Indonesia and North America will enlarge the concentrate surplus during the next few years.

In 1992 and 1993, the surplus of copper contained in concentrates will approach 400,000 tons per annum. In 1994 and 1995, the surplus will decline slightly but is forecasted to exceed 300,000 tons. These estimates assume that the Bougainville mine will remain closed. The second half of the decade is expected to show a declining surplus as old mines become depleted -- yet the surplus will continue until 1999. Experience has shown, however, that exploration will add new projects and mines to the current inventory (Figure 7).

THE SMELTER BOTTLENECK

The expansion of the world copper industry resumed in the second half of the 1980's. This has been a very lopsided expansion, with the bulk of investment directed toward mine and mill development. A substantial part of the mine investment was aimed toward SX-EW development. However, the great majority of investment has been, and continues to be, directed to traditional operations producing copper concentrate which require treatment: smelting and refining.

Smelter and refinery investments during the period have been largely retrofits focussing on environmental improvement. These improvements in smelting technology usually included some expansion of capacity. For example, the replacement of the reverberatory furnaces by an Outokumpu flash furnace at Magma's San Manuel smelter allowed an increase in production by about 50 thousand tons copper. Nevertheless, the sum of these smelter improvements has been far below the increase in world mine production. The inevitable consequence has been the development of substantial incremental volumes of copper concentrate without a balancing increase in capacity to smelt and refine this product.

There are several reasons for the lag in investment in new smelter projects:

- high capital cost
- low profitability
- environmental concerns

These reasons appear to be no different from mine investment considerations. Many mining operations require large capital investments, and all are targets for environmental advocates. The profitability of mining has been notoriously volatile in the past and the outlook for the future contains no guarantees. Copper smelting and refining differs from mining in two major aspects:

> the cash costs of smelting and refining in modern installations are fairly consistent worldwide; and

> ore grades are the key to mining operations, while economies of scale and the efficiency of technology are of corresponding importance to smelting and refining.

Moreover, smelters generate waste products which pose difficult disposal problems, particularly in populated areas. Permits for new plants face complex procedures which often may delay projects for years or kill them altogether.

The capital cost of a new smelting/refining complex is a formidable barrier. This cost forces an unacceptably low rate of return even under optimistic assumptions about the outlook for revenues to be received for the plant services, the TC/RC. The financial analysis of new smelter projects -- even with low operating costs -- invariably produces internal rates of return well below most investor thresholds. Even mines which have large, low-cost and high-grade ore bodies have failed to participate in smelter projects, e.g. Escondida and Freeport Indonesia. The sponsors of the various smelters proposed for development are well aware of this problem. It can only be resolved by obtaining substantially higher project revenues, i.e. higher treatment and refining charges.

The investment problem is compounded because small plants suffer from unfavorable economies of scale. The replacement cost for a large (250,000 ton), modern smelter/refinery, is about US$ 630 million (Figure 8), or US$ 2,500 per annual ton of copper production. The cost increases to US$ 3,500 per annual ton at the smaller end of reasonable smelter/refinery investments; a 100,000 ton per year plant will require US$ 350 million (Figure 9).

The sum of these factors is a considerable deterrent to new smelter development. In the absence of such development the copper industry will continue to see a strong surplus of copper concentrates, with the inevitable consequence of high treatment and refining charges.

Expansions are usually more economic than greenfield investments; and, this is certainly the case for smelters and refineries. Expansions could cost less than US$ 1,000 per ton of production (Figure 9), and produce a rate of return that is very favorable under almost any economic scenario. The opportunity for such expansions is limited and is expected to yield only a small portion of the capacity needed to match the increases in output of copper concentrates resulting from the mine investments.

TREATMENT AND REFINING CHARGES

The consequences of the smelter under-investment are already visible in the copper industry, and can be expected to persist well into the 1990's. We are currently seeing a large surplus of concentrates accumulating at smelters and at loading and unloading facilities. Copper concentrate stored in the open can deteriorate and the accumulation of huge concentrate inventories requires financing which makes storage a very unattractive alternative. The inability to treat all of the concentrate that is expected to be produced may require the reduction of output by marginal producers. The Western Pacific Rim (Japan and Korea) TC/RC has reached record highs; however, these levels are still below the charges levied in the 1970's after removing the effects of inflation (Figure 10). It appears inevitable that the TC/RC will continue to increase. Such increases add to the cost of copper production.

This situation is likely to persist into the mid-1990's, and possibly longer, in large part due to the time required for new smelter construction: at least four years for environmental permitting, design, investor approval, construction and start-up. Financing such a project may add to the time, indicating that new plants which may come under consideration in mid-1991 are unlikely to be in production before 1996. Smelter start-ups can pose numerous problems, as in the case of the current Port Kembla and Naoshima developments, and the recent La Caridad, San Manuel and Chuquicamata start-ups. Smelters failing to achieve designed rates of production even with well-proven technology are all too common, e.g. La Caridad in Mexico, Zlatna in Romania, Georgi Damayenko in Bulgaria and others. Such problems should be expected to affect at least some of the new and modernized plants. It is apparent that existing smelters will be in an enviable situation during this time, a reversal of the relative bargaining positions which favored the concentrate producers during most of the past decade.

A separate consideration affecting the treatment charges is the expected reduction and eventual elimination of the Japanese tariff on refined copper. The indirect impact of the import duty has been to raise the Japanese domestic price of refined copper. This tariff has had the effect of protecting the profit margins of Japanese smelters, while discouraging smelter investment elsewhere. The cost of this tariff is passed on in higher prices to the Japanese copper buyers, which in turn affects their costs and competitive positions. In 1990, for example, the Japanese domestic price was nine percent higher than the LME price. Although other factors, such as transportation costs, may contribute to the price difference, the import duty appears to have a significant impact. Recently the Japan Electric Wire and Cable Makers' Association has brought strong pressure on their government to eliminate the tariff. These moves coincide with similar pressure from foreign copper producers during the Uruguay round of GATT negotiations. For example, the United States has offered reciprocal ("zero-for-zero") reductions in tariffs in an attempt to further open the Japanese market. Any reduction in this tariff will necessarily put upward pressure on the treatment and refining charges.

A parallel pressure arises from the increasing strength of the Japanese currency. With dollar-denominated concentrate purchases defining their revenues and yen-denominated costs, the Japanese smelters have been squeezed on their operating margins. The increasing strength of Japan's currency during the second half of the 1980's has severely eroded treatment and refining charges in their home currency. It appears unlikely that the yen will show any significant reversal in strength and it can be expected that the Japanese smelters will strongly resist declines in treatment and refining charges.

During the past two decades treatment charges in the Western Pacific have been in the range US$ 40-100 per ton (constant end-1990 US$), with refining charges in the range 5-11 cents per pound. The combined effect of a concentrate surplus, a strong Japanese currency, and the probable elimination of the refined copper tariff in Japan is expected to keep the TC/RC toward the higher end of the historical range during the current decade. A tightening of the concentrate market toward the later part of the 1990's may bring about a softening and reduction in the charges. Such softening is conjectural and will depend on the rate of mine expansion compared to the installation of new smelter capacity during this decade. It is very probable that the additional costs imposed by environmental regulations will significantly escalate the floor for future TC/RC's. Any such increase in charges is likely to be supported by the capital requirements of new investments in these treatment plants.

During the mid-1980's the Western Pacific Rim TC/RC was as low as $35/ton and 4.5 cents/pound (in nominal dollars), equivalent to about 10 cents per pound of refined copper. The TC/RC for spot lots is very volatile and currently has exceeded the equivalent of 30 cents per pound of refined copper. Treatment charges may once again reach the level of 40 cents per pound; this is the level reached in the 1970's (in constant 1990 US$). The new smelter projects need more than the current high charges to deliver an acceptable return to the project investors. The conclusion is that over the next decade the TC/RC will exceed the current high level in real terms. These higher charges will substantially increase the cost of refined copper production -- perhaps by as much as 20 to 25 cents per pound.

Higher smelting and refining charges may force some high-cost mines, which are sellers of concentrates, to close or curtail production. A sustained period of very high TC/RC's may also revive interest in alternative methods of recovering copper from sulfide concentrates, such as leaching. Hydrometalurgical methods for recovering copper from sulfide materials have not yet been attempted on a large, commercial scale, however.

CONCLUSIONS

The North American copper industry has had a remarkable renaissance, which has brought about a rationalization in copper trade flows. The renaissance was based in large part on achieving a substantial reduction in the cost of production through greater efficiency and productivity, allowing North American producers to reach parity with the rest of the world.

Change in North American consumption patterns, combined with the Canada-US free trade agreement, has induced western US copper producers to export to the rapidly growing markets of the Far East. These exports have been successful even in the face of high tariff barriers, such as those encountered in Japan. It is expected that such tariffs will be reduced and eliminated during the next few years, in large part due to demands by domestic copper consumers, as well as to bilateral and multilateral trade agreements.

The shortage of copper smelting and refining capacity has been due in part to the economic distortions introduced by the Japanese tariff on refined copper. This tariff has had the effect of protecting the profit margins of Japanese smelters, while discouraging smelter investment elsewhere. As a buyer and seller of copper -- a key material for the growth of modern industry -- Magma strongly supports free trade on a level playing field. Free trade promotes the most efficient use of scarce economic resources and permits consuming industries to extend the benefits throughout the economic system.

Figure 8. REPLACEMENT VALUE
"GREENFIELD" SMELTER & REFINERY
250,000 MT/Y REFINED COPPER

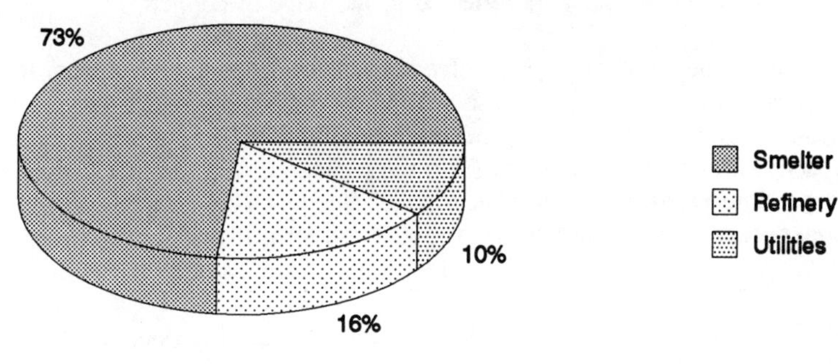

Smelter cost includes oxygen & acid plants,
concentrate storage and handling.

Data from Magma Copper Company

Figure 9. CAPITAL COST FOR SMELTING/REFINING

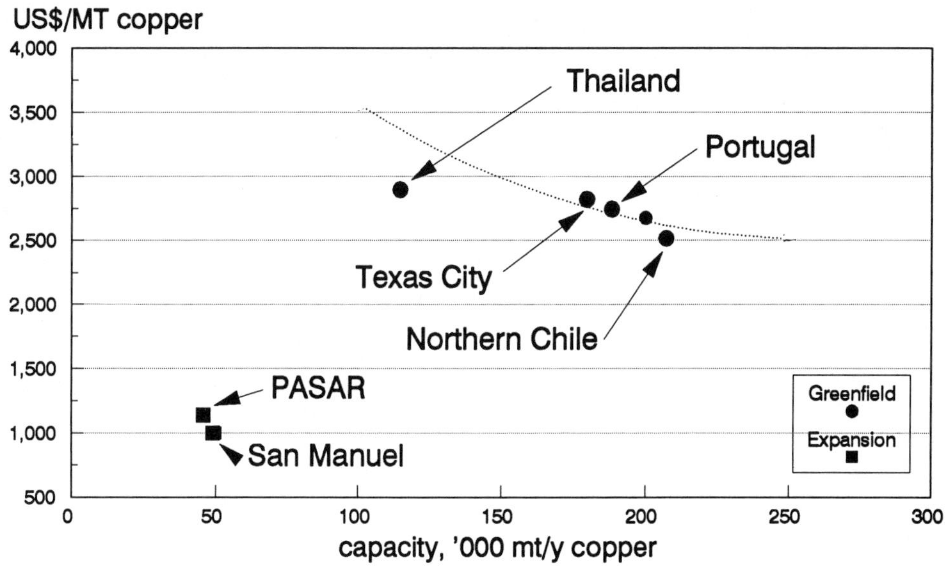

Data from Magma Copper Company

The unfavorable economics of smelter investment will severely constrain the development of new capacity during the next five years. The resultant surplus in copper concentrates is expected to force the TC/RC to record levels of 40 cents, adding as much as 20 to 25 cents per pound to the cost of copper production from the mid-1980's cost levels. The increases in the TC/RC may force some high-cost mines to close or curtail production and may revive interest in processes for leaching sulfide concentrates. To assure the necessary investment in mining, the additional TC/RC should eventually be reflected in the price of copper.

Existing copper smelters will receive significant economic rents from this expected increase in processing charges. This increase will offset the losses posted during much of the 1980's and encourage investment in new production facilities. With approximately 22% of the US smelting capacity, Magma has benefitted from the recent gains in the TC/RC and is currently examining the possible expansion of its San Manuel smelter. The company is well positioned to receive copper concentrates from mines in the southwestern US and is making arrangements to efficiently import from Mexico, Canada, Europe, South America and the Pacific Islands.

Organizationally, Magma views its smelting and refining operations as a separate, stand-alone enterprise which processes concentrates from Magma's mines as well as from outside sources. Smelting and refining revenues will generate a significant percentage of Magma's earnings throughout this decade. Magma is taking the necessary steps to anticipate market trends, to modernize and expand its facilities and to maximize the profitability of its smelting and refining business.

Figure 10. WEST PACIFIC COPPER TC/RC
CONSTANT US$ (end-1990$)

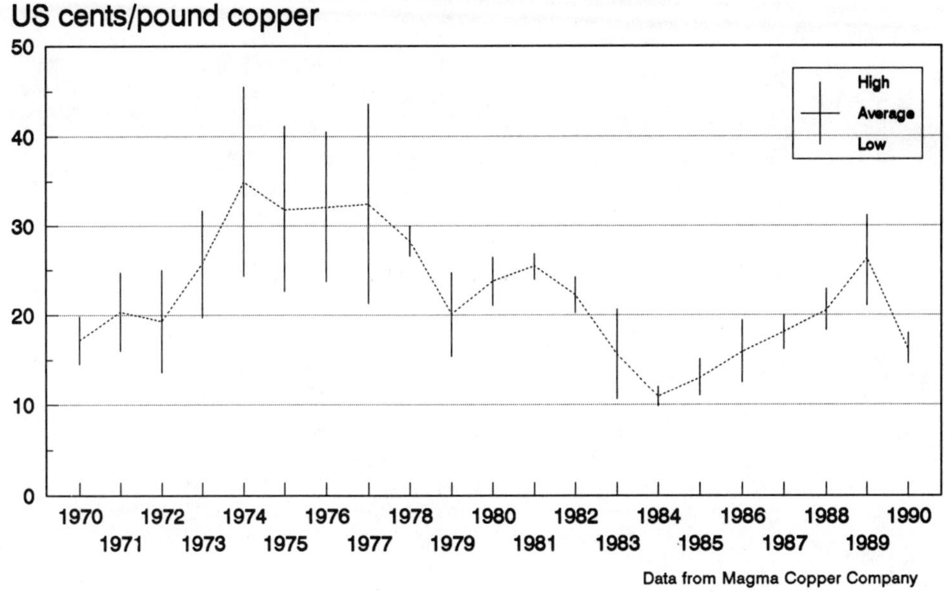

JAPAN COPPER PRICE

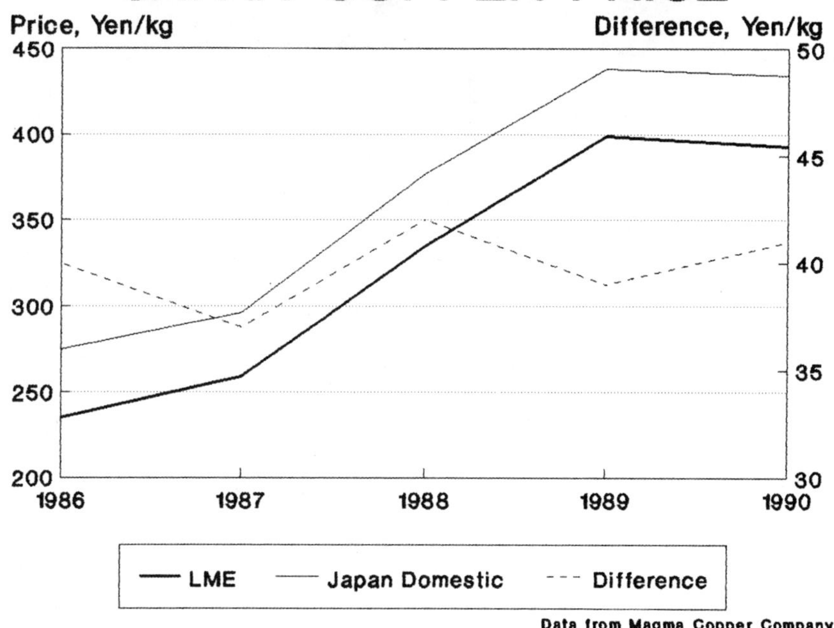

Competitive strategies among leading copper producers and structural changes in the industry's organization: a view from Codelco

I. Marshall, E. Silva and A. González
Codelco-Chile, Santiago, Chile

ABSTRACT

The copper industry passed from a collusive behavior managed by a few companies, to a period of lower concentration, which translated into greater competition and rivalrousness among leading players. This led to trade conflicts between US producers and Codelco in 1978 and 1984. After major cost-cuts in the Northamerican industry, conditions are again right for a more cooperative relation between major players. In this context, Codelco's strategy aims at deepening its competitiveness while keeping its output level. It also favors stronger collaboration with other players in promotion and joint ventures, that support a more cooperative long term development of the industry.

INTRODUCTION

This paper presents an interpretation of the structural changes that took place in the copper industry's organization over the last decades, and describes how these conditions have affected the competitive strategies of the leading copper producers.

Up to the early seventies, the industry was dominated by a few companies, the majority of which were of U.S. origin. These companies had international investments in copper mining abroad and many were vertically integrated. Under these conditions, the leading copper companies were able to manage the copper industry in a relatively collusive fashion. This translated into actions that tended to stabilize market fluctuations.

A series of events that took place in the sixties and early seventies -including the nationalization of copper mines in several LDC´s, the opening of new copper mines in the U.S. and elsewhere by smaller independent producers, and the entry of new investors into the copper industry (such as oil producers)- changed the industry's structure. It became less concentrated and its vertical integration diminished. Traditionally dominant companies, such as Anaconda and Kennecott, lost their main international interests and LDC copper companies emerged as major players in the industry.

This new industrial structure created the conditions for aggressive competition among copper producers, and traditional U.S. producers -whose competitiveness had declined- sought the application of protectionist measures. As protection was not forthcoming, U.S. producers were forced to undertake a drastic restructuring of their companies (and industry) in order to survive and regain competitiveness. They did so successfully, and have now entered a stage of long-term commitment to the industry and of seeking new international investments in copper.

Given the competitive environment of the seventies and eighties, Chile favored a strategy of free competition, in the belief that its position as a low-cost producer would work to its advantage. Though Codelco has continued to be one of the most competitive copper companies and the industry's largest producer, its cost advantage has eroded with respect to companies that applied strong streamlining measures to their operations in the past decade. Although

Codelco grew in terms of output and made important technological improvements, it did not undertake some of the tougher streamlining measures that other firms had to implement in order to survive.

In the atmosphere of greater industrial collaboration that can be foreseen for the coming decade, Codelco´s approach is a twofold strategy. First, it will carry out a number of actions enabling it to increase its cost advantage and leadership within the industry, retaining its output level. Second, Codelco will give high priority to a long-term commitment to the industry´s stable evolution. In order to achieve this goal, it is prepared to continue investing in promotion. It also wishes to strengthen its ties with other copper producers and with consumers, including joint ventures in projects in mining or processing, with major copper companies that share its commitment to the industry´s development.

CHANGES IN THE COPPER INDUSTRY´S STRUCTURE OVER THE LAST DECADES

The Emergence of a more Competitive Industrial Structure and Greater Rivalry between Industry Participants

From World War II up to the seventies, the copper industry was dominated by a small group of transnational companies, mostly of U.S. origin. The strategy applied by these companies was aimed primarily at "regulating" the market. These companies attempted to stabilize the market by adjusting production when demand diminished. They also established the copper producers´ price 1/, that showed greater stability than metal exchange prices.

In the course of the sixties and seventies, and during the eighties, the international copper industry entered into a phase of increasing de-concentration and vertical disintegration on the production side. This process can be partially explained by the fact that companies which had been the industry´s leaders until then, lost their more profitable mines as a result of the nationalization of operations in developing countries such as Zaire, Chile, Zambia and Peru; the opening of new copper mines in the U.S. and elsewhere by independent producers; and the entry of new investors into the copper industry (such as oil companies).

Simultaneously, the market suffered a long recession that made evident that these companies had competitive

---

1/U.S. copper producers´price was established by the producers themselves, and was applied mainly to copper transactions on the U.S. market. Throughout the fifties, sixties and seventies, this price was lower than the price quoted on metal exchanges during periods of upswing or relative improvement in the market; however, it was higher when the price on the London Metal Exchange was low. In 1978, U.S. producers were forced to abandon this pricing system because a significant portion of the consumption of domestic copper was being substituted by imports. At present, the U.S. copper producers´price is very similar to that of the exchanges.

problems in their ongoing operations, expecially those located in their countries of origin. This lack of competitiveness was due to natural conditions (low grade mining resources) and to organization and management problems in these companies, particularly a management style involving heavy spending. This was compounded by a lack of flexibility in the area of human resources and labor relations, which stemmed primarily from the considerable negotiating power held by labor unions and company workers, and from the prevailing industry culture that preferred to smooth out labor conflicts through economic concessions.

The length and strength of the market´s crisis gradually gave rise, in these companies and in the industry in general, to a pessimistic outlook of the industry´s future prospects and the future of copper as a product.

Simultaneously, the fact that the industry had experienced high growth rates in the past, supported by confidence that consumption would increase naturally, had led copper producers to be mainly production-oriented, without realizing the need to carry out marketing activities and to promote copper consumption.

The situation described above, coupled to the process of vertical disintegration in the industry and the entry of companies that lacked a long-term commitment to it (as was the case with the oil companies), help to explain the fact that copper producers became increasingly unconcerned about the industry´s medium and long-term development, and a short-term view in management prevailed. At the same time, the industry entered into a phase of aggressive competition, and greater rivalry and confrontation between the companies involved.

U.S. companies, which had dominated the industry in the past and were now being displaced, initially attempted to ensure their survival and their share in the industry by asking the U.S. government to adopt protectionist measures. Copper producers set in motion several initiatives seeking to limit copper imports into the U.S. market, through the establishment of quotas or higher tariffs. By requesting these measures, these companies adopted, in general, a confrontational stance vis-à-vis lower cost companies, and particularly with respect to Codelco.

On the other hand, the assessment made by Codelco and some other companies was that the industry had entered a more competitive situation, which made them feel that the most suitable strategy was that of free competition. By definition, this approach was non-collusive.

This strategy emphasized Codelco´s advantageous position in relation to the possession of higher quality copper reserves and byproducts, its lower costs and higher profitability, and the availability of a competitive infrastructure.

The above justified an increase in the company´s market share and made it unsuitable to "subsidize" higher-cost producers by reducing production. It should be noted that this strategy was not accompanied by a serious effort to promote copper consumption over a period of several years.

Although Codelco was interested in opening a dialogue with U.S. producers, the opportunity was not forthcoming at the time, and no important initiatives took place. In addition, the large cost differentials between Codelco and U.S. producers hindered the emergence of a shared view of the industry, its outlook and the establishment of a joint strategy.

Confrontation within the industry came to a critical point in 1984 when eleven primary copper producing companies in the U.S. asked the International Trade Commission (I.T.C.) to establish quotas for imports of blister and refined copper, requesting special measures against Chile because the latter had increased its copper exports to the U.S. significantly. Six years earlier, in 1978, another attempt to request this same type of restrictions had been made by U.S. producers. These cases serve to illustrate the confrontational stand adopted by U.S. copper companies. They also make evident Codelco´s lack of foresight and sensibility to the external environment since, in both cases, the company did not anticipate the threat of commercial conflict to which it was exposed.

Restructuring of the Copper Industry in the Eighties and its Consequences

The failure to make the U.S. government adopt protectionist measures forced U.S. companies to choose a different strategy. This strategy was initially aimed at ensuring their survival in the industry by means of strong cost-reducing measures in order to enable them to become profitable once more.

Once this objective had been achieved and confidence in the industry´s medium-term feasibility had been recovered, U.S. companies, set in motion several other strategies, in addition to maintaining competitiveness, for development and expansion both in the U.S. and abroad.

Several actions were undertaken. First, numerous streamlining measures regarding the companies´ organization and management schemes were implemented. These included significant personnel lay-offs and wage cuts both in production and in management; reductions of the size of head offices; training and development programs to increase worker productivity; reductions in the number of hierarchical levels; and modified incentive schemes.

Furthermore, there was substantial investment in technology, in an effort to modernize operations and introduce innovations both in processes and in equipment used in all productive areas. For example, new systems for transporting material by conveyor belts, use of semi-mobile crushers in the mines, high-tonnage mining equipment, semi-autogenous grinding, the use of large flotation cells, the use of autogenous smelting processes, the use of oxygen in smelting, process automatization, cheaper sources of energy supply, and the like were put in use.

Also worthy of note are measures related to streamlining and greater flexibility in the planning of mining operations. This led to the definite closing down of high cost operations and the temporary closing down and later re-opening of mines, under new methods for evaluating reserves and with more flexible production programs.

Finally, one must mention investments geared to increasing copper production through methods based on solvent extraction and electro-winning. This technology was quickly adopted throughout the industry, particularly in the U.S., and it brought about significant reductions in operating costs.

In the remainder of the industry, in some other geographical areas or countries (i.e. South Africa, Australia, Indonesia and Canada) there was also important progress in cost reductions over the eighties. However, other important copper producers, such as Mexico, Zaire, and Peru, did not register positive changes in their competitive situation (See Table I).

Table I - Direct Operating Costs * in Selected
Western World Countries (in 1990
US$ cents per pound of fine copper)

Year	United States	Zaire	Codelco-Chile
1982	87.2	57.3	43.6
1983	81.9	72.9	39.9
1984	74.3	29.6	35.2
1985	67.8	36.8	34.4
1986	66.4	49.3	35.2
1987	63.1	47.8	39.4
1988	60.1	76.1	41.5
1989	58.3	68.7	43.8
1990	58.1	89.0	46.6

*Includes costs covering up to the obtainment of refined copper, net of byproduct credits and not including depreciation and financial expenses.

Throughout this same period, Codelco increased its production, maintaining its position as the major copper producer with practically the lowest costs in the industry. This was the result of a strategy that included three main actions: cost reductions, technological innovations to increase productive efficiency, and production increases. With regard to technological innovations and greater productive efficiency, Codelco followed a path that had much in common with the process followed by U.S. producers and by others that also managed to make greater progress in cost reductions.

However, towards the mid-eighties, this strategy began to show symptoms of exhaustion and the trend toward cost reduction in Codelco began to decline.

This reversion can be partially attributed to technical and operational problems (poorer mineralogical features of the ore deposits, mine deepening, etc.), to deficiencies in the process of mine planning and to the lack of an adequate exploration policy. But, additionally, the company faced problems caused by its system of organization (excessive centralization, top-heavy management and unstable executives) and administration of resources, particularly human resources (lack of adequate training and career programs for human resources and a style of labor relations based on confrontation between the company and its workers).

Towards Greater Collaboration in the Industry

The trends described above resulted in the following two situations.

First, the cost differential between Codelco and other traditionally high cost producers -such as U.S. producers- began to decrease. While at the same time cost differences between

Codelco and other producers, such as CIPEC countries, remained the same or increased.

On the other hand, the restructuring of the U.S. copper industry led to the disappearance or loss of importance of some companies (such as Anaconda and Kennecott) while others grew, became stronger and were willing to adopt a long-term commitment to the industry. In addition oil companies left the copper industry since they were not meeting the expected profitability, which had been their motivation to invest their surplus funds in purchasing copper mining operations.

Acknowledging the fact that the main strategic concern of the copper industry was to expand future demand, this long-term commitment was translated, initially, into greater expenditures in promotion and in research with respect to new uses for copper. Efforts were made to increase collaboration with other producers in this endeavor. Practically all the major companies in the industry, including Phelps Dodge, RTZ, Cyprus, Noranda, Freeport, Inco, Magma, Outokumpu and Codelco, adopted this approach.

This long-term commitment was also present later in the strategy of internationalization and of association with other companies in the industry, focused especially in searching for and implementing copper-related projects located in the lower quartiles of the industry´s cost distribution. The industry´s major companies thus attempted to secure and expand their position as copper producers.

This contrasts to the strategy, broadly applied in the late seventies, of diversifying activities towards other industries and even abandoning the copper industry altogether. Such a strategy was adopted by oil companies that had invested in copper.

The strategy of internationalization and association with other companies has not been limited to U.S. companies such as Phelps Dodge, Cyprus and Freeport; it has also been adopted by highly competitive companies outside the U.S., such as RTZ and BHP.

In short, the onset of a stronger market cycle increased the availability of resources that could be spent in expanding the copper mining business. In addition, the rise in copper prices, coupled with lower costs, improved the optimism of the copper companies with respect to the industry, and made them take a greater commitment with the industry´s medium and long-term development.

Thus, a new atmosphere was created in the industry, which differed from the one prevailing in the previous stage, one that has been marked by the possibility of collaboration and strategic alliances between producers in areas such as copper promotion and expansion of its demand, and the implementation of new joint ventures.

The possibility for collaboration between the industry´s major companies was favored by the fact that the cost gap between them had become narrower, which favored a more similar outlook regarding the industry´s future.

Table II - The Top Ten Primary Copper Producers

| | 1959 | | | 1968 | |
	Production (Thousands tonnes)	Participation (%)		Production (thousands tonnes)	Participation (%)
Anaconda	520	14.1	Anaconda	548	12.2
Kennecott	510	13.8	Kennecott	501	11.2
Anglo American	356	9.6	Anglo American	365	8.1
Union Miniere	300	8.1	Union Miniere	325	7.2
Phelps Dodge	234	6.3	Amax	308	6.4
Amax	227	6.1	Phelps Dodge	215	4.8
Asarco	159	4.3	Asarco	176	3.9
Newmont	154	4.2	Newmont	168	3.7
Inco	140	3.8	Inco	148	3.3
Noranda	74	2.0	Noranda	80	1.8
Total	2,674	72.3	Total	2,834	62.6

| | 1978 | | | 1988 | |
	Capacity (Thousands tonnes)	Participation (%)		Production (thousand tonnes)	Participation (%)
Codelco	880	11.5	Codelco	1,125	16.3
Gecamines	537	7.0	Phelps Dodge	489	7.1
Zimco	429	5.6	Gecamines	450	6.5
Kennecott	376	4.9	RTZ	357	5.2
Asarco	359	4.7	Asarco	317	4.6
Phelps Dodge	346	4.5	Zimco	286	4.1
Newmont	329	4.3	Anglo/Minorco	270	3.9
Anaconda	193	2.5	Cyprus	227	3.3
Anglo American	168	2.2	Magma	178	2.6
Inco	166	2.2	Noranda	130	1.9
Total	3,783	49.4	Total	3,829	55.5

Source: CRU, Copper Studies for 1959, 1968 and 1978; Brook Hunt and Associates for 1988.

Table III – Corporate Strategies in the Copper
Industry in the Nineties

Company	Production Expansion	Cost Reduction	Mining Exploration & Development	Forward Vertical Integration	Backward Vertical Integration	Internationalization	Diversification	Strategic Alliances within the Industry
Codelco-Chile	X	X	X	X		X		X
Phelps Dodge	X	X	X			X		X
RTZ	X	X	X			X	X	
Asarco	X	X					X	
Cyprus		X	X	X		X		X
Noranda		X	X				X	
Freeport	X	X	X					X
BHP	X	X	X			X	X	
Outokumpu	X	X	X	X	X	X		X

CODELCO´S DEVELOPMENT STRATEGY FOR THE NINETIES

Within this new industrial scenario, Codelco´s development strategy for the nineties emphasizes two fundamental objectives. On the one hand, the company will undertake actions enabling it to recover the competitive position which has eroded over the course of the past ten years, seeking to maintain and strengthen its leadership in the industry.

In addition, the company´s strategy is commited to the industry´s medium and long-term development, and also seeks a greater interaction and collaboration with other industry agents. Through this interaction, Codelco wishes to implement a strategy that is consistent with and reinforces the industry´s more cooperative mood, and that takes the best advantage of this mood.

Consolidation of Productivity and Cost Leadership

In order to achieve the first objective, Codelco is currently in the process of consolidating the production expansions, implemented during the past decade, and shall continue to do so in the coming years. Essentially, the idea is to develop new projects and actions geared at reducing the company´s overall costs, rather than reducing unit costs by increasing production.

For this purpose, Codelco is planning to make considerable investments in mining development and in replacing installations, equipment and systems for material transportation, in order to solve the technical and operational problems it faces. One example of this is the project of a conveyor belt to transport waste, which started operating this year at Chuquicamata mine and involved an investment of US$ 91 million.

Codelco has given the highest priority to finding a solution to the geo-mechanical problems that have affected El Teniente for several years, in particular the Sub-6 level of the mine. For this purpose, a team of experts has been working on the subject and the necessary resources and technical assistance has been made available to them. The team has already made some progress in identifying the root of the problem, in proposing possible solutions, and in searching for safer exploitation methods. In order to compensate production losses originated by these problems, El Teniente plans to increase the extraction of mineral in other sectors of the mine.

In the area of human resources, Codelco shall continue working on a new type of relationship between the company and its workers, based on mutual commitment and cooperation. Likewise, it is currently implementing an action plan aimed at overcoming some of the major problems created by lack of flexibility prevailing in this area. Some of the high-priority actions included in the plan are those geared to achieve a greater degree of commitment and a higher managerial spirit among the company´s supervisors and middle level officials; to broaden and increase the flexibility in workers´job descriptions; and to develop incentive mechanisms that will help to solve problems such as absenteeism and lack of work motivation.

Likewise, Codelco´s centralized organizational structure prevents a more agile and flexible decision-making process, a more strategic-oriented management and a greater decentralization of decisions. The action plan prepared includes studying the possibility of establishing a management scheme based on performance agreements which would constitute a formal commitment regarding objectives and results, to be negotatied between the company´s headquarters and its operations. Such agreements, which would be subject to periodical controls, are considered important achieving higher productivity and efficiency.

A third key element in Codelco´s strategy to increase its competitiveness troughout the nineties, is to continuously include technological innovations and improvements in the various stages of the production process.

The above implies taking advantage of a number of opportunities that exist through the company, in order to achieve short term improvements in efficiency, techonology and processing with only small investments. These opportunities were not fully taken advantage of in the past mainly because the previous strategy emphasized the development of large projects requiring high investments for their implementation, rather than marginal efficiency improvements.

Production Expansion

Codelco´s strategy to consolidate previous production expansions plus the depletion of its ore deposits implies that, over the next three years, the company´s production levels will be less than those achieved in the past few years.

Consequently, the company is speeding up the assessment of several highly profitable projects, in order to compensate for the expected decrease in production.

Among the projects under consideration, the most interesting are those that involve producing copper through solvent extraction and electro-winning which, as already mentioned, is a lower cost procedure than traditional production methods (See Table IV).

Table IV - Codelco´s Main Mining Projects [*]

Project	Division	Characteristics	Investment (US$ million)	Production (thousands tonnes of fine copper)
Chuqui-Norte	Chuquicamata	Heap Leaching Sx-Ew	280	150
Ripios 2nd.Stage	Chuquicamata	Dump Leaching Sx-Ew	50	65
Low Grade Sulphide Quebrada	Chuquicamata	Heap Leaching Sx-Ew	30	15
El Teniente	El Teniente	New Mining Area	70	50
Leaching in Crater	El Teniente	Leaching Sx-Ew	45	27
Expansion	Andina	Mine/concentrator	220	65
Oxides	Salvador	Heap Leaching Sx-Ew	13	8

[*]Estimated figures

Geological Exploration and New Mining Projects

Besides production expansion and consolidation, Codelco is
stressing its mining exploration activities, as part of its
strategy to increase competitiveness. Exploration will enable
the company to renew its depleting mining resources base and add
new high grade low cost resources. Special emphasis is placed on
this aspect since one of the weakest elements of Codelco´s
previous strategy was the lack of an appropriate exploration
policy.

Table V - Exploration Expenses

Year	Expenditure in Codelco (US$ million)
1986	1,8
1987	1,9
1988	2,7
1989	2,0
1990	5,8
1991	10,0 *
1992	18,0 *

Comparison of Exploration Expenses
Annual Averages 1986 - 1989

Chilean Middle-Sized Mining Co.	International Mining Co.	Codelco-Chile
US$1-2 million	US$10-90 million	US$ 2 million

*Budgeted figures

For this reason, Codelco places high priority in investments that will increase its knowledge about the geological characteristics of current deposits in exploitation and adjacent areas. Obtaining new information will allow better mine planning and more flexibility in mine development within the company's operations.

Codelco's exploration strategy goes beyond its current operations. It also covers other geographical areas in Chile, placing special emphasis on the search for copper-bearing porphyry and precious metal deposits.

This policy is supported by the fact that Codelco holds title to a significant portion (around 30%) of Chile's permanently established mining property. Only a small part of this property corresponds to the four deposits currently under exploitation and to other known projects that are being assessed. Therefore, Codelco intends to take advantage of this fact by searching for, assessing, and developing new mining prospects or projects that may be located on this property.

Codelco has limited resources to carry out an aggressive assessment of this property. The mechanism that has been developed for overcoming this restriction is through joint-ventures with other companies, either Chilean or foreign, thus increasing available resources and combining the abilities of Codelco and those of other companies.

Under Chilean law Codelco must obtain legal approval to dispose of its mining property (excluding the four deposits currently under exploitation), and to enter into joint ventures with third parties in order to carry out explorations and new mining projects. This legal modification is being discussed in Congress.

Environmental Commitment

Codelco's competitive strategy also takes into consideration the environmental impact caused by its mining operations, a concern that is consistent with objectives and policies established by the present Chilean government. Although in the past the company developed a number of projects to this end, it intends to strengthen environmental protection activities in the future.

For this purpose, Codelco is adopting several measures, as well as committing significant investment resources to adequately counteract the impact of its operations on the natural and human environment, both within the production installations as well as on the wider surroundings in which such operations take place.

Efforts to reduce contamination are mainly oriented to controlling pollution in the smelting area, mainly in Chuquicamata and El Teniente. A new sulfuric acid plant recently started operating in Chuquicamata, a fourth acid plant will be constructed at the same Division over the next few years and Chuquicamata will replace its reverberatory furnaces, that are much heavier polluters.

Simultaneously, an international tender has been put out for the construction of an acid plant at El Teniente.

Regarding de-contamination efforts carried out within the installations themselves, one should note several programs that are being implemented to improve the collection of fugitive gasses in every smelting plant. Also, measures have been taken to improve dust collection in the different operations.

<u>Commitment to Market Development and Strategic Alliance with other Companies</u>.

As already mentioned, Codelco is interested in increasing its interaction and cooperation with other agents in the industry, as well as in committing itself to improving efforts that contribute to the dynamic development of the copper industry in the medium and long-term.

To this end, Codelco is stressing actions and financial contributions that, in collaboration with other major producers, will support promotion activities and research on the uses of copper, both through ICA and local promotion centers. Additionally, the company intends to play an increasingly active role in promoting these efforts (See Table VI).

Table VI - Expenditure in Copper Promotion and Research
(thousands of US dollars)

Year	Codelco Expenditure
1986	905
1987	1,408
1988	1,624
1989	2,037
1990	2,250
1991	3,400 *

* Budgeted figures

The demand for copper in the past few years has increased significantly, contrary to the opinion that the crisis experienced by the market during the past decades would be permanent. However, the danger of a slow-down in copper consumption is an issue that has lately recovered importance, mainly due to the current fear, throughout the developed world, of a decrease in economic activity. Furthermore, the risk of a fall in the use of copper is a permanent threat to the industry. This, in turn, emphasizes the importance of the strategies implemented by the industry to increase copper demand.

Codelco´s strategy is to increase its influence and leverage within the industry, and to support marketing and promotion activities that allow the company to reach end consumers. To implement this strategy, Codelco is planning to enter into strategic alliances and joint ventures with other leading companies in the industry both at home or abroad. This measure has the additional advantage of enabling the company to benefit from the other companies´ contribution in management skills and technological know-how. These alliances can take the form of joint investments in semimanufacturing abroad or in copper mining or processing in Chile and other countries.

Codelco´s interest in doing business with third parties does not focus only on adding company strengths through their association. Codelco also wishes to favor future interaction and collaboration within the copper industry, and support the more cooperative phase that the industry has entered.

CONCLUSION

In brief, it can be said that, upon completion of the copper industry´s restructuring during the eighties, the industry has evolved towards a phase characterized by an increase in mutual trust, commitment, interaction and collaboration among the major participating companies, without involving an impairment to their individual strategies to increase competitiveness.

Thus, interesting possibilities have opened up for leading companies in the industry -such as Codelco- to implement actions that will help to reinforce the companies´ strength in the medium and long-term. Such actions are related mainly to research on copper uses and on increasing copper demand; increasing transparency in providing information within the industry; and establishing strategic alliances to implement new projects or enterprises.

In this context, Codelco´s development strategy for the nineties is guided by two main objectives. On the one hand, the company aims to strengthen its competitiveness and leadership within the copper industry by reducing costs, consolidating the expansions done over the past decade, developing new profitable projects and placing more emphasis on exploration. Additionally, Codelco is interested in entering into associations with other companies in the industry, in support of its strategy of horizontal development and international business, and as a means of strengthening the future development of the industry.

REFERENCES

1.   Annual Reports of Copper Producing Companies, Phelps Dodge Corp., RTZ Corp., Asarco Inc., Cyprus Minerals Company, Freeport - Mc Moran, BHP, Outokumpu Oy, 1989.

2.   J. Bande, "Key Strategic Issues for the Development of Codelco in the 90´s", Paper presented at <u>Metals Week Copper Conference</u>, Orlando, U.S.A., January 7-8, 1991.

3.   Brook Hunt & Associates Limited, <u>Western World Copper Costs. 1988 Edition,</u> England, 1988.

4.   Codelco-Chile, "Objetivos y Estrategias Corporativas dentro de la Industria Internacional del Cobre", ("Corporative Objectives and Strategies in the International Copper Industry"), Subdirección de Planificación, Dirección de Planificación y Desarrollo, 1990.

5.   Codelco-Chile, "Análisis de la Posición Competitiva de Codelco-Chile dentro de la Industria Internacional del Cobre", ("Analysis of Codelco-Chile´s Competitiveness within the International Copper Industry"), Subdirección de Planificación, Dirección de Planificación y Desarrollo, 1990.

6.   Codelco-Chile, "Base de Datos", ("Data Base"), Subdirección de Planificación, Dirección de Planificación y Desarrollo, 1990.

7.   CRU Copper Studies, "The Trend Industry Concentration", Vol. VI, Nº 8, December 15, 1978, 1-5.

8.   CRU Copper Studies, "Oil Companies in Copper", Vol. VIII, Nº 8, December 18, 1980, 1-10.

9.   CRU Copper Studies, "Oil Company Involvement in Copper Production", Vol. XII, Nº 4, October 1984, 1-8.

10. C. Díaz Alejandro, "International Markets for Exhaustible Resources, Less Developed Countries and Multinational Corporations", Research in International Business and Finance, Vol.1, JAI Press, 1989.

11. J. L. Mardones, I. Marshall, and E. Silva, "Chile y CIPEC en el Mercado Mundial del Cobre", ("Chile and CIPEC in International Copper Market"), Estudios Públicos, Centro de Estudios Públicos, Vol. Nº.15, Winter 1984, 5-38.

12. T. Moran, Multinational Corporations and The Politics of Dependance, Princeton University Press, Princeton, N. Jersey, U.S.A., 1977.

13. A. Noemí, "Exposición en el Seminario Internacional sobre Competitividad en la Industria del Cobre: Presente y Futuro", Presentation at the International Seminar on Competitiveness in the Copper Industry: Present and Future, organized by CESCO, Santiago de Chile, December 10, 1990.

14. E. Silva, I. Marshall, and J. L. Mardones, "Estructura de la Industria del Cobre y Políticas de Acción en el Mercado", ("Copper Industry Structure and Market Action Policies"), Estudios de Economía, Universidad de Chile, No.19, Second Semester 1982, 121-143.

The future competitiveness of U.S. copper

M. Rieber
Department of Mining and Geological Engineering, University of Arizona, Tucson, Arizona, U.S.A.

ABSTRACT

Long-run mineral competitiveness depends on grade and tonnage, but the timing and extent of the relative decline and recovery of U.S. copper competitiveness over the past decade is not explained by differential geology, wage rates, or environmental factors. The first two differentials have existed at least since the 1950's; the last has by now been largely capitalized. The factors analyzed here to explain past and anticipated U.S. copper competitiveness are exchange rates, by-product credits, sovereign loans, leach-SX/EW, state-owned mines, and embodied copper imports.

As world-traded copper denominated (primarily) in dollars is produced for local currency, shifts in exchange rates alter the effective prices for buyers, sellers and U.S./non-U.S. operations providing the basis for the historical trade analysis, suggesting future trade patterns. The related problem of unaccounted copper embodied in manufactured imports shows the competitive impact of the reduction from potential base domestic demand from U.S. mines.

U.S. comparative advantage is low cost of capital. This is dissipated by the granting and/or renegotiation of fungible sovereign loans, particularly where mines are state-owned or controlled, at less than full risk-adjusted market rates, reducing foreign mine borrowing costs to less than those in the U.S. Statist mines are wage-bill maximizers, not profit maximizers; they need not be efficient or respond to prices.

The inclusion of by-product credits to determine relative competitiveness, leads to anomalies in mine rankings as the credits vary over time, independently of copper.

U.S. responses emphasized higher cut-off and mining grades, capital substitution and, where possible, shifting towards leach-SX/EW to reduce weighted average costs. The first reduces reserves; the last is subject to special depletion of oxide and non-pyrite sulfides.

INTRODUCTION

Any explanation of U.S. copper competitiveness must be sufficiently robust to explain simultaneously and consistently: (1) the dollar price of copper, annual production rates, and inventory levels over the past decade; (2) why the burden of the supply reduction fell primarily on U.S. producers rather than on all producers; (3) the opposing interest of U.S. copper producers and manufacturers with respect to a protective tariff; and (4) perhaps most important, the timing of the first three observations. The general arguments are presented here followed by a more detailed discussion. The discussion is simplistic; other factors are analyzed in this paper, but it covers both the observed facts and their timing.

It can be argued that U.S. ore grades are low relative to those in Chile, Zambia, Zaire and Peru; that by-product credits are low compared to those in Canada, Sweden and Australia; that wage rates are high compared to everyone. However, this has always been true. It was true in 1979, when optimists argued that copper prices would reach $1.50/pound and pessimists were satisfied with $1.25. The oil companies which purchased domestic copper producers did not think they were buying dogs; and they were well aware of the problems of air pollution control, safety and reclamation.

To examine U.S. competitiveness it is useful to eliminate geotechnical considerations as far as possible. It is not that they are unimportant. In the long run, they are decisive; but for explanation, it is helpful to have as few variables as necessary. The direction of causality may even be reversed. It is less useful to point out that Zambian cut-off grades exceed Chilean mining grades than to ask what must the ore grade/tonnage relationship be if Zambian mines are to operate, if Escondida is to be feasible or, if Ertsberg did not exist, would the existing ratio permit the feasible operation of Grasberg? Besides, since 1979 the geotechnical variables have been constant, though the price of by-products has changed. Arizona and Sonora, Mexico, provide a reasonable comparison. Here it should be noted that if more attention is paid to Mexican-U.S. examples than to those of other countries the purpose is not invidious comparison. Rather, it is the overall similarities that permit a useful concentration on the results due to the differences.

Copper deposits in Sonora and southern Arizona are neighbors in the same copper-belt. They do not differ significantly; ore grades, stripping ratios, and by-products are all similar. Significant wage differentials have existed for decades, partially offset by greater U.S. productivity and, more recently, by domestic wage reduction as well as sharp productivity increases, some of which is due to leaching. On the other hand, until recently both capital costs and the cost of capital were lower in the U.S. than in Mexico, in part leading to the greater U.S. productivity. Air pollution control is an added cost in the U.S. (far less than in Japan, where APC regulations and penalties are more stringent than ours), but some of the cost is offset by increased smelter productivity and decreased fuel use. Furthermore, some of the APC costs were undertaken prior to the recent fall in prices. APC hurt, but it did not cripple U.S. producers. The local problem is its differential application.(1) It is the uncertainty of environmental mandates, the forcing of an industry to hit a set of moving targets which, like projected changes in the mining law and mineral land taxation, creates in the U.S. a country risk premium, not unlike that in the LDCs, which must be added to the familiar project, commodity and market risks of project feasibility analysis. Existing environmental requirements affect cost levels; it is variability and uncertainty that affects risk.

The following is a precis of a continuing, two-year project that was
begun on the basis of a very small U.S. Bureau of Mines grant through the
Arizona Mining and Minerals Resources Research Institute and for which the
author is extremely grateful.

THE ROLE OF EXCHANGE RATES

Only two arguments are required to explain the four observations cited
above. The first is the change in the trade-weighted dollar over the period;
the rise in the dollar against the major European and Japanese currencies
through February 1985 and its fall after February 1988. The second is an
adjustment lag which is longer for supply than for demand. On the production
side this occurs because facility closures and reduced output are related to
prices exceeding average variable costs, while expanded ouptut and reopening
of facilities depend on prices exceeding average total costs. Arguments concerning the need to continue production in order to generate foreign exchange
or to preserve jobs may be contributory, but are not required. They are
further discussed below.

The exchange rate argument is conditioned by two factors: (1) the
reserve currency status of the dollar, and (2) the denomination of primary
(but not manufactured) copper sales in dollars for purposes of world trade.
The distinction of reserve currency status cannot be sought; it is foist upon
a nation. However, it has important implications for U.S. industries competing in world markets, even if those industries do not export, but only face
import competition. Reserve currency status is exemplified by, among other
conditions, a dollar-denominated U.S. external debt and an open economy - the
buyer of last resort, especially for producers with weak currencies.

Similarly, the practice of pricing world traded copper through the
refined stage in dollars, irrespective of the countries of origin and destination, which is common for many commodities, is a matter of trader convenience.
It cannot be a national policy and, in general, it is not followed for
manufactured products. However, only U.S. production costs are denominated in
dollars; those in other nations are counted, primarily, in local currencies.
For primary producers, this may be altered in the long run as mines must be
expanded to maintain ore grade and additional or replacement equipment is
purchased.

For the general argument, only four groups of participants need be identified: U.S. and foreign primary copper producers and U.S. and foreign copper
manufacturers. In the U.S. there is little vertical integration beyond the
refining stage. To 1985, the dollar rose against the currencies of both
foreign copper producers and manufacturers. For manufacturers in, say, West
Germany and Japan, this constituted a price increase in terms of their own
currencies and a consequent decrease in demand. The price increase was sustained, but at decreasing levels, until the later, resultant fall in the dollar price of copper exceeded the fall in local currencies relative to the
dollar. Dollar-denominated copper prices increased for foreign producers as
the value of the dollar rose relative to their currencies and as the revenues
received were converted to their own currencies. Depending on their inflation
rate (and as compared with that in the U.S.) and their indexing of domestic
prices, their real costs in local currencies remained constant or fell. In
local currency terms, profits increased, justifying an increase in production.
Furthermore, local currency profits could be maintained while granting buyers
a lower dollar price for copper.

Over time, supply exceeded demand at the pre-existing dollar prices; therefore, the dollar price of refined copper fell and copper inventories increased. This result would be expected to continue until: (1) the value of the rising dollar stabilized, (2) purchasers achieved a new equilibrium at which the lower dollar price of copper was matched by the higher price of the dollar in local currency terms, and/or (3) non-U.S. primary copper producers found that their local currency price of the lower dollar-denominated copper price was less than their average variable costs in local currency terms. This last condition implies a lag. The time lag is extended by debt renegotiations which reduce the implied local currency outflow burden of a debtor nation.

In general, U.S. copper manufacturers are no longer integrated with domestic primary producers. Because copper semi-manufactured products are not dollar-denominated, even though the dollar price of refined copper fell for both foreign and U.S. buyers, as the value of the dollar increased a U.S. manufacturer could purchase, say, Yen or Deutschmark copper for fewer dollars. As a result, copper product imports rose, reducing the market for U.S. primary copper producers. This also resulted in the establishment of the divergent tariff position in the U.S. industry; primary producers wanted a tariff, manufacturers did not.

The U.S. primary copper producers faced a decline in U.S. manufacturers' demand due to imports, a decline in foreign manufacturers' demand at least until the dollar price of copper fell sufficiently to overcome the effect of the rising dollar, and at least a lagged response by foreign primary copper producers. The dollar price of copper fell, but as U.S. costs are in dollars, prices fell below average total costs and losses were incurred. U.S. output was reduced. As price fell below average variable costs at some mines, they were closed. Output was reduced in the U.S. rather than elsewhere. This situation persisted until the dollar copper price fell sufficiently so that average variable costs for primary copper could not be covered even in local currencies (e.g. Zambia, Zaire and the Philippines).

From February 1985, the trade-weighted dollar fell; it depreciated against the currencies of the European and Japanese copper buyers and manufacturers, but continued to rise against the more rapidly depreciating currencies of the primary copper producing nations. Foreign manufacturers found that the local currency cost of copper inputs fell with the dollar and rose, later, only slightly as the dollar price of copper increased. Their demand increased. Foreign copper producers faced with lower dollar revenues (on equal sales volumes at pre-existing prices) were still profitable in local currency terms due to their currency depreciations. Profits were lower than before so that supply increased more slowly, but they were competitive. The result was that demand increased faster than supply, copper prices rose in dollar terms, and inventories fell.

U.S. copper manufacturers faced a rising dollar price of copper and foreign copper imports which, denominated in foreign currency, required more dollars to buy. The dollar price of copper rose for all manufacturers, but the U.S. manufacturer faces competitive problems as their import competition now has lower copper input costs in local currency terms. The U.S. primary copper producer became more competitive and expanded output due to a rationalization of dollar production costs made to offset the dollar price of copper decline and the increase in the dollar price of copper; profits were again being made. U.S. manufacturers' competitive problems led to a wider opening for domestic refined copper sales. Supply increased, but with a lag leading to inventory problems which, by their nature, are short-run.

It has been argued that exchange rates used for comparisons should be adjusted by the relevant purchasing power parity. Presumptively, in the long-run, exchange rates equate the values (prices) of internationally traded goods. This is the purchasing power parity (PPP) concept: that rate of exchange which equates the price of a market basket of traded goods and services in country A with a similar basket in country B. Alternatively, a dollar should buy as much in the U.S. as a dollar's worth of pounds does in England. Unfortunately, the concept works well only among roughly similar economies. Difficulties arise if the non-traded (domestic) goods sector is large. If a large portion of the economy is non-market, the concept is almost irrelevant. This distinction is clearest when comparing developed with less developed countries, but it is also apparent between market and socialist economies. If automobiles are a part of the market baskets of both Germanys, it remains that a Trabant is not a VW, yet its price does not reflect the differential. PPP will vary depending on which currency is used, the dollar or the Kwacha (Zambia). It will also vary depending on which market basket is used. Commodities generally used or available in the U.S. (Zambia) may not be so in Zambia (the U.S.) with distorting effects on prices in local currencies. Therefore, even relative PPP (inflation adjusted) is unlikely to be meaningful among dissimilar nations. A 20% inflation rate in Zambia versus a 5% rate in the U.S. is unlikely to lead to a 15%/year fall in the K/$ rate. Similarly, a doubling of the Zambian money supply is unlikely to result in a halving of the rate.

For U.S. primary copper producers, all of this suggests that even in the long run, where adjusting exchange rates by divergences from PPP might be justifiable, relative costs and prices adjusted by exchange rates are useful for comparisons only among other developed country primary producers (e.g. Canada, Spain) and with respect to the trade to and among OECD nations of refined, semi-, and manufactured copper products. In the short run, where exchange rates are driven more by capital transfers responding to interest rate differentials than by balance of trade surpluses or deficits (the former are overwhelmingly larger in the balance of payments), PPP is largely irrelevant to the assessment of exchange rate impacts on copper trade or comparative costing at all production levels and among all competitors.

As the foregoing is somewhat abstract , it is useful to add some detail and glance at the future. Unfortunately, the primary copper industry cannot influence, but can only anticipate and react to, exchange rate changes.

Both dollar price denomination and the role of the dollar as the reserve currency are changing, but incompletely and slowly. Over time, more copper will be denominated in such other currencies as the yen and the ecu (if the latter finally becomes a fungible currency). There is some evidence that the role of the dollar as a reserve currency is being increasingly shared with the D-mark and the yen. Nevertheless, the contrapuntal dance between gold and the trade-weighted value of the dollar (its generalized exchange rate) around a common trend demonstrates that both are considered international monetary reserves.

The strong dollar limited U.S. copper exports, but with an open market led to increased imports. As neither the Bureau of Mines nor the Department of Commerce collects data on the export or import of semi or manufactured copper, much less make estimates of the trade in copper embodied in manufactured assemblies, net imports cannot easily be documented from existing data. However, domestic manufacturing firms increasingly used imported refined copper while increased imports of cars and electrical equipment further limited the U.S. market for domestic primary producers. Unfortunately, downstream manufacturing capacity was reduced, losing assured market advantages of vertical integration. A recent analysis (2) shows that copper embodied in net imported finished and semi-manufactured goods has been increasing since 1975 and more rapidly since 1980. From 1970 to 1980 it averaged 8.8% of U.S. apparent copper consumption but, by 1985 when net embodied copper imports exceeded 577 thousand short tons, it was almost 25% of apparent consumption. Forecasted embodied copper net imports were 28% and 41% of 1990 and 1995 domestic copper consumption respectively.

The mid-1980s fall in the trade-weighted dollar should have led to an increase in domestic exports, a reduction in imports, and a rise in all dollar-denominated mineral prices, including copper prices. Of the last, about one-third can be attributed to the dollar's decline against major industrialized nations' currencies. Against the currencies of the major copper producers, including Mexico, until recently the dollar rose, not fell, in large part due to the relative inflation rates of those countries. Existing mines' expansion required large infusions of capital. Presumably, this capital was borrowed or was fungible with capital that was borrowed by governments, especially in the case of government owned companies. Ordinarily, as the peso rate of dollar exchange became worse, the Mexican inflation rate higher, and as the fixed bond interest rates and principal are dollar-denominated (if not in some even higher-valued currency), repayment would become onerous and losses would be made on the new investments, including the LaCaridad smelter and acid plant, the Cananea concentrator, and the expansion of both mines. Mexican mines should be contracting. Mexican copper costs should be high in both dollar and peso terms. Until recently, this was the case.

The recent decrease in the value of the dollar again does not directly affect domestic costs. The indirect effect is through better capacity utilization and the capitalization of prior cost cutting; but the decrease may have slowed the pace of copper imports, both overt and embodied. Two caveats must be noted: (1) While foreign and domestic-owned U.S. copper concentrate exports may increase, unless OECD copper import trade barriers are reduced, particularly the non-tariff barriers, U.S. exports need not increase. (2) The fall of the dollar is usually measured in terms of OECD currencies, but the major copper producers are not OECD members. Compared to the currencies of Zaire, Zambia, Mexico, Chile and Peru, the dollar has risen. Those exporters still have an exchange rate advantage. Nevertheless, the fall in the value of the dollar as well as competitive cost cutting and despite APC regulatory costs has led to a revival of producer profits.

Regardless of exchange rates, the new domestic profitability can disappear if productivity declines or if wage rate increases exceed those of productivity gains. Both are consistent with an approach to U.S. industry-wide full-capacity utilization. On the labor side, it is marked by a return to former work rules and job classification. It is measured by the premium of mine wage rates and benefits over those for comparable jobs paid for by the state, local counties and private contractors. On the capital side, productivity declines as routes are stretched, equipment is stressed, the mining and processing pace is increased, and lower ore grades are utilized. Finally, to the extent that profitability rests on SX-EW and increased oxide leaching, the results may be short-term. Given the ratio of domestic oxide to sulfide ores, increasing the rate of development of the former is akin to high-grading, a shortening of the production horizon. The cost advantages of SX-EW, though smelter operation and the attendant APC regulations are avoided, are limited by the availability of the proper ores, and whether the ores are already mined (and thus separably depletable) or must be mined and the leach sites prepared.

With rising dollar interest costs and adverse exchange rates, except for the actions of the banking community and the institutional lenders, Mexican copper mining costs would have increased perhaps sufficiently to offset increased revenues and peso profits; mine output would have fallen. Planned new capital investment in copper was delayed.

The banks' and financial institutions' first salvage effort should not be considered lending at sub-commercial rates. The reorganization was similar to the results of a Chapter 11 filing for bankruptcy, with the important provision that neither the assets could be seized nor the debt swapped for equity. To the lender, going concern value was important beyond the consideration of the effect of a formal default. The contraction of Mexican mine output did not take place. U.S. copper output bore the brunt of the impact due to an inflated dollar and the Mexican rescue.

As interest rates subsequently fell, the debt burden on existing dollar denominated capital fell. However, in peso terms the advantage was reduced by Mexican inflation and the continued rise of the dollar, but at a slower rate. If new capital can neither be acquired locally nor by an increase in equity, foreign borrowing is required. With a rising dollar, a falling peso, increasing foreign debt and increasing interest rates, project finance would be unlikely unless the prospects were unique. Peso finance became available because loans to the Mexican government are fungible. They were made because the financial community thought that with additional help Mexico would grow out of its debt, enabling a repayment. Additional help came from the Export-Import Bank in the form of trade credits for the Mexican purchase of U.S. commodities.

Even though interest rates subsequently declined, the adverse exchange rate and the size of the Mexican external debt would preclude financing beyond that required for rescheduling. Nevertheless, with the strength of the weak, money is available on terms unavailable to the U.S. copper industry. Rather than an increase in foreign (Mexican) copper costs over time relative to U.S. costs due to an adverse cost of capital, the reverse is quite possible. (1) An anticipation of loan rescheduling reduces Mexican risk. (2) Sales of government and private debt at the current commercial rate would reduce the burden. (3) Even when financed in dollars, equipment purchases (such as acid plants) should be made in nations with softer currencies (Canada).

STATE AND PRIVATE OWNERSHIP

It has been asserted that copper producers do insufficient research and development. However, technical progress in production (if not consumption) is evident; it is the suppliers who do the R & D. One result is technological competition; the other is rapid world-wide diffusion of successful results. To an economist, as to a copper producer, this simply means that the technical production function, from mine through refining, the array of input factors leading to output, is or can be the same for all (partitioned as necessary between surface and underground mines). Locally, the relative prices of each factor will vary; capital is scarce and expensive here, but labor is scarce and expensive there. Given the geophysical and transport conditions, the technical production function, and relative factor costs, one can develop the costs of production or process costing. As commodity prices change, one can also develop the reaction of the firm to fortune and adversity.

During the early 1980's, with copper prices low, it was claimed that LDC copper producers maintained output, even in the face of losses, to generate scarce foreign exchange, resulting in U.S. producers absorbing the major share of output reductions. One may note in passing that if copper is a major exchange source for an LDC, over time sales below costs can only be sustained by internal subsidy (reducing living standards) or by external subsidy in the form of international commodity adjustment or other forms of loans. If true, neither source is likely to be easily available in the foreseeable future.

It is more likely, however, that the reaction of the LDC producers vis-a-vis U.S. producers was, and to some extent still is, a function of the organization of the producing firm. In general, U.S. companies are profit maximizers; this determines their reaction to fortune and adversity, given their costs as derived from their production functions. Elsewhere, in LDC's, eastern Europe, and even parts of western Europe, to a significant degree firms are organized on statist or labor-managed lines. In general, what is maximized is the return to labor.(3) The differences are important as it serves no useful purpose to analyze or complain about the action of one type of firm on the basis of the principles of another; each is simply doing its assigned job.

It is not necessary to develop the two theories in order to identify the results. To a profit maximizer, in general, output decisions are directly related to changes in profit. The factors of production, labor and capital, are suitably adjusted as is land, in the form of cut-off grades, ore grades, and exploration effort. To a labor-return maximizer, employment is fixed or increasing and offsetting adjustments are made in the other factors, including (now) output. As copper is a standardized product, sold competitively on a world market which cares nothing about the internal organization of enterprise, it is revealing to see how each reacts to a competitive reduction in price, given a drop in excess demand.

The hallmark of a statist firm is an excess of labor, which usually cannot be readily dismissed, leading (usually) to low productivity. If labor returns are to be maintained as price falls, output must be increased. To offset a net earnings decline, land may be sacrificed by high-grading, maintenance will be reduced, and new capital equipment will be forgone if its purchase interferes with the return to labor. Amelioration is possible if subsidies are available or if the statist firm is actually a state firm. There may be no difference between long and short run if subsidies are continuous. If some of this seems familiar to U.S. copper firms, it is only because some of them appeared to be "labor-managed" in the 1970s and a few oil companies provided the subsidies.

A return to a more profit-maximizing orientation in the U.S. copper industry began with the strike against Phelps Dodge which, as a non-diversified, independent copper company had no source of internal or external subsidy. The changes made there, and elsewhere, not only highlight the reactions of profit maximizers, but point to the degree to which the industry had effectively become labor-managed.

Short-run, as prices fall, ore grades and mill-head feed grades rise to reduce cost. Cut-off grades rise and stripping may be postponed, reducing reserves. Maintenance may slip and avoidable expenditures are forfeited. In short, operating costs must be covered, but they can be reduced. In the short run, output will be maintained leading, in the face of decreasing demand, to excess inventories. Given the length of labor contracts and the costs of a strike, wage rates and/or fringe benefits may increase while lay-offs are unlikely. The result was exemplified at Bingham Canyon.

Long-run, if the enterprise is to continue despite low prices, adjustments are made to the productive factors and new production schemes are inaugurated. The labor force, restrictive practices and wage rates, are reduced,
increasing productivity and reducing the wage bill. Physical capital is
replaced, modernized and may be expanded (along with output) if funding can be
obtained and the reduced unit costs can be shown to justify the expenditure.
The expansion of leach-SX/EW to sulfide ores is simply an example of the
implementation of a new production function, one with a capital, not a labor,
bias. What is relevant is its diffusion and its timing. Again, this is very
familiar.

For the U.S. copper industry to survive long term, the consensus seems to
indicate a weighted average cathode cost of $.65-.70/lb. (1990), probably at
the lower end of the range. Leaching helps, but is limited by the oxide
reserves and the amenable sulfides. These, including dumps, amount to a
reserve subject to a special depletion. For some companies, if the SX-EW
cathode is assumed to be $.35-.45/lb., their conventionally mined and pro-
cessed copper exceeds current cathode prices. It is probable that labor-
saving shifts in production or wage rates must yet occur. Clearly, the
mine/mill complex is the most plausible location.

To measure the differential wage rate in U.S. copper mining, data were
obtained on wages and fringes on a job specification basis. These were compared with the wages for the same specification paid by government (city,
country, state) and by private contractors as reported averages. Not
surprisingly, on a job-by-job basis, the mines paid more, even without consideration of such hidden fringes as low cost housing. Mine productivity may be
higher, but is must be demonstrated that it is compensating.

On the other side, the privatization of statist firms will, in the long-
run, increase their productivity and reduce their costs, even more than did
the elimination of the labor-managed firm in the U.S. U.S. firms were never
as overstaffed as, say, Cananea, nor was the infrastructure and physical capital allowed to deteriorate to the extent found in Africa. In the short-run,
which may last quite a few years, the shift from statist to capitalist
precepts will bring labor unrest to many countries, impacting supplies, and
providing at least near-term opportunities for U.S. firms with respect to
price and customers.

SOVEREIGN DEBT

It is almost true that all LDCs are characterized by relatively low-cost
labor while all developed countries exhibit low-cost capital. In trade terms
these are the bases of their comparative advantages. The explosion of sovereign LDC debt since 1975 and the renegotiation of its terms since 1982 has
done much to reduce the comparative advantage in the cost of capital enjoyed
by the primary copper producers in the U.S. It is the relative cost of
capital in the future which, with the role of privatization, is likely to
determine the competitiveness of U.S. copper.

Indeed, were it not for international lending, it is doubtful that the
LDC primary copper producers could have maintained output for as long as they
did in the face of adverse exchange rates, lower prices, and the efficiency of
statist organizations. Without an effective cartel or UNCTAD commodity agreement, a competitive world market cares little for high cost producers. International lending, and the renegotiation of its terms, will continue
irrespective of the impact on those adversely affected.

In competitive terms, for the US copper industry the availability of
international loans, particularly sovereign loans, provides the LDC recipient
with funds not domestically available for projects or sales which might otherwise not be commercially feasible. If the recipient's sovereign interest rate
is below the appropriate (copper) project rates; if the loan terms are
concessionary with respect to market rates, payment periods, or contain a
forgiveness period; and if the loan is (or is expected to be) renegotiated,
LDC competitors' cost of capital is reduced, and may well be less than that of
the US producers' market determined or internal rate. The impact is related
to both costs and the feasible time horizon of a mining project.

Both earned and borrowed hard currencies are fungible assets. Even funds
borrowed for a specified, limited purpose simply frees other funds for expenditure elsewhere: the military, the mining industry, mineral commodity export
subsidies, though there may be internal transactions costs due to the
shifting. Where the government owns (in whole or in part) or regulates a
minerals project the shifting of funds is eased, the transactions costs are
reduced, and the project cost of capital approaches the sovereign borrowing
rate.

India provides a classic example of sovereign loan shifting. Its
$200/capita/yr average income qualifies it as a major recipient of aid at
concessionary interest rates and repayment schedules. With China and Ghana,
it is the largest recipient of World Bank interest free International Development Association loans. Yet it maintains the world's fourth largest military
force at a cost estimated in 1988 at $4-6 B/yr (4) and in 1990 at $9.3 B.(5)
Between 1985 and 1989, its imports of major weapons ($17.35B in 1985 dollars)
was the highest in the world, with Iraq coming second at $12 billion.(6)

The subsidy element, if any, in the cost of capital borrowed by the LDCs
involves initial and effective interest rate differentials from appropriate
commercial bank rates, off-loan costs or conditionality, and added capital
access. Central to these is the role of the international financial institutions (IFI's) including the World Bank, its associated Asian, African and
InterAmerican Development Banks, and the International Monetary Fund (IMF).
Their importance lies less in their share of lending than in their almost
necessary presence before commercial banks will subscribe to a large-scale
loan, sovereign or otherwise, and the terms of their subscription; this is
leverage.

If IFI loans, or domestic loans made possible by the availability of IFI
lending, are made with non-market objectives, projects need not be commercial
per se, might be uneconomic but for the loan, hence lead to excess capacity.
Here the loan is, itself, a subsidy. Excess provisions conditionality,
increases costs, but it is not clear that such conditions are ever fully
carried out.

The basis for an IFI loan implies a subsidy. For example, the Article of
Agreement III 4(ii) of the International Bank for Reconstruction and Development (the major portion of the World Bank) states that the Bank lends for a
project when "the Bank is satisfied that in the prevailing market conditions
the borrower would be unable otherwise to obtain the loan under conditions
which in the opinion of the Bank are reasonable for the borrower." Simply
stated, commercial interest rates and/or loan conditions would be more onerous. The calculation of benefits for infrastructure, beyond that needed for a
project, or which may enhance another project, yields a credit worthiness
greater than that perceived by a lender contemplating financial rates of
return. The result is expanded, not necessarily low cost projects, possibly

excess world capacity and, where private lenders are associated with IFIs, additional project risk.

As an indication of the effect of renegotiation on interest rates, the Mexican experience is revealing. From 1983 through 1986, as Mexican debt problems increased, its interest premium on sovereign debt, as a percent over the London Interbank Offer Rate (LIBOR), fell steadily from 2.25 to 0.81.(7) Recently, in the face of considerable pressure to increase lending, over 45% of the banks holding Mexican bonds elected to drop the interest rate to a fixed 6.25% under the Brady plan, in contrast to the Cete rate (for Mexican bonds akin to 30 year U.S. Treasuries) in early 1990 of about 46% per annum. For comparison with the Brady Plan renegotiated Mexican interest rates, in early February 1990, the one year LIBOR rate was 8.62%, the U.S. prime was about 10%, with 30 year Treasuries at 8.5%. In Japan, 10 year governments yielded 6.62% while in West Germany the same maturities yielded 7.9%. In all these cases the risk was far less than that of any LDC. The differential provides some measure of the subsidy element.

The explosion of debt and its international default is not new. The origin of the recent debt problem began on the lender's side with the necessary relending of petrodollars and, on the borrower's side, with excess demand. Where the use of the funds was consumption, social services, the subsidization of public and private industry, and generally inefficient investment, a surplus for use in repayment was not generated. The import barriers of developed nations (despite GATT) on manufactured rather than raw materials reduced value added, slowed development and skewed trade. So too did LDC trade barriers which raised domestic prices by supporting monopoly.

Debt serviceability, or sustainability, depends on growth rates in GNP or net exports and the real (inflation-adjusted) interest and exchange rates. Where problems arise, creditor responses can be divided into self-interest yielding no new loans, risk or exposure limitation, loan loss reserve augmentation and debt sell-outs or swaps. Cooperation (forced or genuine) implies loan extensions, forgiveness, renegotiation to reduce effective rates, risk limitation by the provision of loan guarantees and the provision of new money. Debt service (amortized or levelized principle and interest) can be sustained even with floating exchange rates, as long as export revenues are rising. As the interest rates (LIBOR +) depend largely on OECD country rates, an LDC takes a long-term gamble. Export dependency on a single commodity (copper), coupled with recession and/or a weak commodity price, will push an LDC to increase exports as price falls, if only to maintain debt service.

To 1980, the growth rate of LDC exports exceeded the increase in the interest rate at which they could borrow. Their credit worthiness was maintained. From 1980-86, the growth in interest rates exceeded that of exports, so that the ratio of debt service to net exports grew. To reduce net exports, a trade surplus was necessary: either increase exports or reduce imports which, for many LDC's, would have created severe hardship. To reduce debt service, it is necessary to renegotiate principle and interest, or reschedule the debt or unilaterally limit or repudiate it, or engage in debt/debt swaps at better terms. Alternatively, as the value of the bonds fall in the secondary market, they can be repurchased by the issuer, as U.S. firms do when their own prices fall, but this requires hard currency which, at times, has been available from later loans. An interest cap is possible, but as lender bank earnings and share prices would fall, banks would limit lending and accept far less risk.

A debt/equity swap, if to local buyers, simply shifts sovereign to private debt, unless the funds were found by non-local, "silent partners" who
provided the bonds purchased at a discount and presented to the government at
face value. A debt/equity swap to non-nationals both reduces debt service and
shifts risk. This is not likely in countries with a statist or Napoleonic
Code tradition. From a foreign banks' viewpoint, added risk, especially for
long-lived projects facing commodity price and exchange risk, may be unacceptable.

It is commonplace to suggest that when in debt one must sell something.
To U.S. creditors, the unwillingness of many indebted LDC's to sell prime
assets, rather than threaten non-payment without renegotiation, appears as
unwillingness to pay. This view arises because there are very few private
U.S. assets that cannot be purchased by anyone. Indeed, the U.S. was developed with foreign funds and considerable foreign ownership. The principle of
service and repayment of sovereign debt was established by Alexander Hamilton.
Individual states have defaulted as, of course, have private ventures. Nevertheless, it was learned early that the hole in the ground, the in-place
resource, could not be exported.

The debt service burden (net interest as a percent of cash flow where the
latter equals pretax profits plus depreciation and net interest) of U.S. mining companies was 18.2%, 56% and 34.3% in 1980, 1982, and 1989, respectively.(8) This is similar to the burden of many LDC's. Even the problems of
repayment caused by recession, falling product prices and loss of
competitiveness are similar. Except for bankruptcy, the solutions of merger,
new buyers, and restructuring with bondholder losses are familiar. Foreign
investment in U.S. copper mining capacity in 1987 for the five major producers
ranged from 100% (Kennecott/Sohio) to about 20% (Phelps Dodge), averaging 39%
for the five.(9)

That debt renegotiation is neither foreign nor uniquely an LDC problem,
one needs only recall Public Service of New Hampshire, Texaco, and Bridgeport,
Connecticut. In all cases, contracts are abrogated, and contractual payments
(including amortization and interest) are reduced. The result is a lower
cost, more competitive entity. LDC debt history, however, is already providing evidence that the benefits gained may be one-time and short-lived. If US
companies can weather the additional capacity problems resulting from the
low-cost capital already provided to LDCs, they should regain their comparative advantage in cost of capital. A few notes are in order: (1) Zambian and
Zairian debt service cannot be sustained by exports. However, as almost
three-fourths of that debt is owed directly to governments, it can be for-
given; the taxpayer will have paid. The remaining debt, mainly to the IMF and
World Bank, can be serviced. (2) Even with Brady Plan guarantees, only some
10% of Mexican bond holders chose to supply new capital at 25% of their exposure. Chile, which has already done some refinancing, does not appear interested as their participation might actually reduce their credit rating. Peru
is ineligible without IMF-approved reforms. (3) Current borrowing by newly
privatized companies in Latin America is from direct investors, including
insurance companies and nationals repatriating capital, but not banks. Even
so, Mexican loans with maturities beyond 1994 are made only with a put option
to sell back to the issuer.(10)

Money-market banks, having insulated themselves against the repercussions
of their existing LDC loans are unlikely to make more junk bond money available. In the absence of another avalanche of petrodollar deposits, they will
not need to make high-risk loans to cover interest payments. If noncommercial
loans to LDCs for balance of payments coverage (sovereign) or project support
are to be make, they will come from the multilateral lenders (the IMF and the
World Bank or its affiliates) and governments.

Money-market bank participation in sovereign loans to traditional LDC borrowers, including the copper producers, is likely to be very limited over the next decade. What lending does occur will probably be traditionally project-oriented, as in the 1950s and 1960s. The U.S. lenders face significant domestic portfolio problems. Both U.S. and European banks will be interested in new, competing opportunities in Eastern Europe. They will be joined by the Japanese in the lending opportunities arising in the Middle East. Little will be left for lending to nations that have conclusively demonstrated both high risk and low returns. The need for U.S. government guarantees for Brady Plan bonds demonstrates that they are not commercially marketable at existing rates. The guarantee, with the taxpayer as the ultimate resource, is itself the manifestation of the subsidy.

COMPETITIVE ORDERINGS

In the end, competitive analyses seek to develop an hierarchical ordering, a copper supply schedule. The common, simplistic use of production costs to rank cathode output by mine and processing units or by country presents some interesting problems. Even if copper is dollar-denominated, all costs are not so denominated. International comparisons require that a stance or viewpoint be taken. Principal and interest may be repayable at various rates over varying periods in several currencies. Exchange rate variations and, where multiple exchange rates exist, choice of the appropriate rate, may alter competitive position over time when viewed in local currency terms. Operating costs, when put on a common dollar basis for comparison, are also subject to exchange rate variation. Adjustments made to reflect purchasing power parity are only useful when the countries involved are roughly similar. Furthermore, in the producer's view it is domestic profits in local currency that counts. The same issue arises periodically when comparing U.S. and Russian military expenditures - which market basket, which currency, whose prices, and what exchange rate; differences in expert estimates appear to vary widely.

The common practice of deducting by-product credits from net copper revenues also clouds competitive evaluations. If the by-products contribute so little to expected revenues that they may be safely ignored in feasibility studies, they are true by-products. If, as in eastern Canada, western Europe, and Australia, the mine is poly-metallic, it may not only be improper to classify a mine as a copper mine, but changes in the market price of the co-products over time may easily alter competitive position to the point of closure.

The use of by-product credits to offset copper production costs is useful for existing mine financial analyses and for feasibility studies when anticipated prices are explicitly stated. As used by the U.S. Office of Technology Assessment (11), or by the World Bank (12), to establish a cost-based hierarchy of copper producing countries, the results are anomalous, especially over time.

If all credits (and debits such as arsenic, acid, bismuth and mercury) moved in a constant relation to copper prices, copper supply hierarchies would be invariant; but as there is no evidence that this is so, the result is shifts in the apparent relative standing of countries in a manner unrelated to copper demand, supply, or price. Not all mines or countries are equally affected. Countries such as Chile have few such credits; others, such as Canada, would be priced out of the market were credits not explicitly included. Within countries, individual mines may differ widely.

Analytical problems arise when by-product prices drive production decisions. For the right price a company chases moly all over the mine, while
producing the associated copper. For the wrong price, the moly circuit is
unused. For political and other reasons, a ceiling on uranium output limits
the copper-gold production at Olympic Dam. In neither case is copper produced
as a simple function of prices, costs and technology. A Canadian mineral
resource publication carried the by-product credit problem to the limit. In
separate analyses of nickel and copper, each was the by-product of the other,
leading to the enviable conclusion of zero costs.

A reasonable solution is to determine which commodities are co-products
and which are by-products, based on whether they were or were not included in
the income (or cost) stream of the latest feasibility study. Then the costing
and pricing of by-products (or residuals) begins at the point of product
stream separation. For co-products it is also necessary to allocate the joint
costs. The result remains a bit arbitrary, but income and cost streams are
separated by commodity.

For practical purposes, commodity separation rather than crediting highlights competitive (low-cost) output from secondary and by-product sources,
points to the potential vulnerability of high-cost sources, and suggests
commodity analyses of alternative production methods. For the last, if leach
SX-EW is proposed for an amenable sulfide ore, what, if any, associated mineral values are lost compared to copper recovery by conventional means? Are
the losses compensated for by the lower copper costs?

Proper treatment of by- and co-product income or cost streams changes no
physical dimensions, but by establishing a reasonable hierarchy of copper supply, it will help minimize the "mining is dead" rhetoric in the U.S., which,
in turn, may lead to a more respectful audience.

REFERENCES

1. M. Rieber, "Acid, Acid Rain and U.S. Copper Competitiveness: The
    Mexican-American Agreement," *Society of Mining Engineers* (AIME), Preprint
    88-56, January 25-28, 1988.

2. K. Al-Rawahi, and M. Rieber, "Embodied Copper: Trade, Intensity of Use
    and Consumption Forecasts," *Resources Policy*, 17(1)2-12, March 1991.

3. E. Domar, "The Soviet Collective Farm as a Producer Cooperative," *The
    American Economic Review*, **56**(4)734-57, Sept. 1966.

4. E. Margolis, "Asia Worries About Ghandi's Military Complex," <u>Wall Street
    Journal</u>, 2 May 1988.

5. Anon., "India's Military Buildup has Sparked Arms Race in South Asia,"
    <u>Wall Street Journal</u>, 20 April 1990.

6. Anon., "Swords not Ploughshares," <u>The Economist</u>, 23 March 1991, p. 50.

7. "Debt Does Us Part," *The Economist*, 3 October 1987 (85-6).

8. "Diffusing the Debt Bomb, *The Economist*, 3 November 1990 (75).

9. L. Sousa, et al., "Foreign Investment Increasing in U.S. Minerals Indus-
    try," *Mining Engineering*, August 1988, Table 2, p. 797.

10. "The Latin Market Comes to Life," *The Economist*, 8 June 1991, pp. 77, 78,
    80.

11. Office of Technology Assessment, *Copper, Technology and Competitiveness*, Congress of the United States, Washington, DC, 1988.

12. K. Takeuchi, et al., *The World Copper Industry, Its Changing Structure and Future Prospect*, World Bank Staff Commodity Working Papers, No. 15, Washington, DC, 1987.

Competitive cost analysis for production of copper

J.C. Agarwal, F.E. Katrak, M.J. Loreth and G.D. Rainville
Charles River Associates Incorporated, Boston, Massachusetts, U.S.A.

ABSTRACT

Cost of producing copper in the noncommunist world is presented for 1981 and 1991 to show the reduction in costs that has occurred in the industry and the factors that have influenced the competitive position of copper producers worldwide. Also, the significance of costs on short-term prices is examined and the implications for long-term prices are analyzed. A discussion of the future trend in copper technologies and costs is also presented.

The competitive position of the United States in world copper production has changed dramatically over the past decade. The transition of the U.S. copper industry from a weak and floundering albatross to an industry that no longer is the "marginal cost" producer is a tribute to the creativity and adaptability of U.S. copper companies.

This paper reviews the worldwide trend in costs during the 1980s and the factors that have influenced the competitive position of copper producers worldwide. Also, the significance of costs on short-term prices is examined and the implications for long-term prices analyzed. We also discuss the impact of new technologies and worldwide environmental regulations.

WORLDWIDE COPPER COSTS: 1981 VERSUS 1991

During the 1980s, the cost of copper production worldwide actually declined in nominal terms and, naturally of more significance, in real terms. Figure 1 illustrates this point. It is a plot of the "cumulative capacity" versus our estimated cash operating costs for 1981 and 1991. It shows the relative differences in "nominal" U.S. dollar-denominated costs. In 1981, 50 percent of the total Western world annual capacity of 7.5 million tonnes had a cash operating cost of about 70 cents per pound of refined copper; whereas in 1991, the cash operating cost at the 50 percent of capacity level is just over 60 cents per pound. A large number of mines/smelters in this group represent production facilities in the United States and Chile.

The changes in cash operating costs (averaged across various mines) in certain major producing regions are compared in nominal terms for 1981 and 1991 in Figure 2. Note the actual decline in nominal U.S. dollars for both Chile and the United States. Other major producing countries such as Canada, Peru, the Philippines, and Zaire have increased in nominal terms. In real dollars terms, however, the cost even in these countries has remained relatively flat or declined for the most part (Figure 3).

U.S. COSTS

The United States deserves special attention because of the excellent job that U.S. companies have done in reducing costs through the 1980s. Some of the cost-saving measures were implemented by sheer desperation, as cost increases led to nearly continuous red ink throughout the early to mid-1980s and resulted in the near collapse of several companies. Some of the actual factors that have reduced U.S. production costs in particular but are also applicable to many countries worldwide include:

1.   Technological and process improvements;

2.   Increases in labor productivity attributed to increases in more automation and computerization;

3.   Incentive-based profit-sharing contracts that have increased labor productivity;

4.   Shutdown or permanent closure of marginal mines despite profitability in other mining divisions of the company;

5.   Export of concentrate under favorable conditions, rather than processing in high-cost smelting facilities;

6.   Capital expenditures on technologically advanced smelting/refining facilities including better pollution control equipment;

7.   Utilization of byproduct sulfuric acid to produce low-cost copper precipitate from dump leaching and solvent extraction and electrowinning (SX-EW) copper from copper oxide ore;

8.   Willingness to change long-run mine plans that may reduce mine life but will lead to lower costs for the reduced operating life;

9.   Merging of production from smaller operations to attain better scale economies;

10.   Willingness to direct capital to facility improvements and modernization in addition to capacity expansion; and

11.   Infusion of capital from overseas sources.

All of the above factors have contributed to lowering the average U.S. cost per pound of finished copper to the low-60-cents-per-pound range. The actual costs would be even lower if it were not for the lower prices of byproduct credits in 1991 relative to 1981.

COSTS OF PRODUCTION IN OTHER COUNTRIES

The improvements in cost in several other countries have also been substantial. Chile is a mixed bag: it has had an extensive restructuring of the mining labor force, but it also lowered its costs during the 1980s by postponing some development and maintenance work. The actual percentage of labor costs from mine to refined copper production as a percentage of total cost has declined from over 30 percent to about 20 percent from 1981 to 1991. This in part has resulted in the more efficient use of labor since the demise of the previous regime and the increasing level of output from private copper producers in Chile. These results have been achieved despite the general decline in ore grades at most operations. Increased mechanization of the mines, improved ore transport, and milling improvements have aided productivity improvements. However, some of the Codelco mines in particular will be penalized in competitiveness during the 1990s as they suffer because of postponed maintenance work. Thus some of Chile's past success in lowering costs was a temporary and ill-advised reprieve that is not sustainable over the long run.

In other countries, the labor productivity and technical improvement have not been as pronounced. This is particularly true of countries such as Zambia, the Philippines, and, to a lesser extent, Zaire, where capital investment must be made to allow the mines, and processing facilities to utilize fully the existing capacity of the operations. This has tended to offset technical advances such as computerization at the mills and improvements of the hydrometallurgical operations.

Labor problems in Peru as well as continued instability have tended to erase productivity gains at the large operations. Particularly significant has been the continued unrest in Papua New Guinea, resulting in the closure of operations that were producing 170,000 tonnes per year of copper.

EFFECT OF COSTS ON PRICES

The years after 1981 proved disastrous for copper prices. The decline of prices into the mid-60-cent-per-pound range for the years until 1987 made copper producers either lower costs or face extinction. The result was that many marginal producers that could not lower costs to compete in the marketplace were forced to close — for the most part permanently. The other producers that were able to lower costs sufficiently remained in production. The effect of eliminating high cost producers and lowering the costs of more efficient producers resulted in a downward shift of the supply curve.

The unfortunate temporary side effect of the cost restructuring in the copper industry was a further decline in prices during the mid-1980s. There is not sufficient demand elasticity in the short run, so the decrease in costs led to a decrease in price. In other words, the quantity demanded in the short run is relatively fixed; cost reductions remove pressure over the long term but not generally over the short term.

With the general demand increasing by 1987, the long-run supply-demand relationship again came into play and more copper was demanded from a rationalized base. The largely maintenance- and development-related production constraints of countries like Zambia, Zaire, and the Philippines (along with continued withdrawal of high-cost capacity) made rapid response to demand pressures difficult to meet in the world copper industry. This left very marginal producers to meet part of expanded requirements and resulted in the high price of copper since the latter part of 1987 to the present. This situation has been exacerbated by unexpected loss of production due to political instability in several countries, and has tended to keep prices high even during the current recession.

The benefit to U.S. producers is that they have improved their cost structure considerably, and are still earning good profits in an economic downturn.

MISCONCEPTIONS REGARDING WORLD COPPER PRODUCTION COSTS

It has been generally accepted that the quality of world ore reserves is deteriorating and will continue to deteriorate. However, improved exploration techniques have resulted in the continual uncovering of deposits of higher grade, particularly in places such as Chile where Sur-Sur and Escondida are prime examples. Also, rethinking of exploration philosophy has opened up new ore genetic environments, with Olympic Dam in Australia and Neves Corvo in Portugal as new sources of non-porphyry copper.

Another misconception is that new supply sources are always more expensive than existing ones. This is not always the case, particularly if one distinguishes between capital cost required to justify a new investment and the operating cost necessary to exploit the deposit after the capital has been sunk. Many of the new mines are often more cost-efficient as they are able to incorporate technological improvements into mine design.

In terms of grade declines at existing deposits, technological improvements in large haulage vehicles, pit crushing, conveyor transport, improved flotation cells, semi-autogenous grinding, improved explosives, enhanced recoveries, and secondary recovery (dump leaching) have allowed for maintenance or improvement of the existing cost levels. New technologies have also allowed for exploitation of previously uneconomic reserves.

ENVIRONMENTAL PRESSURES WILL MAKE U.S. AND CANADIAN PRODUCTION COMPETITIVE

The actual costs of smelting and refining in the United States have declined in real and nominal terms. The improvements in smelter technology and automation at refineries has certainly helped. The investments, however, in acid capture systems have provided "cheap acid" to leach low-grade dump ores and added considerably to the output of many mines, enhancing smelter feed and lowering the per pound cost of copper.

FUTURE TRENDS IN COPPER PRODUCTION COSTS

In copper smelter operations, the future trend to decrease costs will be in the following areas:

1. Oxygen-based smelting: Many process schemes are now available and used to make high-grade matte and convert it to blister copper. The use of oxygen increases smelting rate, decreases the offgas volume, and increases the concentration of sulfur dioxide in the offgases. As a result, the cost of environmental compliance has decreased while increasing the production of relatively inexpensive sulfuric acid.

2. Continuous smelting and converting will reduce the labor requirements while generating excess energy for sale or use in other portions of the copper smelter complex. In-plant emissions caused by batch converting could be eliminated, thereby avoiding large ventilation and sulfur dioxide capture costs.

3. Recovery of copper and other metal values from slag will also become a continuous operation, based on decreasing the oxygen content of slag in electric furnace-based slag treatment processes. The continuous processing of molten slag and recycling of recovered matte will decrease cost and increase yield.

4. Refining operation: Automation of refineries has certainly helped in decreasing labor costs. Additional cost reductions are possible with air agitation at the cathode in the electrolytic cells. This technology will permit higher current densities and efficiencies resulting in lower copper inventories in the refinery and lower operating costs.

5. Solvent extraction-electrowinning: In the last decade, the improvement in this technology has increased U.S. copper mine production by solvent extraction-electrowinning technology to 20 percent of total U.S. output (i.e., about 0.3 million tonnes out of 1.5 million tonnes). The cost of producing cathode-grade copper at an acceptable quality is now lower than that of the conventional smelter-electrorefining route. The oxide ores that were mostly discarded in the past have now become a low-cost source of high-quality copper. With the improvements in the chemical reagent for solvent extraction, it will be possible to replace most of the precipitate copper production from dump-leach solutions. The problem has been that the copper concentration in dump-leach solution has been considered too low (about 0.6 grams per litre) for solvent extraction technology to be economical. In the future, most, if not all, the leach solutions in hydrometallurgical copper operations including *in-situ* mining will be followed by solvent extraction processes to increase the copper concentration, as well as to purify the copper content by improvements in solvent extraction technology. The quality of electrowon copper cathode is now sufficiently high to be acceptable without the need for further refining.

6. *In-situ* mining technology that leaches copper out of ore bodies without moving the ore will become increasingly useful in remote areas or where conventional hydro- or pyrometallurgical processes are uneconomical because of low ore grade or geology. Until now the technology has been successfully applied to oxide ore deposits on a developmental basis. Based on the work done in the late 1970s at Kennecott's Safford, Arizona deposit, it is possible to oxidize the sulfide mineralization by oxygen injection and then leach the copper by *in-situ* solution mining. In the near term, *in-situ* mining will be confined to the oxidized ore bodies first.

In conclusion, the copper industry has made vast improvements in the cost structure, and the near-term future will bring even lower costs. Given the prospects of the industry, more funding to develop and implement technical improvements is likely to lead to exploitation of *in-situ* and other processing options early in the next century.

Figure 1

PRIMARY COPPER CAPACITY CURVE: NONCOMMUNIST WORLD (1981 AND 1991)

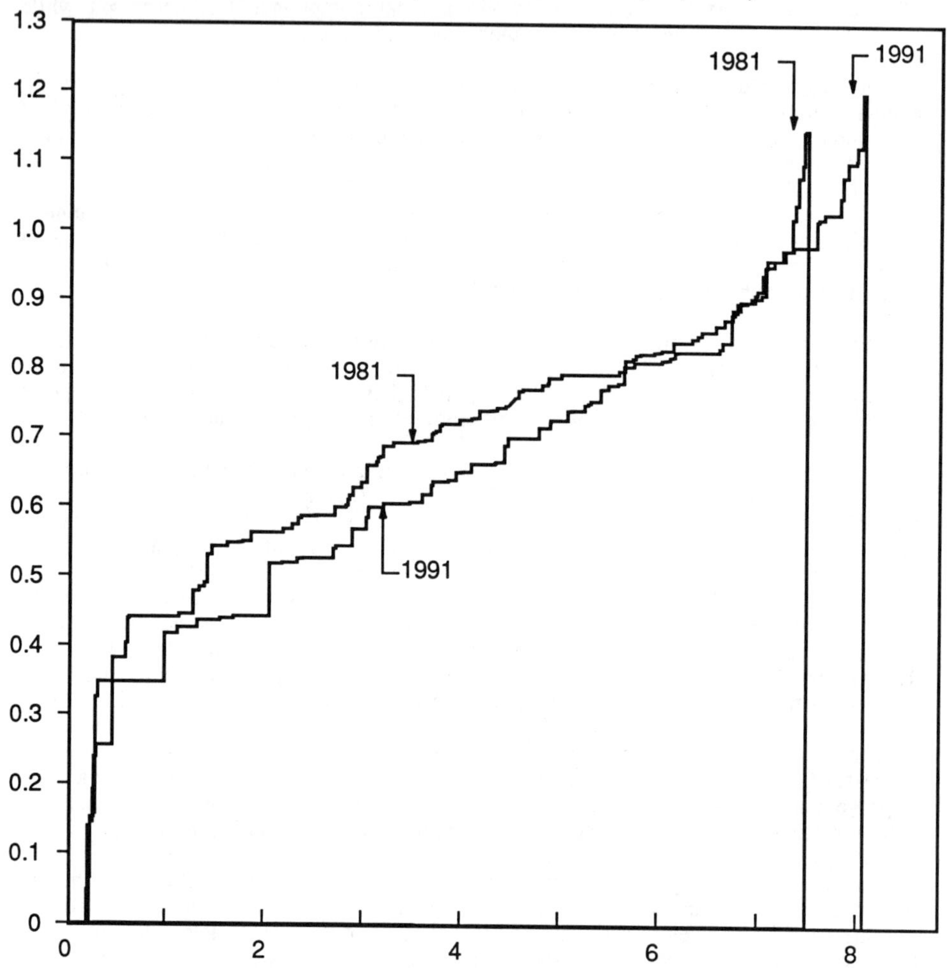

*Does not include depreciation, corporate charges, or finance charges; net cost after credits for all other metals.

SOURCE: Charles River Associates, 1991.

Figure 2

AVERAGE DIRECT CASH OPERATING COST IN NOMINAL U.S. DOLLAR TERMS FOR PRIMARY COPPER PRODUCTION IN SELECTED COUNTRIES

DIRECT CASH OPERATING COST (nominal U.S. dollars per pound)

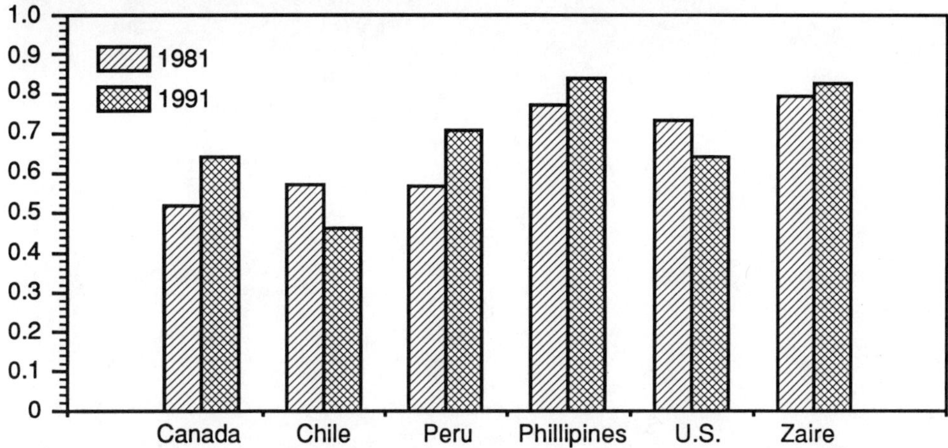

Figure 3

AVERAGE DIRECT CASH OPERATING COST IN REAL U.S. DOLLAR TERMS FOR PRIMARY COPPER PRODUCTION IN SELECTED COUNTRIES

DIRECT CASH OPERATING COST (real 1991 U.S. dollars per pound)

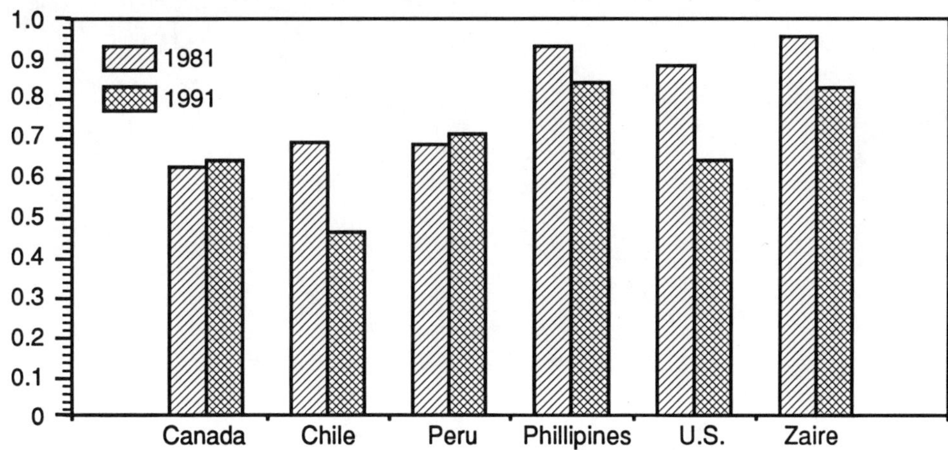

SOURCE: Charles River Associates, 1991.

Copper price predictions: perils and promises

D.L. Morgan and D.H. Brown
Placer Dome Inc., Vancouver, British Columbia

INTRODUCTION

Good copper price predictions can be very useful. Industry often uses price predictions in making decisions about production planning, financial projections, investment and deposit development. The financial sector also uses price estimates in evaluating project financing and share prices.

Since we do not know the future, even the best estimates are subject to prediction error. However, metal analysts face two additional problems. Predictions which are too high can easily lead corporations and investors into making ill-advised investment. On the other hand, predictions which are too low can lead to missed opportunities. Prediction is a risky profession.

Despite the use and significance of price predictions, relatively few papers on prediction methods have been presented in the past decade. Perhaps this is a good time to encourage the development of better methods.

OUTLINE

With this objective in mind, the following is a rough outline of prediction methods and their limitations. For simplicity, the short term will refer to the 0 - 2 year period; the medium term, to a 3 - 5 year period; and the long term, to 10 - 15 years.

SUPPLY AND DEMAND

Supply and demand projection seem to be the most popular method which appeals to conventional economists. This approach is probably best for short term prediction, but does not seem to work very well for medium and long term periods.

Two factors pose inherent problems for metal analysts: (1) statistical limitations and (2) limited accuracy in economic and financial prediction.

The copper industry is very fortunate to have several excellent statistical groups. However, usually their estimates of consumption do not agree due to political and technical reporting problems, especially in the short and medium term. Also, the apparent production vrs. consumption balanced (Table 1) do agree with reported inventory changes (Table 2) on an annual basis. However, the professional persistence of the statisticians clearly shows up in a longer time frame. Over 15 years, the accumulated difference is insignificant (0.01%).

Even if the analysts could get their supply/demand balances correct, they are then faced with the disposition of inventories. The liquidity of stocks and their carrying costs can affect prices significantly. Financial liquidity pressures have limited producer and fabricator capacity or willingness to hold stocks, so the financial burden has fallen on exchanges and merchants. This change in inventory liquidity seems to have influenced price response curve.

The limited accuracy of economic and financial forecasting usually shows up in medium term forecasts. Part of this problem may be that most economists are not interested in or cannot measure factors which influence metal consumption. Another problem is the increased prediction error at economic and financial turning points. We are not really sure why this occurs. Some of the reasons suggested are the tendency toward linear projection, reliance on outdated economic models, shifts in political attitudes of economic and financial control, changes in technology and limited understanding of economics. We suspect that a modified fourier extrapolation of business, capital investment and international trade trends and cycles might improve consumption prediction in the medium term. But, so far, we have not had time to evaluate this idea.

For long term projection, we have concluded that the copper industry tends to restructure every 10 - 12 years. Periodically, this restructuring involves a major change in consumption trends (Exhibit 2). This is usually accompanied by financial instability (Exhibit 3) as well. We suspect the 1972 - 77 shift in trend was caused by increased government spending combined with declining financial liquidity and saturation in major product markets, such as electrical generation, steel production and automobile demand etc.

For many years, analysts used the OECD index of industrial production to forecast refined copper consumption. Since 1970, a better correlation seems to involve capital goods production and non-residential construction (Exhibit 4). Unfortunately, the OECD no longer report some of these useful indices. With metal consumption growing rapidly in Southeast Asia, perhaps this change in OECD data may be a blessing in disguise.

LONG TERM DEMAND

While detailed prediction of supply and demand seems to work pretty well in the short term, this method does not seem to work very well in the medium and long term. Medium term demand, major supply disruptions and shifts in Sino-Soviet trade are all very difficult to predict.

These problems and our emphasis upon feasibility studies have led us to emphasize long term prediction. Our basic approach has been to concentrate on five and ten year consumption trends, major shifts in supply structures and the lagged response of production to medium term trends in deflated prices.

Three approaches seem to be reasonably effective in estimating long term demand trends: (1) technological surveys, (2) evaluating shifts in competitive materials and (3) estimating major economic trends which relate to metal consumption. Technological trends are apparent to executive and chief engineers in charge of research and development at major manufacturers around the world who use significant amounts of copper. These executives usually have a 5 - 10 year planning horizon and are well aware of the technical feasibility, economic viability and social acceptance which are required for good product sales trends.

Trends and shifts in competitive materials usually show up in intensity of usage and market share of materials. For example, copper's share of non-ferrous metals markets (Exhibit 6) declined from 1950 to 1974 and has since stabilized at about 25%. This suggests that copper is now very efficiently used in sound applications.

The third approach is estimating general economic trends which relate to metal consumption. The patterns of capital investment and consumer durable production by region should be related to population and disposable real income. At this stage, these trends should be combined with the technological survey and usage patterns, to assure a coherent evaluation, especially in the shifts from old to new industries which is now ell underway.

LONG TERM SUPPLY

The size and capital intensity of copper production leads to a slow response to price/cost margins. Slowly growing markets are easily over and under supplied. During periods of falling prices, as in 1975 - 86 (Exhibit 7), producers are highly oriented to low capital investment and low cost production. The development of Escondida and SXEW deposits are examples of this natural economic response.

We now probably have an under-invested industry which cannot respond to higher levels of demand. The current emphasis upon smelting capacity is perhaps symptomatic of this factor which is compounded by political instability in some areas. The smelter capacity problem may take five or six years to correct because the political process have slowed down investment decisions and because scrap may not be so plentiful in the future.

While the total scrap ratio (Exhibit 8) has been stable for some years at roughly 37 - 38%, the trend seems to be down. The more efficient use of scrap has made copper more expensive to collect. The trends of total copper consumption and total scrap supply now appear to be diverging (Exhibit 9).

"CYCLES"

In the 1970 - 1975 period, we became quite fascinated with the repetitive patterns in many commodity prices, including copper. We suspected that copper has a regular "family of cycles" of 50 years, 48 months and 52 weeks which seem to reflect major economic and financial behaviour.

However, periods of major capital investment tend to shift demand into a 7 - 11 year pattern and the price peaks tend to shift accordingly. The shift from 4 year intervals between peaks to the current 9-year interval seems to fit this pattern. Of course political events like the Korean War and resulting U.S. stockpile can distort any regular pattern.

Unfortunately, these capital investment "cycles" are not uniform in magnitude and in intensity of copper usage. In addition, central bank and product demand patterns can shift demand in any business "cycle". The so-called "cycle" approach is probably best applied to copper consumption on a regional basis to reflect their different economic structures and on a limited time horizon.

From 1289 to 1946, real interest rates in the leading economies of Europe and North America have fallen to periodic low points about every 51 years or so. Usually, 2 - 4 years after this low interest rate, economic activity starts to accelerate as in the 1949 - 1969 period. If this pattern repeats, the next low in interest rates could occur around 1997 and copper demand could start to grow more rapidly starting about 1999 - 2001 or so. However, we have reservations about reliance upon "cycle" analysis and feel this approach must be combined with other methods to be effective.

ECONOMETRIC MODELS

We have had some temporary success with econometric price prediction. However, every structural shift in the copper market rendered our models obsolete. We only know of one short term system at SGM which has apparently been working fairly well. All the other attempts at modelling which we are aware of, have distinct limitations.

ERROR ANALYSIS

Most competent metal analysts try to quantify the error of their estimates. This is especially important in evaluating the financial risk in a feasibility study because price variability is usually the major factor in economic viability. Perhaps the best way to evaluate this error potential is to run both optimistic and pessimistic scenarios.

SUMMARY

In summary, the combination of several methods outlined above seems to give pretty good results in the short term and in the longer term, but so far we have not found a combination of approaches for the medium term. Following the market in the detail required to predict short term prices is a lot of work and perhaps best left to people who are continuously exposed to copper markets.

While long term prediction also requires thorough analyses, the approach is quite different because it is oriented towards trends and many of the prediction errors tend to offset.

CONCLUSION

Hopefully, the above outline of prediction methods and problems will help you and stimulate the development of better approaches, especially in the medium term horizon. If you have any suggestions for medium term prediction, we would certainly like to hear them.

Good luck in a hazardous, but important profession, and remember that most of your critics won't put their estimates down for public scrutiny, it's too risky.

TABLE 1

REFINED COPPER MARKET STATISTICS
(in 1,000 metric tons)

Year	Production	Sino-Soviets*	Reported Supply	Consumption	Apparent Balance
1975	6,263.4	2.2	6,265.6	5,438.0	827.6
1976	6,644.9	59.1	6,704.0	6,427.4	276.6
1977	6,853.4	17.1	6,870.5	6,874.8	(4.3)
1978	6,903.1	10.8	6,913.9	7,277.8	(363.9)
1979	7,016.1	34.1	7,050.2	7,513.1	(462.9)
1980	7,036.4	41.2	7,077.6	7,101.0	(23.4)
1981	7,349.7	58.0	7,407.7	7,252.0	155.7
1982	7,169.3	46.6	7,215.9	6,760.5	455.4
1983	7,320.5	(347.5)	6,973.0	6,843.4	129.6
1984	7,184.5	(122.9)	7,061.6	7,656.4	(594.8)
1985	7,290.0	(176.4)	7,113.6	7,328.7	(215.1)
1986	7,433.5	7.4	7,426.1	7,674.2	(248.1)
1987	7,631.3	112.6	7,743.9	8,012.3	(268.4)
1988	7,978.3	153.8	8,132.1	8,213.3	(81.2)
1989	8,392.5	147.0	8,539.5	8,646.5	(107.0)
Total				109,019.4	

Notes: * Net exports per WBMS estimates, including 1,060 KMT exports to China in 1983 - 86.

Source: World Bureau of Metal Statistics.

TABLE 2

REPORTED COPPER STOCKS VRS APPARENT BALANCE
(in 1,000 metric tons)

Year End	Commercial Stocks*	Strategic Stockpiles	Reported Stocks	Stock Changes	Apparent Balances	Difference
1974	1,048.7	32.0	1,080.7			
1975	1,743.9	23.7	1,767.6	686.9	827.6	(140.7)
1976	1,828.2	88.6	1,916.8	149.2	276.6	(127.4)
1977	1,985.5	85.9	2,071.4	154.6	(4.3)	158.6
1978	1,564.3	92.4	1,656.7	(414.7)	(363.9)	(50.8)
1979	1,115.6	63.3	1,178.9	(477.8)	(462.9)	(14.9)
1980	1,034.9	27.8	1,062.7	(116.2)	(23.4)	(92.8)
1981	1,133.2	25.0	1,158.2	95.5	155.7	(60.2)
1982	1,639.7	20.2	1,659.9	501.7	455.4	46.3
1983	1,708.9	20.2	1,729.1	(69.2)	129.6	(198.8)
1984	1,195.3	20.2	1,215.5	(513.6)	(594.8)	81.2
1985	1,073.9	20.2	1,094.1	(121.4)	(215.1)	93.7
1986	892.8	20.2	913.0	(181.1)	(248.1)	67.0
1987	521.6	20.2	541.8	(371.2)	(268.4)	(102.8)
1988	585.1	20.2	605.3	63.5	(81.2)	144.7
1989	664.9	20.2	685.1	79.8	(107.0)	186.8
Total						(10.1)

Note: * Includes metal exchange stocks.

Source: World Bureau of Metal Statistics.

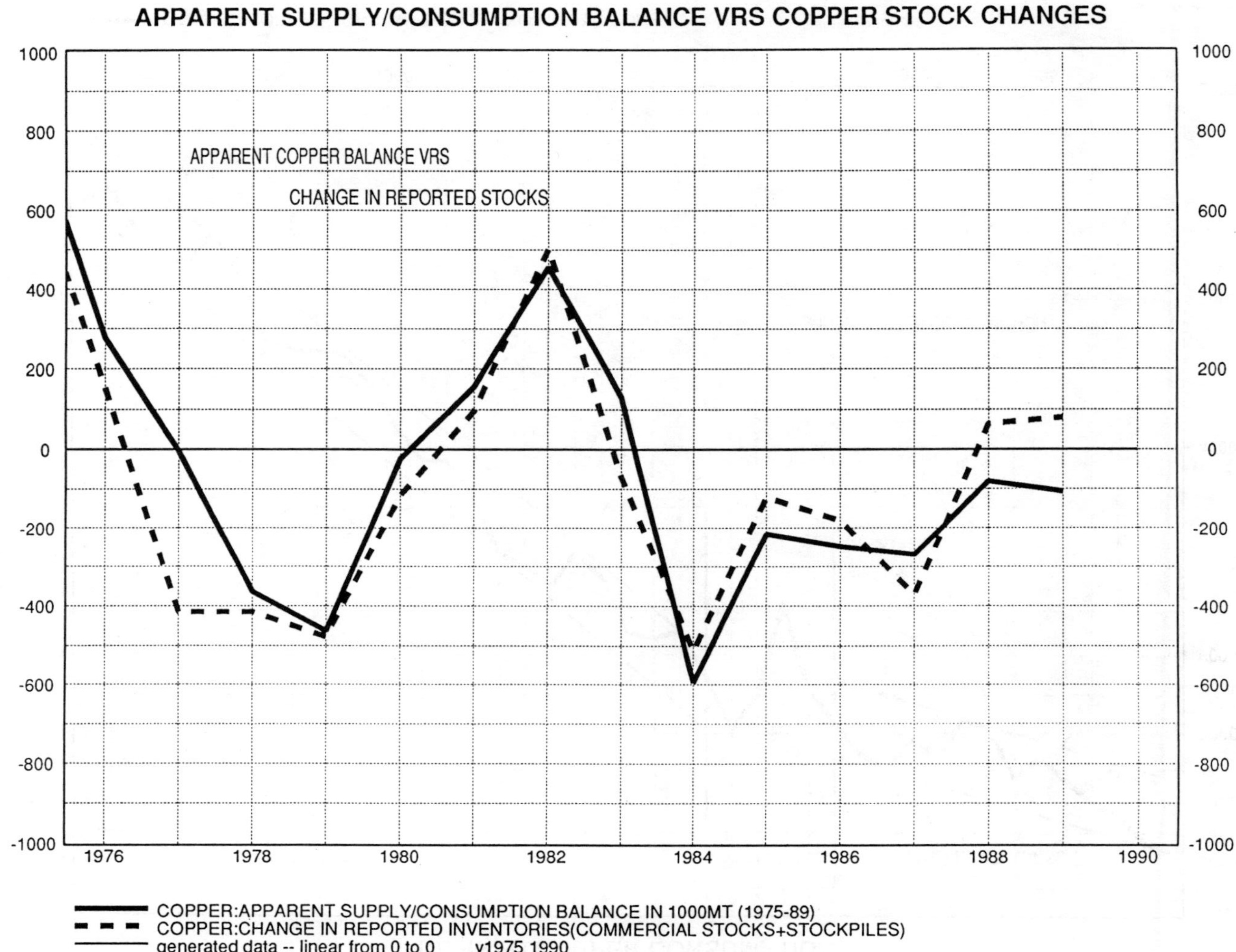
APPARENT SUPPLY/CONSUMPTION BALANCE VRS COPPER STOCK CHANGES
IN 1000 METRIC TONS OF COPPER
IN KMT
EXHIBIT 1
APPARENT COPPER BALANCE VRS
CHANGE IN REPORTED STOCKS
1000
800
600
400
200
0
-200
-400
-600
-800
-1000
1976
1978
1980
1982
1984
1986
1988
1990
COPPER:APPARENT SUPPLY/CONSUMPTION BALANCE IN 1000MT (1975-89)
COPPER:CHANGE IN REPORTED INVENTORIES(COMMERCIAL STOCKS+STOCKPILES)
generated data -- linear from 0 to 0   y1975 1990

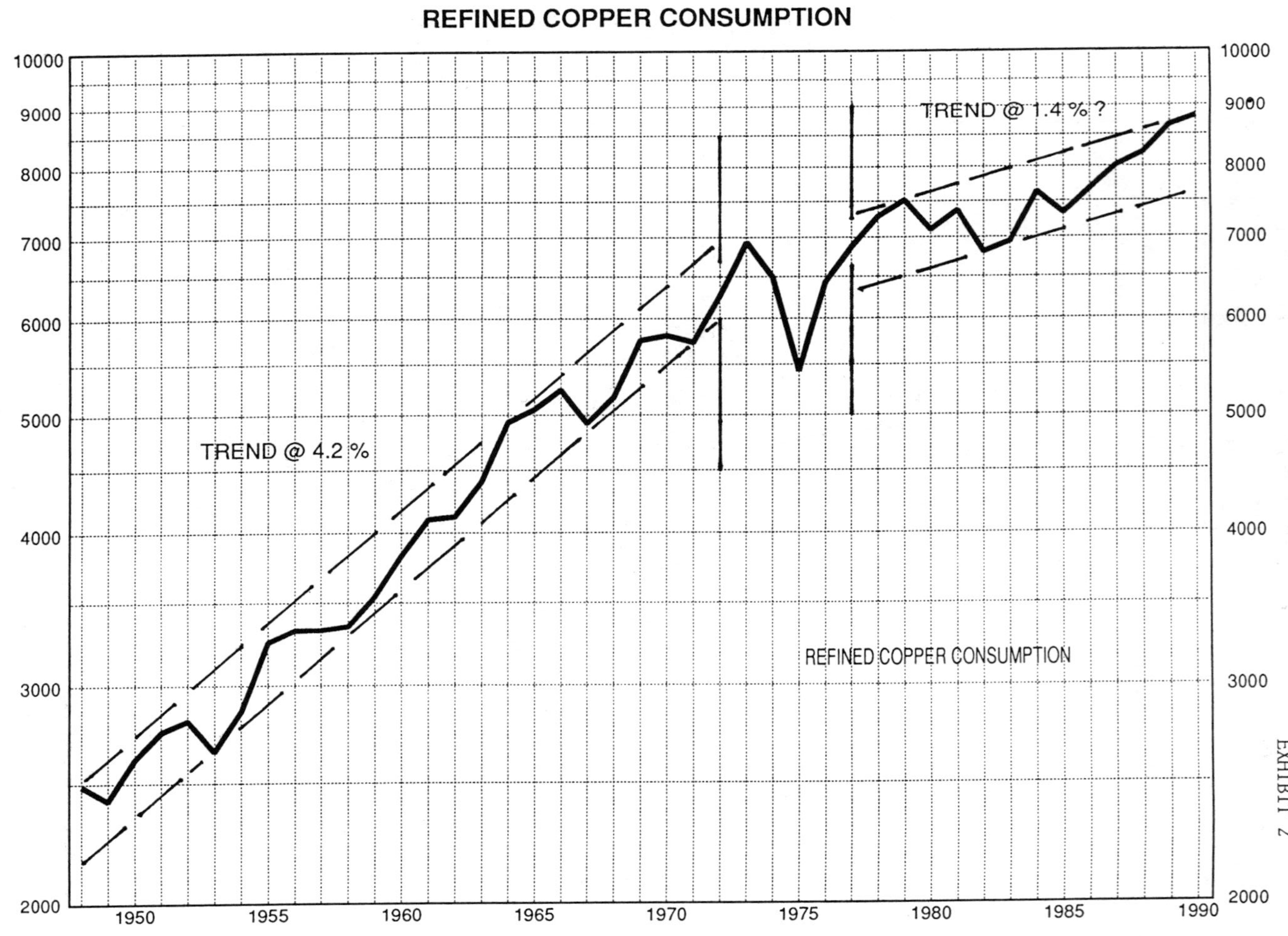
REFINED COPPER CONSUMPTION
IN 1000 METRIC TONS
REFINED COPPER CONSUMPTION IN KMT
TREND @ 4.2 %
TREND @ 1.4 % ?
REFINED COPPER CONSUMPTION
EXHIBIT 2
REFINED COPPER CONSUMPTION IN 1000 METRIC TONS
10000
9000
8000
7000
6000
5000
4000
3000
2000
1950
1955
1960
1965
1970
1975
1980
1985
1990

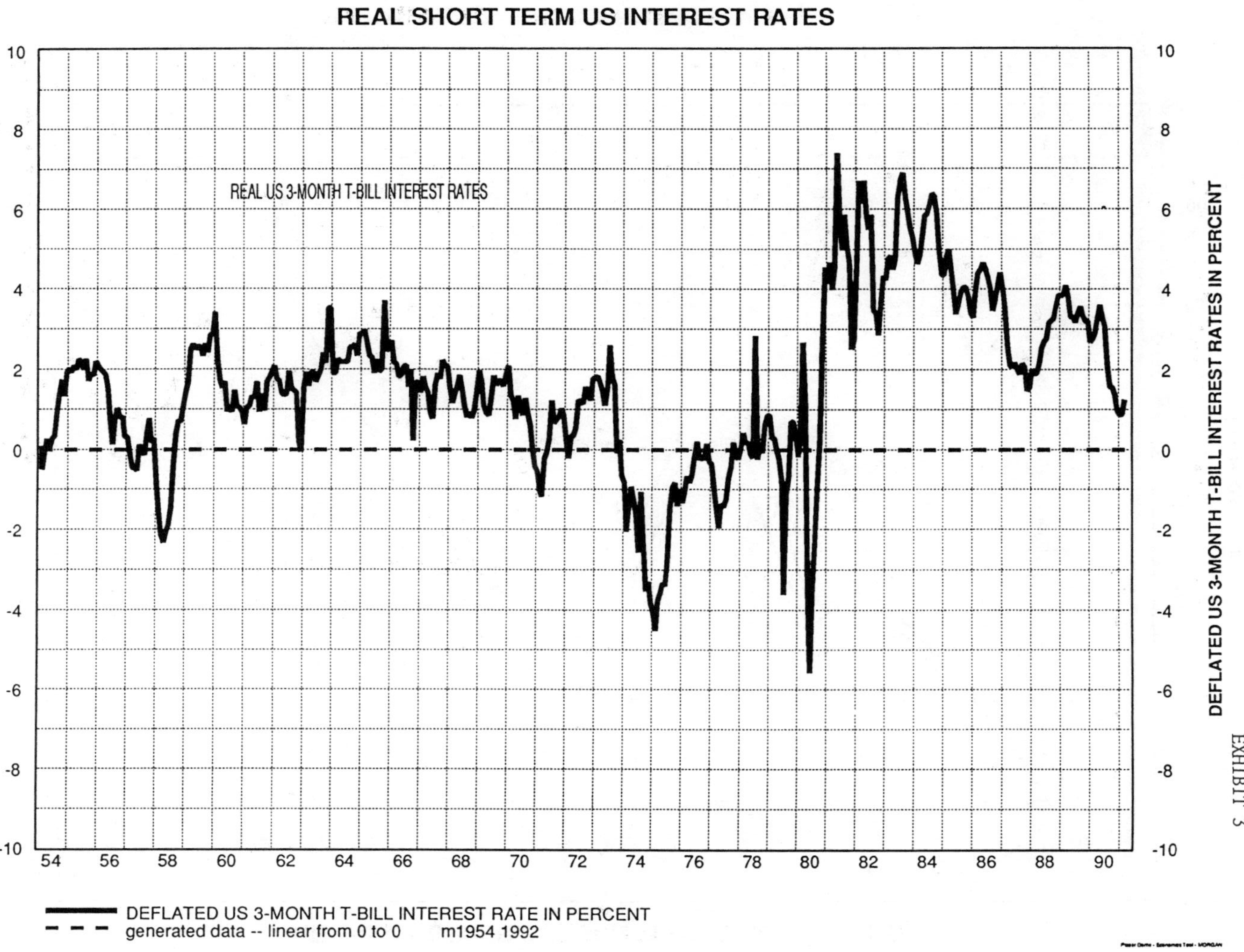
REAL SHORT TERM US INTEREST RATES
INTEREST RATES IN PERCENT
DEFLATED US 3-MONTH T-BILL INTEREST RATES IN PERCENT
REAL US 3-MONTH T-BILL INTEREST RATES
EXHIBIT 3
DEFLATED US 3-MONTH T-BILL INTEREST RATE IN PERCENT
generated data -- linear from 0 to 0   m1954 1992

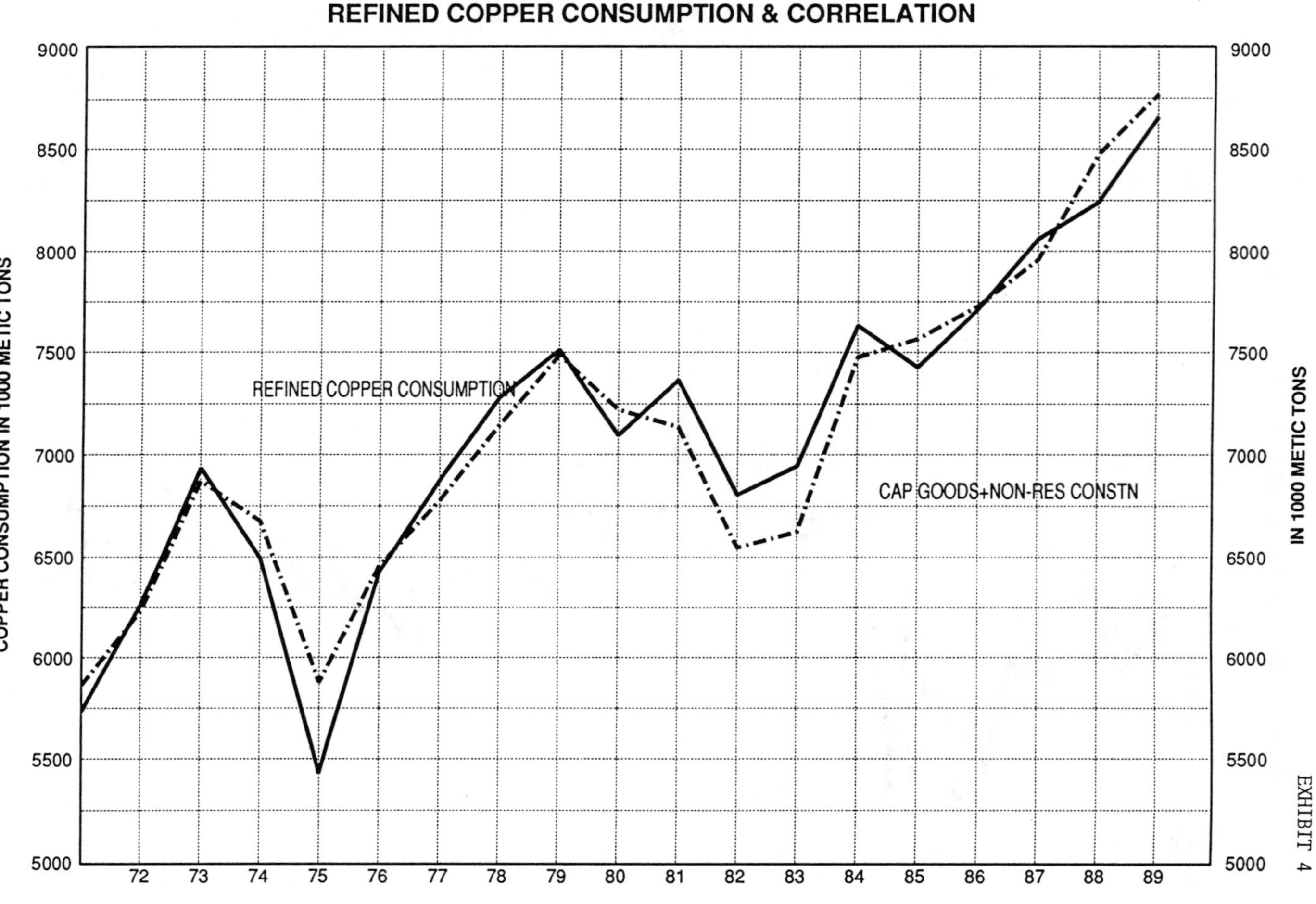
REFINED COPPER CONSUMPTION & CORRELATION
EXHIBIT 4
COPPER CONSUMPTION IN 1000 METIC TONS
IN 1000 METIC TONS
REFINED COPPER CONSUMPTION
CAP GOODS+NON-RES CONSTN
REFINED COPPER CONSUMPTION IN 1000 METIC TONS
-435.04 + 69.3612 * CH1 + 10.6403 * CH5

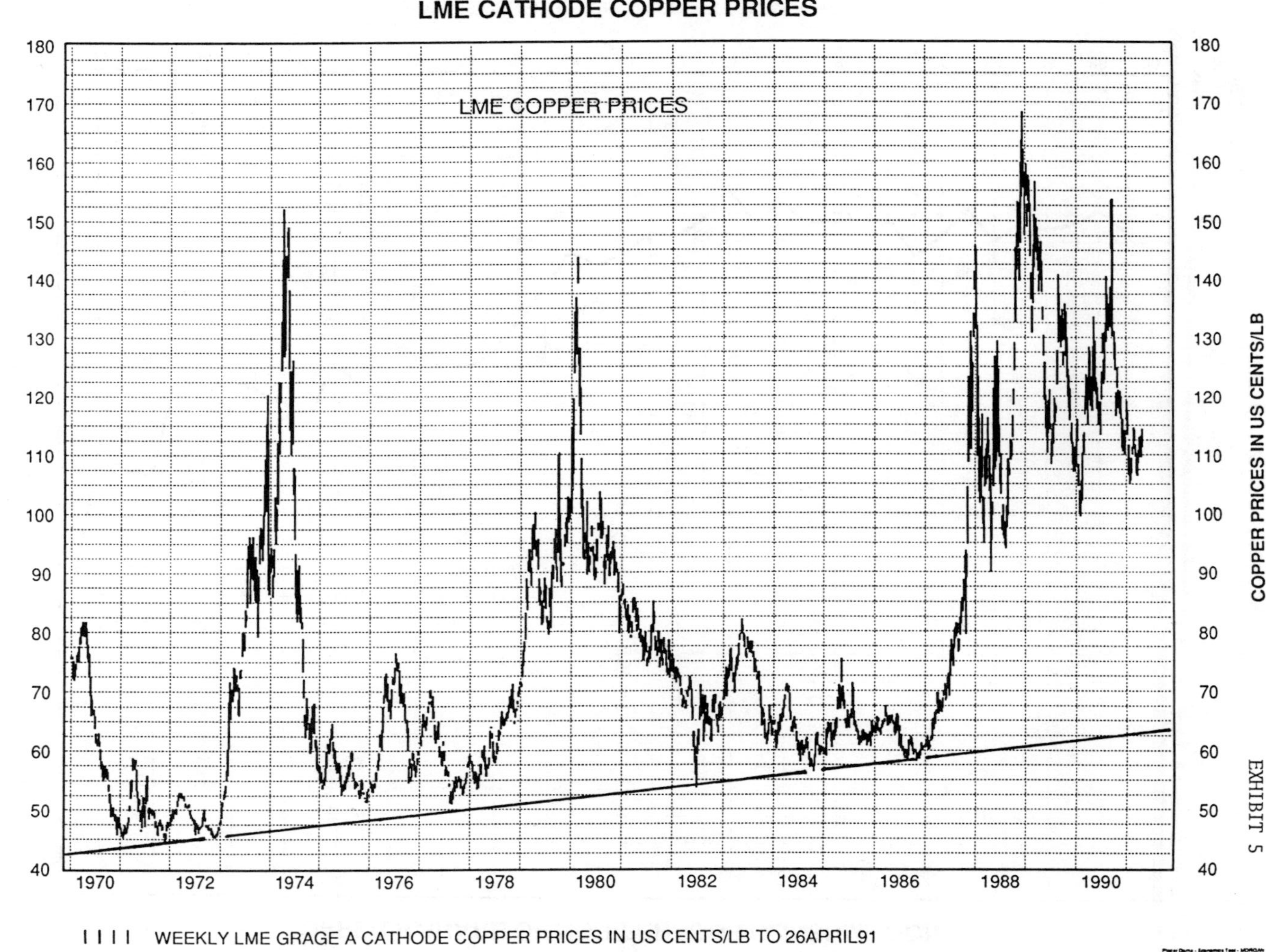
LME CATHODE COPPER PRICES
LME COPPER PRICES
IN US CENTS/LB
COPPER PRICES IN US CENTS/LB
EXHIBIT 5
1970 1972 1974 1976 1978 1980 1982 1984 1986 1988 1990
40 50 60 70 80 90 100 110 120 130 140 150 160 170 180
WEEKLY LME GRAGE A CATHODE COPPER PRICES IN US CENTS/LB TO 26APRIL91

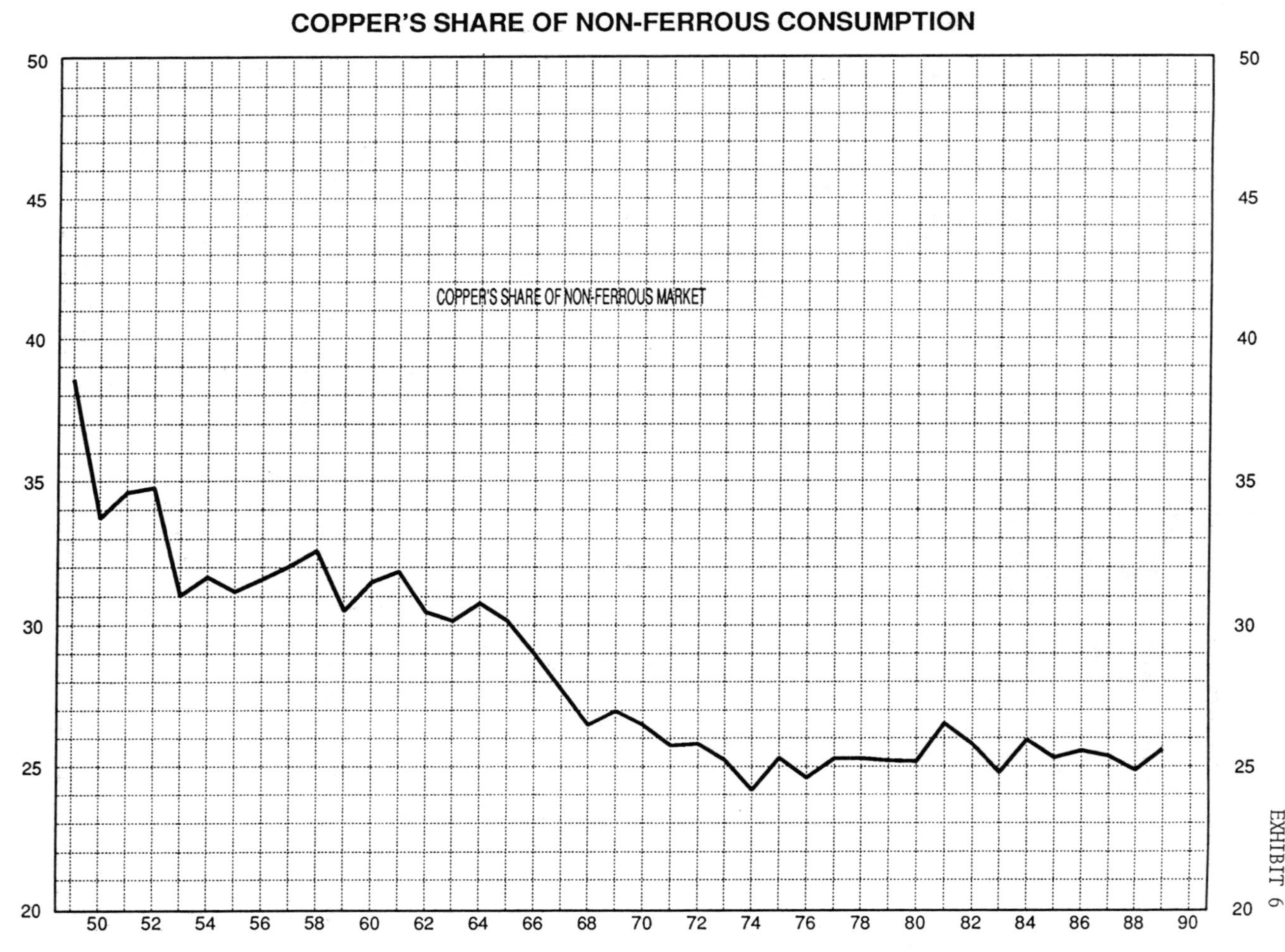
COPPER'S SHARE OF NON-FERROUS CONSUMPTION
COPPER'S SHARE IN PERCENT
COPPER'S SHARE OF NON-FERROUS MARKET
COPPER'S SHARE IN PERCENT
EXHIBIT 6
COPPER'S SHARE OF NON-FERROUS CONSUMPTION IN PERCENT BY WEIGHT

EXHIBIT 7

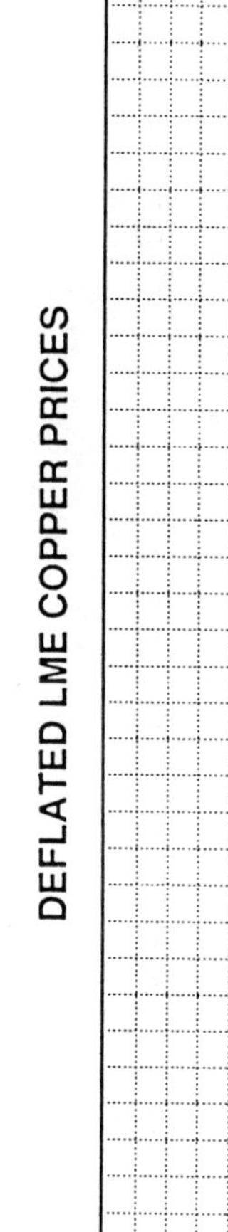

DEFLATED LME COPPER PRICES
DEFLATED LME COPPER PRICES
COPPER PRICES IN 1990 US CENTS/LB
COPPER PRICES IN 1990 US CENTS/LB
0 40 80 120 160 200 240 280 320 360 400
1900 1910 1920 1930 1940 1950 1960 1970 1980 1990
DEFLATED LME COPPER PRICES IN 1990 US CENTS/LB TO 1990

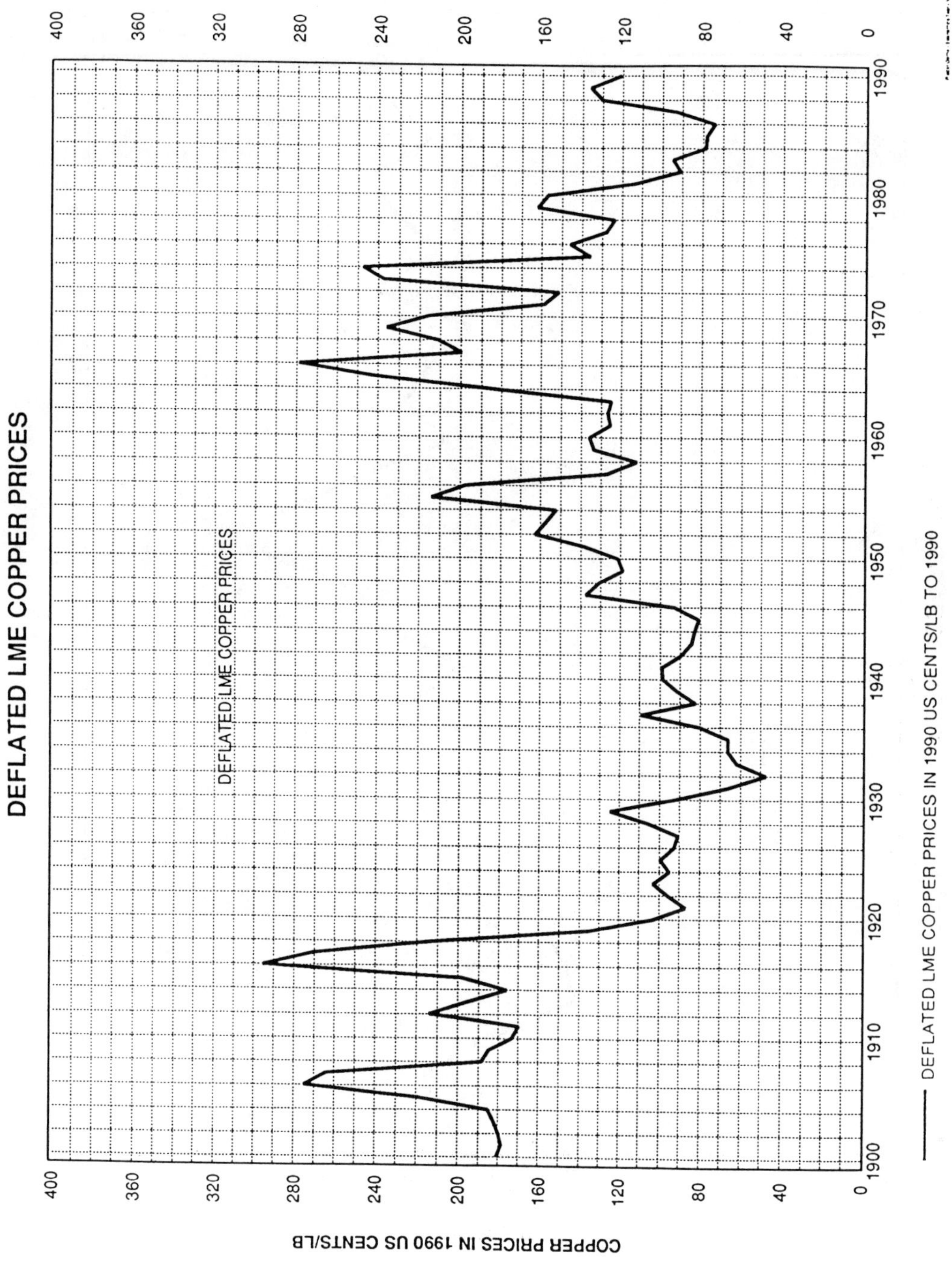

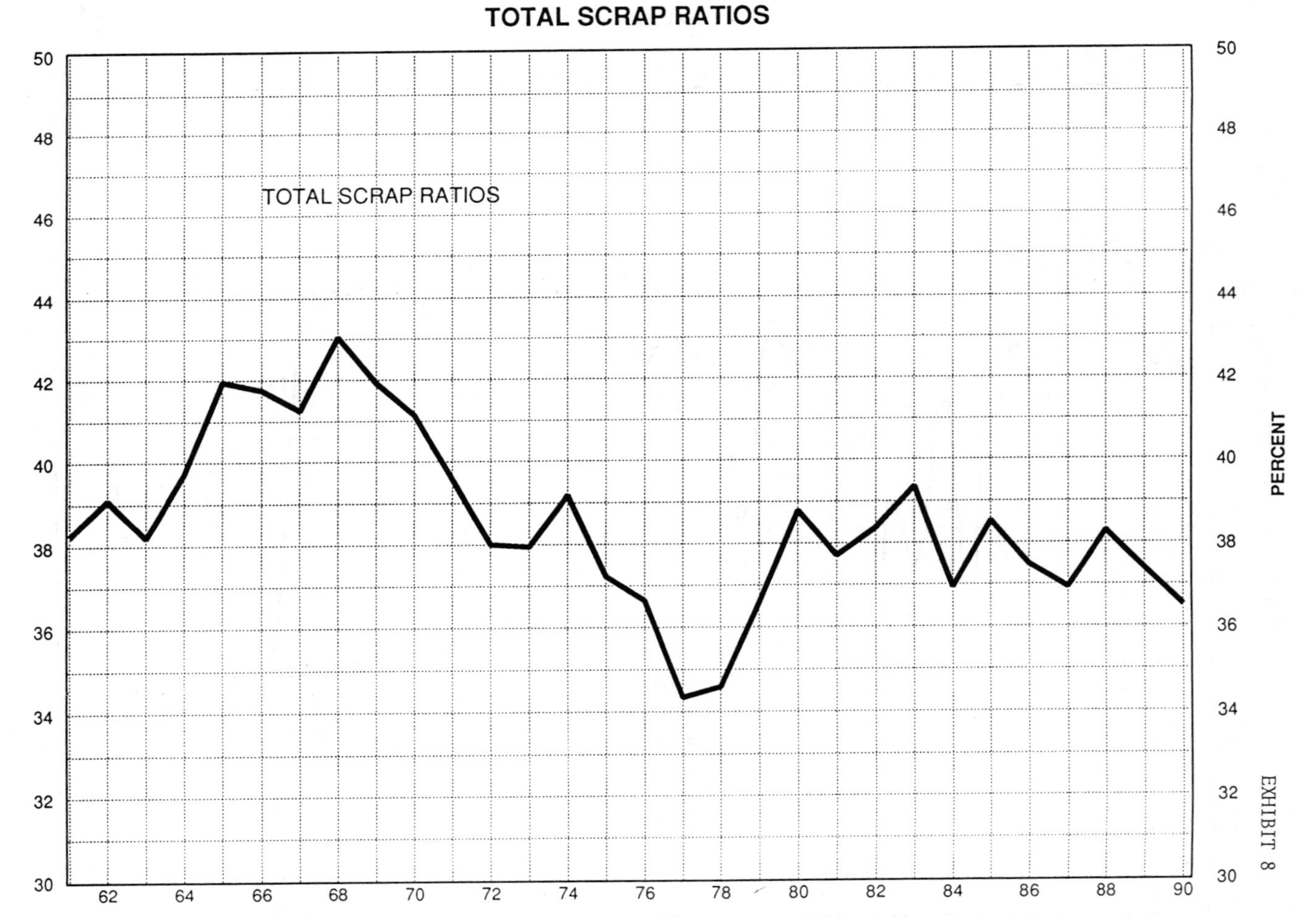
TOTAL SCRAP RATIOS
TOTAL SCRAP RATIOS
PERCENT
PERCENT
EXHIBIT 8
TOTAL SCRAP RATIO(TOTAL SCRAP/TOTAL CONSUMPTION) AS PERCENT

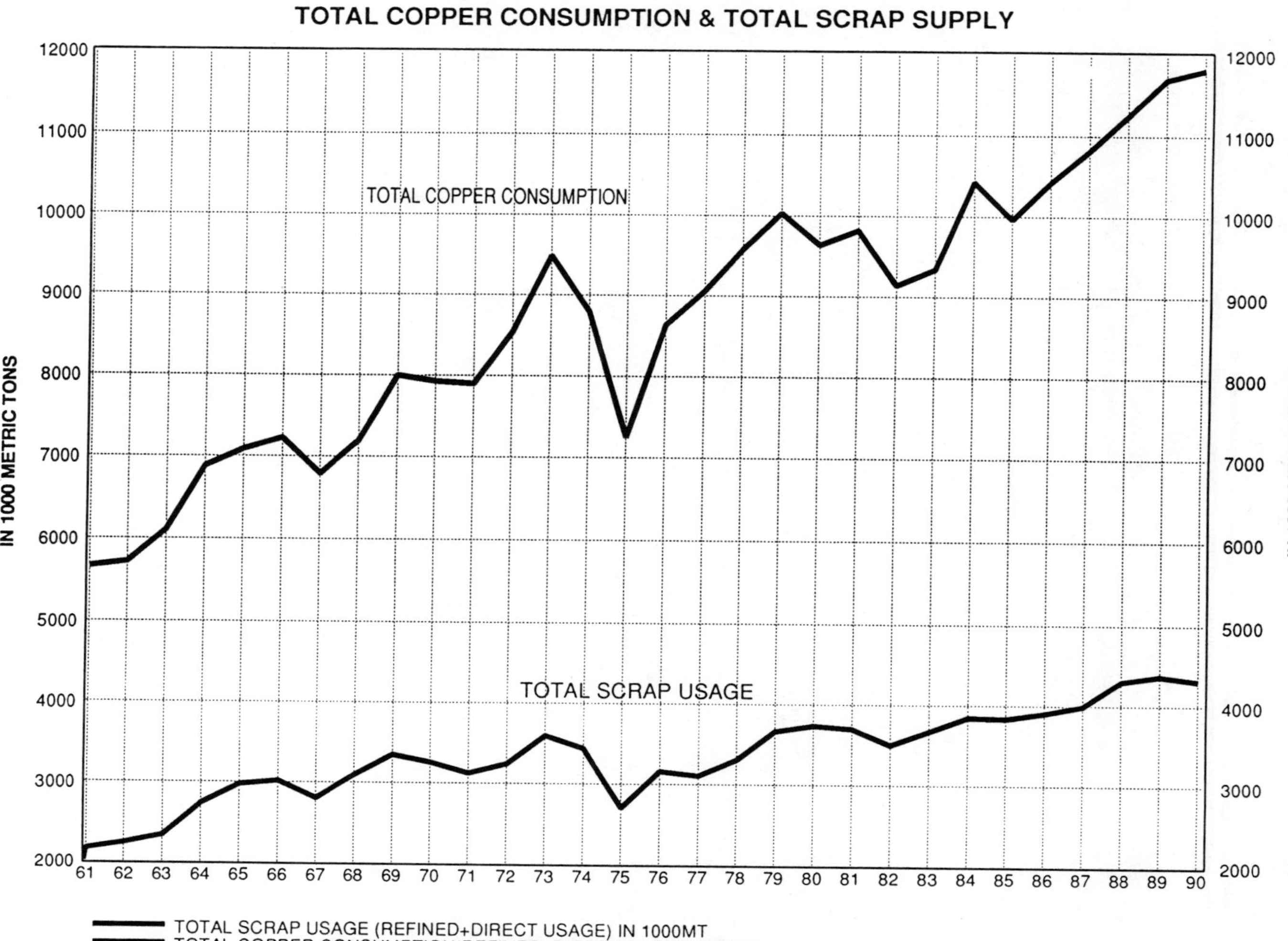
TOTAL COPPER CONSUMPTION & TOTAL SCRAP SUPPLY
IN 1000 METRIC TONS
TOTAL COPPER CONSUMPTION
TOTAL SCRAP USAGE
EXHIBIT 9
TOTAL SCRAP USAGE (REFINED+DIRECT USAGE) IN 1000MT
TOTAL COPPER CONSUMPTION(REFINED+DIRECT SCRAP) IN KMT

Production response to price surprises through optimization of cut-off ore grades: its impact on supply and on mine value. An application of option valuation to a copper mine

J. Luis Mardones
University of Chile, Santiago, Chile

ABSTRACT

We determine the production response implied by the modification of the cut-off ore grade in an open pit copper mine when copper prices change. We assess the value added by the possibility of modifying the cut-off grade, adapting a model of cut-off grade optimization into a simulation model of option valuation or valuation by "reduction to a risk neutral world." The extra value comes from the fact that production optimization in fact controls risks better than a fixed production schedule. This method also gives an estimate of the variable risk profile of a project.

INTRODUCTION

Once capacities are in place, the cut-off grade strategy is already determined, but there usually is some flexibility to optimize it if prices and price expectations change. Flexible production practices and other contingent rules add an extra value to a project and also change the risk profile of its cash flows. Traditional net present value methods are unable to assess this extra value while valuation methods based on option theory capture the value of contingencies. Option theory provides an analytical solution for the value of an option, an instrument that captures contingencies.

We make production of the period flexible to price changes, adapting a model of cut-off grade optimization. The parameters that are optimized in an inter-temporal optimization procedure are the velocity of extraction and the cut-off grades to send material to treatment, to stockpiling for later treatment, or to the waste deposit. Optimal cut-off grades depend on limiting capacities in the successive phases of mining, treatment and refining, and in general they descend over time.

In sum, we use option simulations to estimate the value of a mine which can optimize production and cut-off grades when expected prices change, and the extent of variation of production that may result.

The contribution of this article is the application of the derivative asset valuation (DAV) method, proposed by Jacoby and Laughton (1), to a copper mine using a realistic operating option, such as the flexibility to modify the cut-off grade to govern the tradeoffs between mining, treatment and refining capacities. The cut-off grade model developed by Lane (2) is extended from its application in the planning stage to its application as a flexible response by the mine administrator. It is also extended to work with non-uniform expected prices; with increasing transportation costs as the pit gets deeper; and with the exploitation of more that one mining phase per period of time. The model also includes a temporal stockpiling alternative.

The project is a real copper mine currently under development in Chile. The value of the cut-off grade flexibility is estimated. Also, the variable risk profile over time of the system of mine and processing plants is calculated. This computation permits to support the recommendation to at least value separately the "mine" period and the "stockpile" period, because of the important disparity in risk levels. The "stockpile" period is defined as the period when plants are fed mainly from stockpile because the mine is nearing exhaustion or is already depleted,

In simple net present value (NPV) methods, the expected value of uncertain variables are used to estimate the expected value of cash flows. This procedure is correct only when cash flows depend upon uncertain variables in a linear form. Usually complex projects are characterized by non-linear relations. When relationships are not linear the expected value of an uncertain cash flow must be estimated by Monte Carlo simulation, starting from the full distribution of each uncertain variable.

Still, Monte Carlo simulations in a discounted cash flow framework do not address the effects of non-linearity on the discount rate. When operating leverage changes over time, the risk level of the cash flows also varies over time. The operating leverage is measured as the proportion of fixed to total costs.

Usually, as prices increase, production is also increased, and variable costs are raised, decreasing operating leverage and the risk level of the firm. A firm with high operating leverage is more risky than a firm with low operating leverage. In the extreme, a firm with only fixed costs has no flexibility, while another with only variable costs is very flexible to respond to price surprises. The flexibility to respond in such way reduces the <u>risk</u> of each cash flow and of the project as a whole, because the manager is able to put a lower limit to losses in the event of meager prices, and the range of variation of cash flows narrows.

The flexibility to save costs in response to price developments depends on the relative magnitude of variable costs to fixed costs. In consequence, the risk level is also contingent on the price level. Risk levels vary over time even when there is no possibility of response, but operating leverage is not uniform over time, for example due to exponential decay in production as in oil exploitation.

Traditional valuation methods do not relate the risk level to operating leverage. Usually only one discount rate is calculated and used to discount cash flows of every period. Option methods are able to deal with the effects of non-linearity on the risk of the project, because a non-linear relationship can be represented by an option. A financial option has a variable risk over time, and still it can be valued, although not with traditional discounted cash flow methods. Black and Scholes (3) provided a formula for an option, but a numerical method must be used when the analytical solution does not exist. A real option like a project with variable operating leverage should also be valued using option-based methods.

Boyle (4) suggested an option valuation method based on Monte Carlo simulations, making use of Cox and Ross (5) proposition of "reduction to a risk neutral world". The DAV method of Jacoby and Laughton (1) elaborates on Boyle (4) and applies it to value investment projects; this method captures the variation of risk over time by the simple procedure of discounting risk at the level of uncertain basic variables such as the price, where risk structures are simpler, instead of discounting at the level of cash flows.

The only source of risk considered in this application is the price of copper, which is modelled as a loglinear diffusion model. While the price information model is simple, the copper mine has a complex cash flow structure, because there are inter-temporal relations given by cut-off grades, and also given by variations of the ton-grade distribution for successive phases of mine exploitation. Tax regimes also add inter-temporal complexity but in this exercise we omitted taxes to simplify the presentation.

The contingent rule applied by the mine operator is the optimization of the cut-off grade strategy each time current and expected prices change. In addition to a set of cut-off grades, the grade optimization model which we use to mimic the operating option of the manager of the mine, determines flows from the mine to the treatment plant; to the waste deposit; and to the stockpile for later treatment. Also, it determines flows from the stockpile to the treatment plant; and from the treatment plant to the last phase of refining and marketing. This operating option model is then applied in each simulation step of the option simulation procedure.

The model is written in BASIC for a VAX mainframe, where a complete strategy is solved in two minutes, but it takes almost 11 hours to run 6,000 simulation draws.

METHODOLOGY

The DAV method proposed by Jacoby and Laughton (1) carries out the discounting for risk at the level of underlying uncertain variables, which usually show a simpler resolution of uncertainty. In our case, the underlying uncertainty is the price of copper.

Price Information Model

The information model represents the resolution of price uncertainty in time. Instead of estimating a convenience yield as a proportional dividend, as in Brennan and Schwartz (6), Jacoby and Laughton (1) utilized the concept of a pure discount commodity bond that pays at maturity the price the underlying commodity attains at that date. A commodity bond does not yield a convenience return, and its price at maturity equals the price of the underlying commodity at maturity.

The price process is stated in terms of the expectation of prices in the future and not on the spot commodity price itself. The underlying asset is a copper bond. For ease of exposition, it can be stated in terms of a set of copper bonds, with maturities in each period of the project horizon, but strictly only one copper bond is needed with a long maturity date, longer than the project horizon.

Not many commodity bonds are actually traded in the market. To calculate the value that commodity bonds would have in the market if they were traded, Jacoby and Laughton (1, 7) assumed that the Capital Asset Pricing Model (CAPM) holds. They estimated price trends, and then calculated the implied hypothesized market price of commodity bonds discounting their maturity prices (equal to the prices of the commodity, oil in their case, at each maturity date). The discount rate used in each case is estimated applying the CAPM. Formally, they needed to assume that financial markets are complete, in the sense that existing traded assets span all possible commodity bonds.

The term structure of initial expectations of copper prices is estimated according to market information, when long term bonds are traded in the market, or from experts' estimates. The volatility may also be estimated from the behavior of long term traded copper bonds, from historical values, or from experts' opinions.

The pattern of volatility may be taken as constant. In this case a surprise in the current price results in proportional changes in the expectations of all future copper prices. The price surprise is considered as a permanent change in market conditions. Although not demonstrated unambiguously, practitioners hold that prices in most markets show reversion to a long term central tendency, driven by long run equilibrium forces. This reversion to the mean can be obtained by making the pattern of volatility to decay exponentially over time. The time scale of the decay in volatility determines the time scale of the reversion to the mean in prices, as suggested by Laughton and Jacoby (8).

Valuation

The price process is then used in a Monte Carlo simulation to generate simulated paths of the copper price. The trend in the price is discounted for its systematic risk. Then for each discounted (or certainty equivalent) price path, the cash flows of the project for each period are generated. Then the sample mean of the cash flow for each period is calculated. These are the risk-adjusted expectations of net cash flows. These expectations of cash flows are then discounted by time using the risk free discount rate. The value of the project is the sum of the present values of the expected cash flows.

The steps are the following:
- the underlying assets which are the sources of uncertainty of the project are identified, and the variances of their processes are identified; these assets must be traded in the market;
- the "cash flow" formulas that relate cash flows with the underlying asset prices are specified;
- we discount the trend in asset price processes by the systematic price risk;
- we implement a simulation to obtain the terminal distribution of asset prices (terminal refers to the period in which the relevant cash stream flows);
- using the "cash flow formulas" we get the terminal distribution of each cash flow;
- from each terminal distribution we get the mean or expected value of each cash flow;
- we discount each cash flow by the riskless rate to obtain its present value, and then add present values up to obtain the value of the project.

Equivalent Constant Discount Rate

The risk characteristics of the project can be analyzed by calculating the Equivalent Constant Discount Rates (ECDR) for each period's cash flow, following Laughton and Jacoby (8). The ECDR is defined as that constant rate that will discount a net real expected cash flow (not adjusted by risk) to its calculated (present) value. It is also possible to calculate an ECDR for the project as a whole: it is the constant rate that discounts the series of real expected cash flows (not adjusted by risk) to its calculated (present) value.

This indicator can only be calculated after the fact. Actual discount rates are stochastic. The ECDR may even not exist for all projects because there may be no real root to the internal rate of return equation used to calculate it.

The ECDR series for the cash flows of a project is the implicit structure of discount rates that would have given the same project value using an NPV framework. The exercise to obtain them described at length is:
- an NPV type simulation is carried on, in which trends in asset prices are not replaced by the riskless rate;
- the terminal distribution of asset prices and of cash flows are estimated, and cash flow means are calculated;
- ratios between cash flows obtained by NPV simulation and cash flows obtained by DAV simulation are measures of risk, because those rates applied in an NPV framework to discount each period cash flows, give by construction the same present value as the DAV method.

The Operating Option Conundrum

The possibility of changing the cut-off ore grade in the mine when prices have unanticipated changes implies that production levels can be altered with respect to the originally planned levels. Accordingly, variable costs also change.

In a dynamic programming method each chance node of the decision tree represents a possible movement in the copper price, and the objective function is valued according to a discounted cash flow evaluation of the flows of the project at that stage. Each possible state must be valued, starting by the last period, and going back through the selected branch of higher value. This procedure requires enormous amounts of computation time.

An optimizing model such as Lane's (2) searches directly for the optimal decision at each stage, maximizing the contribution to value of each period's exploitation. The contribution to value includes the net cash flow of the period plus an opportunity cost term, which represents future possible states. This opportunity cost is composed by the return that the capital tied up in the operation would have in alternative investments of the same risk level, plus a term representing the change in value of the mine over time due to capital gains if prices would improve.

This maximization of the contribution of the exploitation of resources in each period determines an optimal cut-off grade for the period. However, to perform this maximization it is necessary to estimate the value of the capital tied up in the operation (the value of the mine at that stage), which is one of the elements in the opportunity cost of each period's contribution to value as described above. It is necessary then to calculate the value of the mine at this stage, to represent the capital tied up in the operation.

To do it in a manner consistent with option valuation would require to run a Monte Carlo simulation at this step; however, in each step of this interior simulation, the value of the mine has to be estimated again, giving rise to new Monte Carlo simulations, and so on, <u>ad infinitum</u>. In conclusion, the value of the mine for the purpose of the cut-off grade model optimization cannot be estimated in a manner completely consistent with option valuation.

In general, DAV simulation models cannot solve contingent control rules in a consistent form with option valuation. Only closed analytical solutions provide a control rule. However, only simple operating options have closed analytical solutions. Brennan and Schwartz (6) solved the control rule and valuation problem for a simple copper mine that has the flexibility to close down and to reopen.

In our example, the contingent rule is much too complicated to be solved analytically. Its advantage is that it is a much more realistic decision rule, applied customarily in the planning stage of every mine, and applied again when large price movements occur. Managers apply a model based on NPV analysis, and not on option valuation anyway. In comparison, we represent more realistically the operating option in an open pit copper mine, but we have to simplify the task by using a valuation method in the interior of the cut-off grade model that is based on NPV analysis. This bias is probably small, and in any case it should probably undervalue the operating option contribution.

The Inter-temporal Dimension of the Operating Decision

We describe here the determination of a cut-off strategy in the planning stage, a procedure that we then repeat every time there is a price surprise, as described in the next section.

The value of the mine at each stage is needed in the optimization cut-off grade model. But the value of the mine is exactly what the exercise is about, and it depends on future grades selected. The approach is to use, for the internal workings of the cut-off grade model, an approximation to the present value of the mine based on NPV concepts, and then to iterate.

The iteration starts with an initial estimate of the present value of the project under the optimal strategy to be determined. Then in each period the current value of the mine is adjusted; it decreases as the resource is being depleted. The effects of time and of use of the resource are considered to adjust the value of the mine from period to period.

At the point when the mine is depleted, the current value of the project should be zero, or a positive figure accounting for the residual value of the equipment. If the value at termination date (when the mine is depleted) is different from the residual value, then the initial estimate used is corrected by the present value of this difference between the termination value obtained and the expected residual value.

The present value of the mine is the result of the iteration until this terminal value difference is negligible. The inter-temporal optimization process is based on the optimization principle of dynamic programming (although it is an iterative optimizing procedure and not a dynamic programming one). The principle of optimization holds that an optimal strategy has the characteristic that for any state and initial decision, subsequent decisions must be optimal.

Summing up, the cut-off grade decision for the current stage depends on an estimate of the current value of the mine. To estimate the current value of the mine, it is necessary to estimate the complete cut-off strategy until the mine is depleted.

Valuation of the Flexible Mine

In each simulation step, the cut-off grade model is run as if the mine had to be planned again, given that a change in prices will make it preferable to choose a different cut-off strategy.

The cut-off grade selected for the period determines, depending on the ton-grade distribution of the phase of the mine in exploitation, the quantity of material that complies with the requirement of grade and also the average grade of the material to be sent to the treatment plant. When the cut-off grade is high, the selection of material will be strict, sending little material to the treatment plant, but with a high average grade. Most of the material will be sent to the waste deposit (or to stockpile for later treatment). Also, if mine capacity permits, the next pushback will be exploited earlier to send more material to the treatment plant, shortening the mine horizon. Operating costs are also dependent on the cut-off grade, because it determines flows of materials between the different stages.

Value of Flexibility

The value added to a project by an operating option can be measured by comparing the DAV valuation of the project without applying the operating option, and the DAV valuation of the same project, in this case applying in each step the operating option model, which implies that the mine is sequentially "re-planned" for the future based on the new expected price path.

The DAV valuation without applying the operating option is done by running the cut-off grade model only once, and using for all simulation steps the cut-off grades determined in that only run.

The risk characteristics of the project should improve with the application of an operating option, making it less risky. This result can also be obtained by comparing the Equivalent Constant Discount Rates of the two DAV simulations indicated above.

THE MINE

The project is a copper mine and treatment and processing plants currently under development in Chile. The project is planned for ten years of open pit production. The exploitation pattern selected calls for extraction of 1.6 million tons of copper oxide ores per year, with an average grade of 1.5% of copper, yielding approximately 20,000 tons of fine copper per year in the form of electrowinning cathodes of high quality (Grade A).

A block model of the mine gives a spatial representation in blocks. Each block is assigned its average ore grade. Each expansion (or pushback) is a fraction of the ore body, geographically defined according to mineral homogeneity and exploitation logistics, composed of several blocks. Its statistical representation is a histogram of tonnage - ore. The production plan considers seven "expansions" or partial pits, which represent the geometrical division of the ore body and the global sequence of exploitation. In each period, the production plan combines minerals from different "expansions", in such manner that the stripping ratio and grade are maintained as constant as possible. These combinations are called "increments," "equivalent phases" or simply "phases." The speed of exploitation of equivalent phases is a variable of control in the cut-off grade model, but the sequence of exploitation is considered fixed. This is a reasonable assumption, although in extreme cases of changes in price expectations, it may pay to review the exploitation sequence. The same is true for capacities, which in this exercise are considered fixed.

Typically the material will be extracted from the mine at full capacity, although prices may dictate a slower pace. There is an additional cut-off grade decision that determines the destiny of a rejected material: the waste deposit or stockpile for eventual later treatment. In the stockpile the material is mixed in one pile, and when the mine is nearing exhaustion a decision has to be made on how much material, if any, is sent from the stockpile to the plant.

The plant will process 4,500 tons of ore per day. The treatment process consists of a conventional crushing plant, with a primary crusher, and then secondary and tertiary crushers. The crushed mineral goes to conditioning and sintering with concentrated sulphuric acid. The product is then stockpiled on a non-permeable floor to continue processing with conventional leaching techniques, using diluted solutions of sulphuric acid in sea water. Rich solutions of copper sulfate are accumulated and sent to the electrolysis tank. Copper is electro-deposited, producing highly purified copper cathodes as final output. These operations are grouped for the purpose of the cut-off grade model in three stages according to the base of costing: mine (costs per ton of mineralized material); treatment (cost per ton of ore); and refining (cost per ton of contained copper metal).

THE OPERATING OPTION

The Cut-off Grade Model

The optimum cut-off grade in every period is selected from a set of six grade alternatives, three "economic cut-off grades" and three "balanced cut-off grades". The six cut-off grades are the result of maximization of value under different capacity constraints, and in this sense they all are economic grades.

An "economic cut-off grade" denotes the material that at the margin just pays for downstream processing. For example, material with a quality that is barely the mine economic cut-off grade just pays for treatment and refining. If the price goes up, a higher unit cost would be covered; and it would be convenient to process materials with even lower grades. This argument presumes that there is no limitation in treatment and refining capacities.

A "balanced cut-off grade" is one that uses a pair of capacities up to the full, and it depends on the respective capacities of the pair of stages, and on the quality of the material as represented by the tonnage-grade distribution. The balanced grade does not change with price or cost alterations.

A well designed operation should use all capacities to the full. The balanced grade for treatment and refining will probably be selected, as these two stages are the more capital intensive ones. However, price surprises may dictate the convenience of changing from a balanced grade to an economic cut-off grade, leaving some capacity unused.

Current and expected prices determine which one of the six grades is selected as optimal for each period. The set of cut-off grades for the whole mine lifespan is the <u>optimal cut-off grade strategy</u>.

Cut-off Grade Formulas

We present the cut-off grade formulas with explanations that are necessary to introduce the results obtained. The definition of variables is the following:

<u>Price:</u>

p : price, per unit of metal;

<u>Variable and Fixed Costs:</u>

k : variable cost of refining per unit of metal;
h : variable cost of treatment per unit of ore;
m : variable cost of mining per unit of mineralized material;
f : fixed costs per year;

F :   opportunity cost; it is composed by the cost of the capital tied up in the project, and the capital gain that could be obtained from one period to the next if the resource is not consumed and economic conditions improve.

<u>Ratios</u>:

x :   proportion of the mineralized material classified as ore;
y :   yield of mineral in the treatment process (recovery typically in the range of 80% to 90%)
g':   ore ratio, or average grade of the ore as a mineral, expressed in terms of copper content in the ore;
g :   mineral ore ratio, or cut-off grade applied to mineralized material to define ore;
q :   quantity of mineralized mineral of grade **g**;

<u>Economic Cut-off Grades</u>:

g_m :   mine economic cut-off grade;
g_h :   treatment economic cut-off grade;
g_k :   refining economic cut-off grade;

<u>Balanced Cut-off Grades</u>:

g_{mh}:   mine/treatment balanced cut-off grade;
g_{hk}:   treatment/refining balanced cut-off grade; and
g_{mk}:   mine/refining balanced cut-off grade.

Table I - Three Component Stages of the Mining Process

STAGE	BASIC THROUGHPUT	QUANTITY	VARIABLE COST (per unit of throughput)	CAPACITY (throughput / year)
Mining	Min. material	1	m	M
Treatment	Ore	x	h	H
Refining	Metal	$x \cdot g' \cdot y$	k	K

Table I shows that **x** represents the technical coefficient to convert mineralized material into ore. The ratio **g'·y** represents the coefficient that transforms ore into metal. And finally, the ratio **x·g'·y** is the coefficient to transform directly mineralized material into metal. Assuming for ease of exposition that there is one unit of mineralized material, then the coefficient **x** represents the quantity of ore, the ratio **x·g'·y** represents the quantity of recoverable fine copper, and the ratio **g'·y** represents the recoverable fine copper <u>expressed in units of ore</u>.

The mine economic cut-off grade depends inversely on price and directly on treatment and refining costs. The derivation of these formulas can be seen in Lane (1) and with more detail in Mardones (9).

$$g_m = \frac{h}{(p-k)\cdot y} \tag{1}$$

The treatment economic cut-off grade depends also (and directly) on fixed costs **f** and on the opportunity cost **F**:

$$g_h = \frac{h + (f+F)/H}{(p-k)\cdot y} \tag{2}$$

The refining economic cut-off grade depends on the same variables as g_h but in a somewhat different form. Note that the recovery yield is not present:

$$g_k = \frac{h}{p - k - (f+F)/K} \tag{3}$$

Balanced cut-off grades

The mine/treatment balancing cut-off grade is given by the equality between the percentage of mineralized material classified as ore, and the quotient between treatment capacity **H** and mine capacity **M**. These two components will be used up to their capacity when the mine moves **M** units of material per year, which converts into **H** units of ore per year. In consequence, the quantity **x** of ore arising from one unit of mineralized material must be in the proportion **H/M**:

$$x = \frac{H}{M} \tag{4}$$

This balanced value of **x** determines a balanced cut-off grade g_{mh} between mine and treatment capacities, using the relationship between **x** and **g** given by the cumulative grade distribution of the **increment** to be processed.

The relationship shows that the percentage of mineralized material classified as ore is the sum of the percentage quantity of mineral for each grade over the cut-off grade:

$$x = \int_g^\infty q\cdot dg \tag{5}$$

and is a descending line in the plane **g** - **x**.

The exact formula relating the balanced cut-off grade g_{mh} to the variables that determine it can be deduced when the ton-grade distribution function $q = f(g)$ follows a theoretical distribution such as the lognormal one. However, this function in practical cases is an empirical distribution.

The mine/refining balanced grade shows that the recoverable mineral above a corresponding cut-off grade, expressed per unit of mineralized material $(x\cdot g'\cdot y)$, must equal the relationship between refining and mine capacities.

$$x\cdot g'\cdot y = \frac{K}{M} \tag{7}$$

The relationship between the cut-off grade **g** and [x·g'·y], the fine copper contained in mineralized material (given **g**), is a descending line, indicating that the amount of copper contained in one unit of mineralized material descends with an increasing cut-off grade. The slope of the curve is **-q·g·y**.

The recoverable mineral above a corresponding cut-off grade, expressed per unit of ore (g'·y), must equal the relationship between refining and treatment capacities:

$$g' \cdot y = \frac{K}{H} \tag{8}$$

The relationship in this case is an ascending line with decreasing slope, given by q·(g'-g)·y·x.

Patterns of Cut-off Grades

Economic Grades: Decreasing Grades in Time

With an uniform or decreasing expected price profile, the resulting <u>economic</u> cut-off grades for mine, treatment and refining limits present a strictly decreasing trend (except mine economic cut-off grades which for constant expected prices are also constant). This result (shown theoretically in Mardones, 9) reflects that the time value of money in an inter-temporal optimization, with only one limiting capacity, forces a quick treatment path.

However, with an increasing time structure of expected prices, <u>economic</u> cut-off grades may display rising values. The reason is that with increasing expected prices, the opportunity cost **F**, or change in the value of the mine with time (if no exploitation took place, that is, the value with the same reserves but one period ahead) may increase. This happens when the positive effect on value of an expected price rise is more important than the negative effect of the discount rate from period to period.

When this opportunity cost rises in time, then the economic cut-off grade also increases, as can be seen in equations (2) and (3). These valid responses are not usually considered in the literature, which is limited to simple cases of uniform expected prices.

Balanced cut-off grades

Even if economic grades were decreasing, choosing a <u>balanced</u> cut-off grade as the optimum cut-off grade instead of an <u>economic</u> grade, may interrupt the monotonic trend of the optimum cut-off grades.

The profile of balanced grades is not uniform because it depends on the ton-grade distributions of phases in the exploitation sequence (defined above) which exhibit some differences among them. This profile does not change with any price vector, because it depends exclusively on the capacities of the process' components, and these capacities remain unaltered during the lifetime of the project.

It is interesting to note what happens with very low expected prices. Economic grades are high, and the planned balance between capacities is ruined, and <u>balanced</u> grades are not selected as often. One or the other limiting capacity becomes operative more frequently. The observed result is that selected grades are higher and that there are jumps from period to period when a different limiting capacity or pair of limiting capacities become the active restriction. In the last periods the selected cut-off grade usually goes up, reflecting the fact that it may not even be economical to exploit lower grades.

Lifespan of the Project

A higher average price profile determines a longer lifespan of the project, with and without the stockpile option. The reason is that with low prices the opportunity cost of postponing the exploitation of high grade minerals is not significant. Given a low price, the model imposes high cut-off grades to obtain benefits enough to cover unit costs. A high cut-off grade dictates the extraction for each period of high quantities of material from the mine, because the material that satisfies the high degree of desired mineralization is scarce.

<u>Adaptations of the Model</u>

When the cut-off grade model is applied in a simple NPV valuation, the cost of capital used to determine the "opportunity cost" is the discount rate of a project with the same risk level as the undertaken project. This discount rate used to discount the model's cash flows must include a premium for time and risk, when an NPV evaluation framework is applied.

When used in option valuation, this cost of capital must be the risk free rate perceived by the manager of the mine, because all calculations are being made in certainty equivalent terms.

COPPER PRICE MODEL AND ECONOMIC PARAMETERS

The copper price model used here is a simple logarithmic diffusion. It has been left for subsequent research the application of a model with <u>reversion to the mean</u> characteristics. In particular, it will be of interest to observe the behavior of cut-off grade strategies under a reversion to the mean price model.

We assumed in NPV valuations a "beta" for a copper project of 0.75 for a well diversified foreign investor, a risk-free rate of 5% per year, and a market premium of 4%, yielding a discount rate of 8% for real U.S. dollar flows.

In DAV valuations we assumed that the "beta" for the copper price was 0.75 for a well diversified investor; the risk-free rate and the market premium were taken at the same levels of 5% and 4% respectively. With these assumptions the implicit risk profile of the copper project is for every period somewhat higher than the 8% discount rate used in NPV exercises.

RESULTS

<u>Effects on production</u>

The cut-off grade strategy determined before the first year of production is for almost every period the balanced cut-off grade for treatment and refining, for both constant expected price profiles, at 70 ¢/lb and at 100 ¢/lb. Only in years 10 and 11, the two last years of the mine, the the economic grades for treatment are selected instead of the balanced grade for treatment and refining. The economic grades for treatment are different for the two expected price profiles.

This result shows that the cut-off grade strategy is very robust to price variations in the case of this project. For much lower expected prices the balance of capacities gets distorted and the cut-off grade strategy gets very different. Figure 1 shows the initial cut-off grade strategy for a constant expected price profile of 70 ¢/lb.

Being the refining capacity a limiting capacity in almost every period, production levels in metal content are not much affected by changes in prices, at least in the range of 70 ¢/lb to 100 ¢/lb. Nevertheless, the lifespan of the project is longer for higher expected prices. The mine is exhausted in 11 periods for both price profiles considered, but production only from the stockpile takes seven years with expected prices of 100 ¢/lb, but only one year for the 70 ¢/lb profile. Total metal production in 18 years is 243,256 metric tons for the higher price, and 212,667

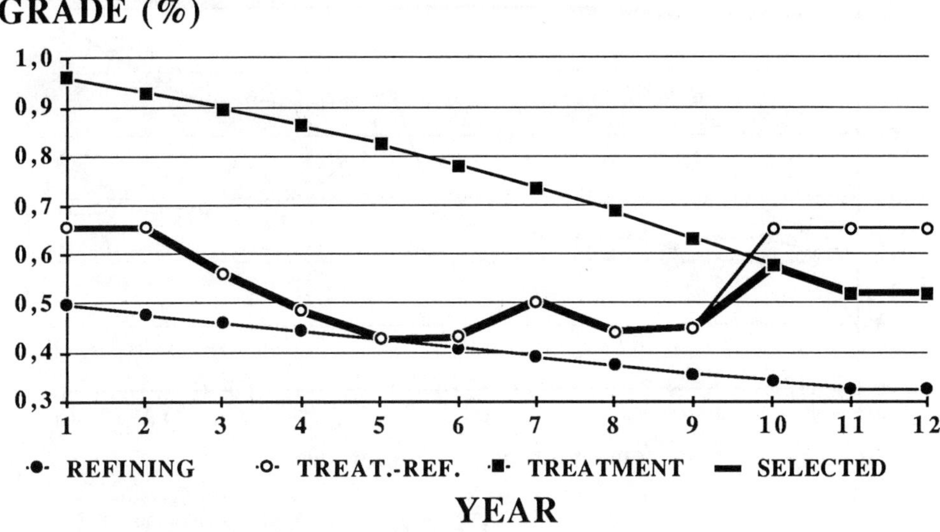

Figure 1 - Cut-off grade strategy for a constant expected price of 70 ¢/lb

metric tons for the lower one.

In each simulation draw the price profile may be much higher or lower than the expected profile as a consequence of simulated price surprises. In those cases the model registers important changes in cut-off grades and production levels with respect to expected levels.

Measurement of Bias in Simple NPV valuation

Simulation methods correct for the bias in the estimation of expected cash flows present in simple NPV estimations due to Jensen's inequality. The bias is measured as the difference of present values estimated by simple NPV and by NPV simulations. The bias for a constant expected price profile of 70 ¢/lb the bias is 6.14%, for a standard deviation of copper changes of 0.2. This result shows that a low price profile generates non-linearities in the cash flow model, and also that a higher percentage results with a low NPV associated with low expected prices. The simple NPV is $ 47,993,000, while the simulated NPV reaches to $ 50,398,000, a 6.14% difference.

For a constant expected price profile of 100 ¢/lb the bias is smaller, of 1.37% of value, for a standard deviation of copper price changes of 0.2. The simple NPV method gives a value of US$ 141,366,000, while using simulations the result is US$ 143,310,000. For higher standard deviation of copper price changes, this difference would probably be higher also.

Value of Flexibility

The value of flexibility can be estimated by running the option simulation model with and without the application of the contingent rule. Our simulations give a substantial value of flexibility for this particular mine, of 6.34% for an expected price profile of 70¢/lb, but only of 2.36% for an expected price profile of 100 ¢/lb.

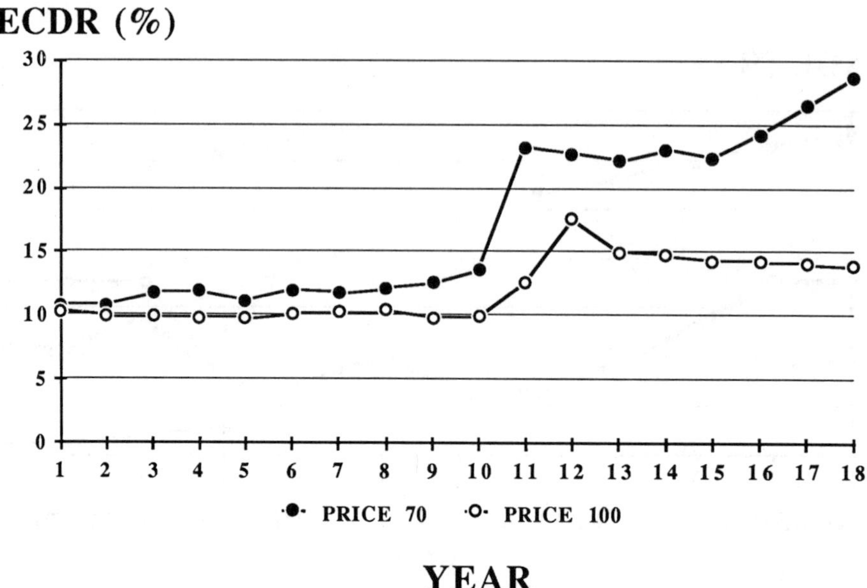

Figure 2 - Equivalent constant discount rates (ECDR) for expected prices of 70 ¢/lb and
100 ¢/lb (standard deviation of copper price changes at 0.2)

DAV valuation results

The DAV valuation gets a higher value of almost 2% for a standard deviation of copper
price changes of 0.3, compared to a standard deviation of 0.2. This result illustrates that firms
with flexible options get more value in a more volatile environment.

The values for the project are not comparable to those obtained under NPV valuations
reported above, because the discount rate structures are not compatible. In fact, in our case the
implicit discount rates (Equivalent Constant Discount rates) found for option valuations were
different from those we used in NPV valuations.

Risk structure

We follow Laughton and Jacoby (8) in the procedure to calculate the Equivalent Constant
Discount Rate (ECDR), which are a measure of the implicit risk profiles of the project's cash
flow. The ECDR is obtained by making equal the present value of the NPV cash flow, with the
present value of the option cash flow for the same period. The idea is that the ECDR is the
discount rate that in an NPV simulation would give the same result as an option simulation.

Option cash flows are discounted at the risk free rate. The rate of discount that applied to
the NPV cash flow gives the same present value as the option NPV discounted at the risk free
rate is the ECDR of each cash flow, estimated as a geometric average yield.

The ECDR profiles for the two price profiles that we are analyzing are shown in figure 2.
Both are very similar, but as expected the ECDR's are higher for the lower prices. We analyze
now the particular form of risk profiles.

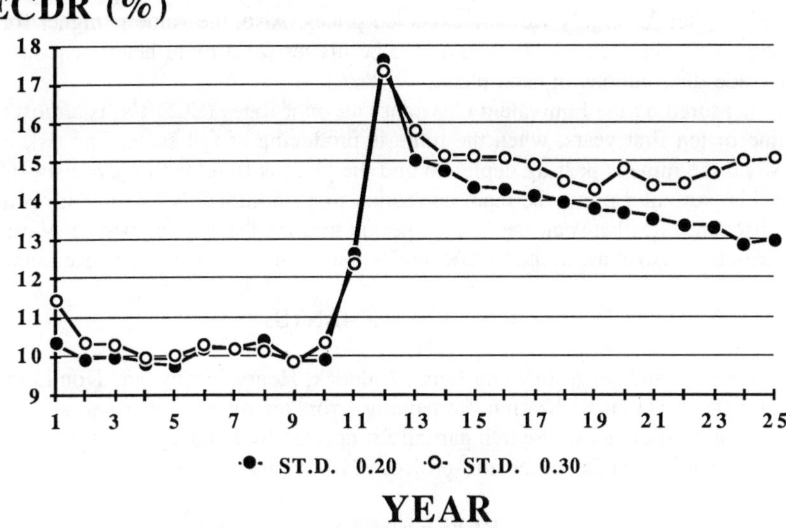

Figure 3 - Equivalent constant discount rates (ECDR) for expected prices of 100 ¢/lb and standard deviation of copper price changes at 0.2 and 0.3

Mine vs. Stockpile

After year ten the mine is almost always depleted, and processing from stockpiled ores starts. This phase is much more risky than processing from the mine. The reason is that ore grades are lower, and actual processing takes place only when prices are relatively high.

Period 11 is the last one with production from the mine in simple NPV cut-off grade models for both price profiles, although with lower production than before, as the mine is nearing exhaustion. Period 12 shows an important increase in risk, which coincides with a drastic drop in production, as the mine is already exhausted and production comes only from the stockpile. Of course these production characteristics are the expected ones, but there are wide distributions of results around them. The period of production from stockpile is the most sensitive one to prices.

Ton-grade Distribution

Figure 2 also shows that there is a term structure of risk (measured as ECDR yields) during the lifespan of the mine, that is, the first 10 years, which is not flat, although ECDR's are not very different. This structure is attributable to the characteristics of each period, which differ mainly on the ton-grade distribution of the sequential phases in the mine. For this particular mine, differences are very small.

Variability of the copper price

Figure 3 shows that the risk profile measured by ECDR's is higher for a higher variability of the copper price, even when the systematic risk of the price has not been changed. This result shows that there are relevant non-linearities in the project under consideration.

Summing up, the project presents some non-linearities as measured by the bias or difference in present values using simple NPV and simulation NPV methods. The value of the project's flexibility is small for high and uniform prices, but higher for low expected prices, and presumably still higher for highly variable expected prices. Also, the value is higher for a price process showing a higher variance. It should also be higher for a more heterogeneous mine in terms of ton-grade distributions in each phase.

The risk measured by the Equivalent Constant Discount Rates (ECDR's) is almost constant during the nine or ten first years, when the mine is producing at full scale. The risk increases importantly when the mine is nearing depletion and the plant is filled with previously stockpiled material. For this particular mine, the main correction to a traditional NPV method would be to differentiate discount rates between the "mine" period and the "stockpile" period. Variations in ton-grade distribution also show in the ECDR profile, but are not of great relevance in this mine.

ACKNOWLEDGMENTS

I thank comments and suggestions by James Paddock, Henry Jacoby, and Don Lessard, and the research assistance of Andrés Kettlun. Remaining errors are mine. This paper is based on my Ph.D. thesis and on post-doctoral research partially supported by Fondecyt, the Chilean Fund for Scientific and Technological Development, project 1214 of 1991.

REFERENCES

1.   H.J. Jacoby and D.L. Laughton, "Project Analysis Using Methods of Derivative Asset Valuation", Working Paper MIT-EL 87-001WP, 1987, M.I.T. Energy Laboratory.

2.   K. Lane, The Economic Definition of Ore, Mining Journal Books Ltd., London, U.K., 1988.

3.   F. Black and M. Scholes, "The Pricing of Options and Corporate Liabilities", Journal of Political Economy, Vol. 81, 1973, 637-59.

4.   P. Boyle, "Options: A Monte Carlo Approach", Journal of Financial Economics, Vol. 4, 1977, 323-338.

5.   J.C. Cox, and S.A. Ross, "The Valuation of Options for Alternative Stochastic Processes", Journal of Financial Economics, Vol. 3, 1976, 145-166.

6.   M.J. Brennan and E.S. Schwartz, "Evaluating Natural Resource Investments", The Journal of Business, Vol. 58, 1985.

7.   H.J. Jacoby and D.L. Laughton, "Uncertainty, Information, and Project Evaluation," Working Paper MIT-CEPR 90-002, 1990, Center for Energy Policy Research, MIT.

8.   D.L.Laughton and H.D. Jacoby, "A Two-Method Solution to the Investment Timing Option", Document 90-023WP, 1990, MIT Center for Energy Policy Research.

9.   J.L. Mardones, "International Investment Contracting with Financial Market Segmentation: Option Valuation of a Copper Mine", Ph.D. Thesis, The Fletcher School of Law and Diplomacy, 1991.

Applications of Copper

The hardening of complex aluminum bronze by heat treatment

S. Lu
*Department of Material Science, Shanghai Jiao-tong University,
Shanghai, People's Republic of China*

A. Wu
*Attached Machinery Factory of Shanghai Jiao-tong University,
Shanghai, People's Republic of China*

S. Zhou
*Department of Material Science, Shanghai Jiao-tong University,
Shanghai, People's Republic of China*

ABSTRACT

A complex aluminum bronze with nominal composition Cu−11 wt pct Al−5 wt pct Ni−5 wt pct Fe is treated by water quenching at 950°C or thermo−mechanical treatment at 980°C, and followed by tempering in the range of 350~450°C. Its hardness increases remarkably. The original microstructure of massive pro−eutectoid α −phases and coarse granular and lamellar κ−phases are transformed into that of the dispersed distribution of fine κ−phase precipitates in the α−matrix. It may be the cause of hardening.

INTRODUCTION

The complex aluminum bronzes containing iron and nickel are generally used for components working in heavy load condition and marine environment due to their good mechanical property and corrosion resistance. The composition, microstructure and property of these bronzes has been studied by different authors. (1−7) Now these bronzes are generally used in as−cast or forged state, hence many references concentrated on the investigation of the morphology, composition and structure of the phases in as−cast microstructure of the bronze and their formation sequence, (1, 2, 5, 6) According to the vertical section of $Cu-Al-5Ni-5Fe$ equilibrium diagram shown in Figure 1, (8) the bronze with nominal composition $Cu-11$ wt pct $Al-5$ wt pct $Ni-5$ wt pct Fe is in the β−phase region at high temperature in equilibrium state, and the α−phase and κ−phase precipitates successively during cooling process. Jahanafrooz, Lloyd and Hasan et al (1, 2, 6) suggested that the κ−phases precipitated in the as−cast microstructure of these bronzes may be classified into κ_I, κ_{II}, κ_{III} and κ_{IV} on the basis of the various morphologies and their precipitation sequence. The κ_I phase looked like bulky dendrite or rosette precipitates from liquid alloy. The κ_{II} phases appeared as coarse grain (diameter about $5-10$ μm) precipitate from β−phase at high temperature, and the κ_{IV} phases are the final precipitates of fine particles (diameter less than 1 μm) from α−matrix. All of them are iron−rich phase based an the $FeAl_3$ compound. But the κ_{III} phases formed by eutectoid decomposition are the lamellar nickel−rich phase based on the NiAl compound.

In as−cast and forged microstructure of these bronzes, the κ−phases are generally coarse and non−uniformly distributed. But a dispersed distribution of fine grains of κ−phase can be obtained by adopting quenching at high temperature and followed by tempering, and the hardness of bronze rises remarkably. The tensile strength and proof strength of the bronze also increases with a corresponding decrease in ductility. (9) For example, the bronze with nominal composition $Cu-11$ wt pct $Al-6$ wt pct $Ni-6$ wt pct Fe was treated by water quenching at 925°C and followed by tempering at 400°C, its hardness rised and reached HV365. (10)

Brezina (9) also reported that the higher corrosion resistance of these bronzes can be gained by quenching and tempering due to the dispersed distribution of fine κ−precipitates in the microstructure amd their uniform composition.

In addition, the strength of the aluminum bronzes can be raised about 15 percent by the thermo−mechanical treatment owing to the fine microstructure. (9, 11)

In the present work the effect of quenching and tempering or thcrmo−mechanical treatment on the hardening of complex aluminum bronze containing iron and nickel has becn studied, the optimum parameters of heat treatment and the variation of microstructure of the bronze are also investigated.

EXPERIMENTAL

The composition of the bronze investigated, as determined by wet−chemical analysis, is given in Table I . The casting samples (20mm × 12mm × 120mm and 34mm × 12mm × 120mm for the test of hardness and tensile property respectively) were heated at 950°C then forged on a crank press with a capacity of 1.6MN, the reduction was 40 pct in a single operation. The average hardness of six tests on the as−cast and forged samples is listed in Table II .

Table I − Chemical Composition of the Bronze. wt%

Element	Al	Fe	Ni	Cu
Amount	10.6	4.4	4.5	balance

Table II − Hardness Determined on the Bronze at Original State

State of Material	As−Cast	Forged
Hardness (HRC)	22.5	27.4

The forged samples were heated for quenching and tempered in electric furnace.The cast samples for thermo−mechanical treatment were also heated in electric furnace, after forging on the crank press the samples were quenched into water immediately. And then tempered. All the furnaces were controlled by the silicon controlled temperature regulators. The determination of hardness were carried out on a Rockwell hardness apparatus type 69−I. The test bars for tensile test were machined in accordance with the 21 K bar (d_0=10mm,l_0=5d_0)in the national standard GB228−76, and tested on a tensile test machinc type ZDM−30T.

For metallographic microscopy the specimens were polished using conventional procedures and etched in a solution of 10 pct ferric nitrate in water, and were photographed on the optic microscope type Neophot II and the scanning electron microscope type S−520.

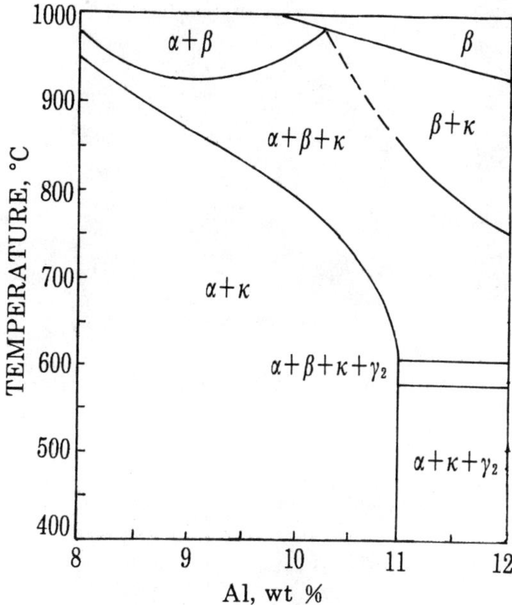

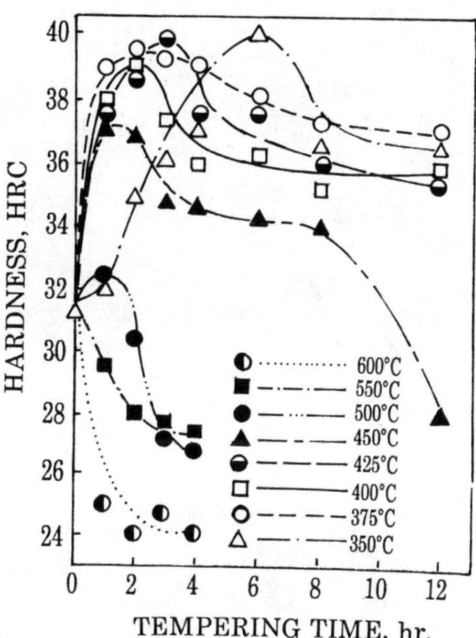

Figure 1−Vertical section Cu−Al−5Ni−5Fe equilibrium diagram, after Cook et al. (8)

Figure 2−Variations of hardness of the bronze quenched at 950°C and tcmpered at different temperatures

RESULTS

Effect of Quenching and Tempering

Effect on Hardness

The variations of hardness of the bronze quenched at 950°C and followed by tempering at different temperatures in the range of 300−600°C are shown in Figure 2. All the data of hardness illustrated in the figure are the average of 5−6 tests. The hardness of bronze tempered in the temperature range of 350−450°C rises obviously and reaches HRC34∼39. The hardening effect of tempering in the range of 375−425°C increases more remarkably and the peak hardness is achieved after tempering for 2−4 hours. The tempering time reached the peak hardness shortens gradually with the rise of the tempering temperature, and the peak hardness reduces correspondingly.While the specimens are tempered in the range of 500−600°C, the hardness doesn't increase obviously, and even the peak value disappears.

Variations of Microstructure

The as−cast and forged microstructure of the bronze are shown in Figure 3 and Figure 4 respectively. The massive pro−eutectoid α−phases exist in the as−cast and forged microstructure of the bronze, although the size of α−phases in forged microstructure became smaller. And in the as−cast and forged microstructure of high magnification by scanning electron microscopy, the κ_{II} and κ_{III} phases appear in large granular and lamellar respectively.

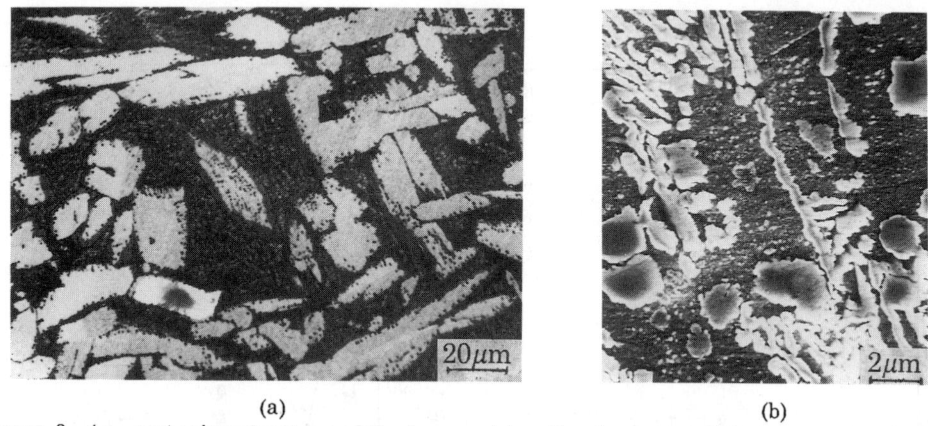

(a)   (b)

Figure 3−As−cast microstructure of the bronze: (a) optic microscopy,etched in water solution of ferric nitrate, (b)scanning electron microscopy.

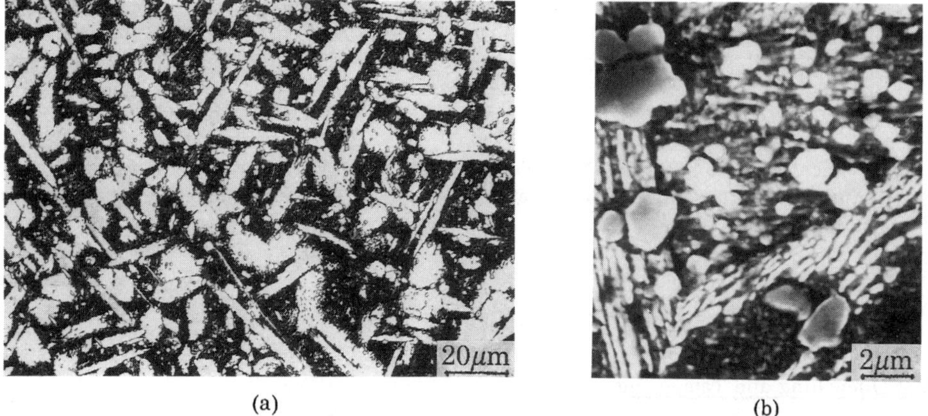

(a)   (b)

Figure 4−Forged microstructure of the bronze: (a) optic microscopy, etched in water solution of ferric nirate, (b)scanning electron microscopy.

The microstructure of the bronze water qucnched at 950°C is shown in Figure 5. The white slender stripes of the α' phase are the supersaturated solid solution by aluminum,and distribute as a Widmanstätten structure. The martensites β' locate between the α' stripes.This mixture structure is namely the "Coarse Bainite" reported by reference (9). In this figure we may find the fine particles of κ−phase still precipitate in the martensite after quenching. The microstructure of the bronze quenched at 950°C and tempered at 400°C for 2 hours is shown in Figure 6. The optic microscopic structure is similar to that of quenching state, but in the high magnifying microstructure by scanning elctron microscopy the volume fractions of κ-phase precipitates increase. In the scanning electron micrograph of the bronze quenched at 950°C and tempered at 450°C for 2 hours shown in Figure 7, the particles of κ-phase are further coarsened.

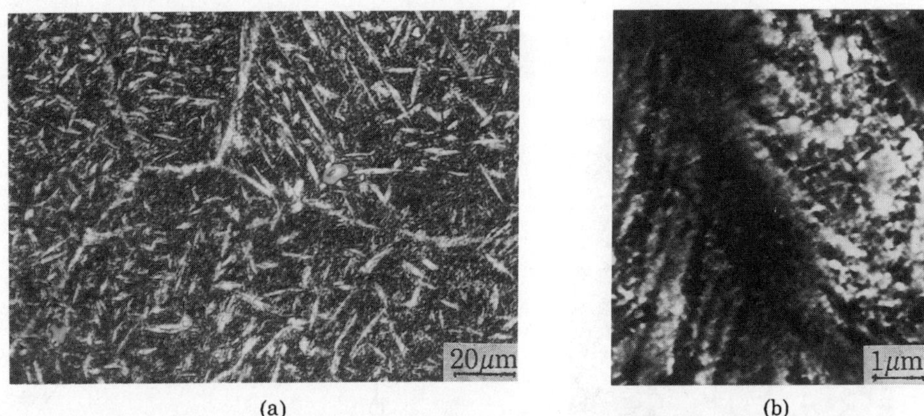

(a)   (b)

Figure 5 — Microstructure of the bronze quenched at 950°C: (a) optic microscopy, etched in water solution of ferric nitrate, (b) scanning electron microscopy.

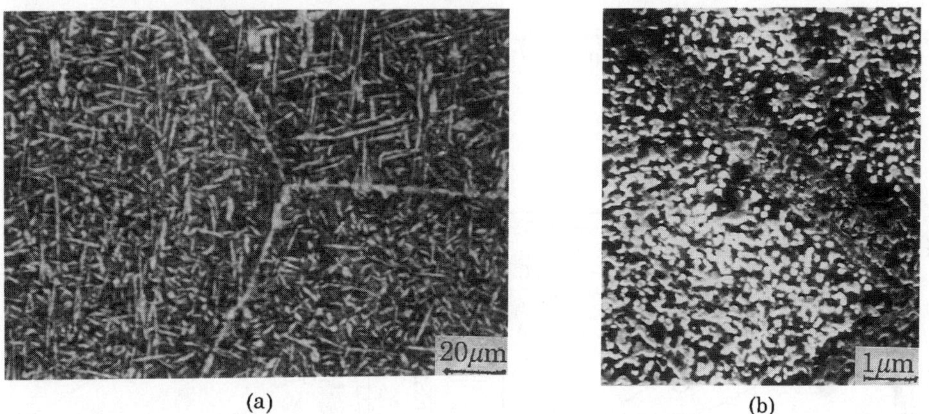

(a)   (b)

Figure 6 — Microstructure of the bronze quenched at 950°C and tempered at 400°C for 2 hours: (a) optic microscopy, etched in ferric nitrate, (b) scanning electron microscopy.

Effect of Quenching Temperature

The variations of hardness of the bronze quenched at 950°C and 850°C then followed by tempering at 400°C are shown in Figure 8. In comparison with the quenching state, the hardness of the bronze quenched at 950°C then tempered at 400°C rises obviously regardless of the tempering time. But the hardness of the bronze quenched at 850°C then tempered is lower. Its hardening effect descends, especially while the tempering time is prolonged.

Effect of Thermo — Mechanical Treatment

Effect of Tempering

The specimens for thermo—mechanical treatment were forged at 980℃ then quenched into water at once, and then were followed by tempering. The variations of hardness of the bronze treated by TMT at 980℃ and tempered at different temperatures are shown in Figure 9. The hardness of the bronze treated by TMT at 980℃ and then tempered in the range of 350—450℃

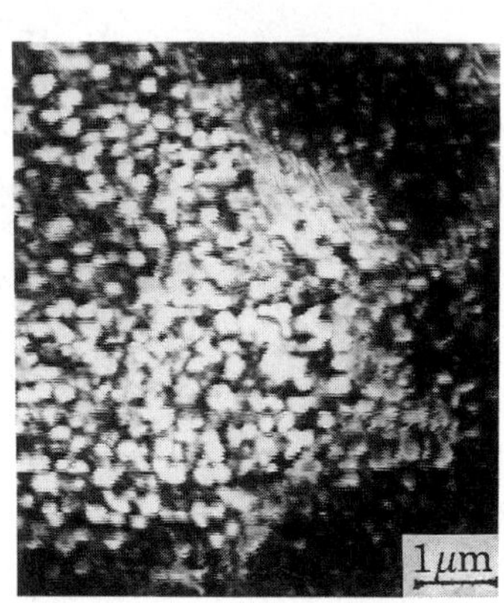

Figure 7—Scanning electron micrograph of the bronze quenched at 950°C and tempered at 450°C for 2 hrs

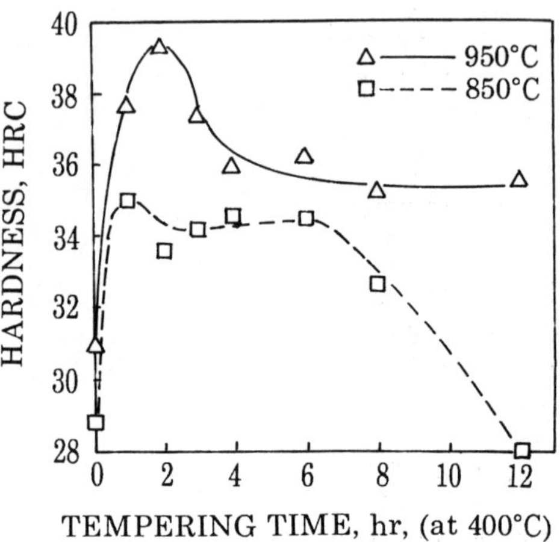

Figure 8—Variations of hardness of the bronze quenched at different temperatures then tempered at 400℃

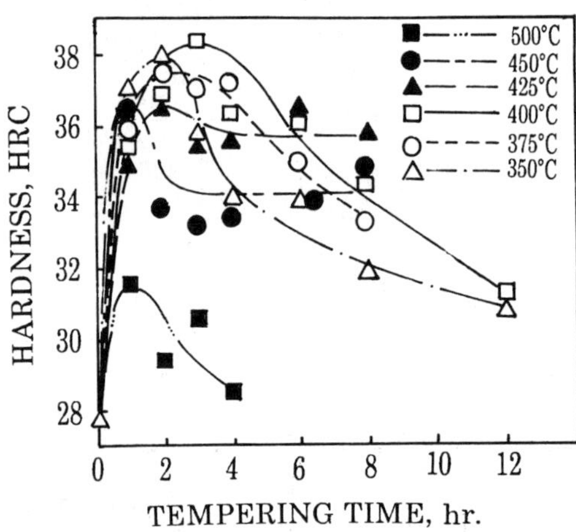

Figure 9—Variations of hardness of the bronze treated by TMT at 980℃ and tempered at different temperatures

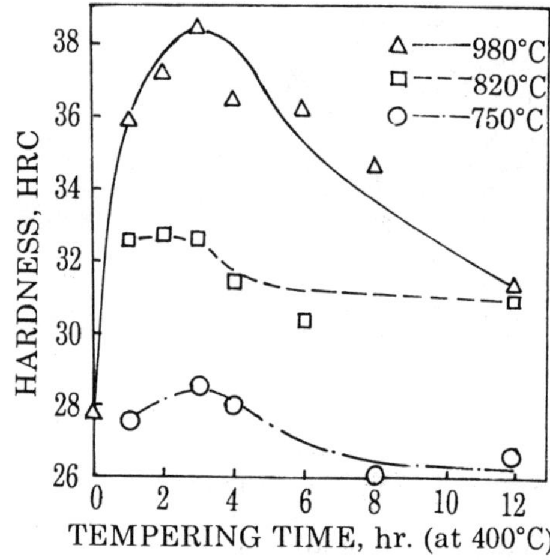

Figure 10—Variations of hardness of the bronze treated by TMT at different temperatures and tempered at 400°C

increases obviously and reaches HRC32~38. In comparison with Figure 2, the variations of hardness of the bronze are similar. But the difference is that the tempering time reached the peak hardness for the bronze treated by TMT and tempering is shortened to 1~2 hours. Similarly the hardening effect decreases while the tempering temperature is higher then 450℃.

Effect of TMT Temperature

The variations of hardness of the bronze treated by TMT at 980℃, 820℃ and 720℃, than followed by tempering at 400℃ are shown in Figure 10. The hardness of the bronze treated by

TMT at 980℃ and tempered at 400℃ for 1−6 hours rises remarkably But the hardening effect of the bronze treated by TMT at 820℃ and tempered decreases obviously, and there is no obvious hardening effect in the bronze treated by TMT at 720℃ and tempered at 400℃.

Variations of Microstructure.

The microstrueture of the bronze treated by TMT at 980℃ is shown in Figure 11, and is similar to that of the same bronze at quenching state shown in Figure 5. Because the

(a)

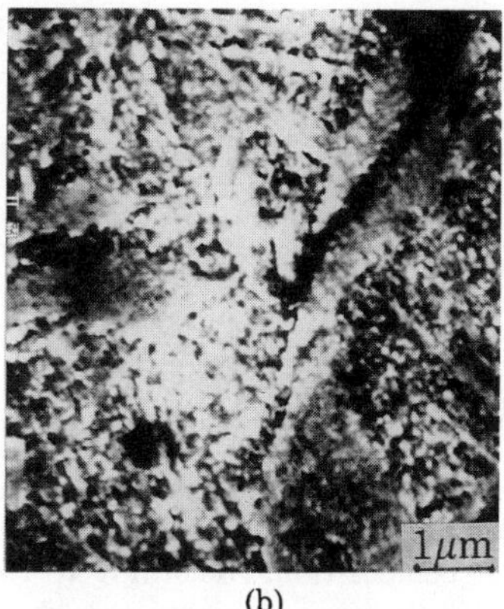

(b)

Figure 11 − Microstructure of the bronze treated by TMT at 980℃: (a) optic microscopy, etched in ferric nitrate, (b) scanning electron microscopy.

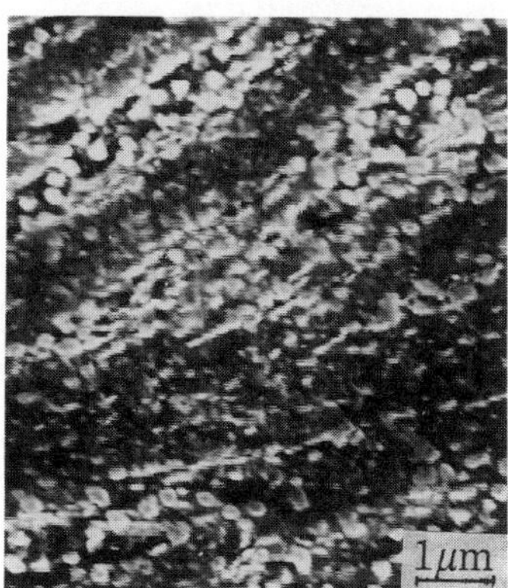

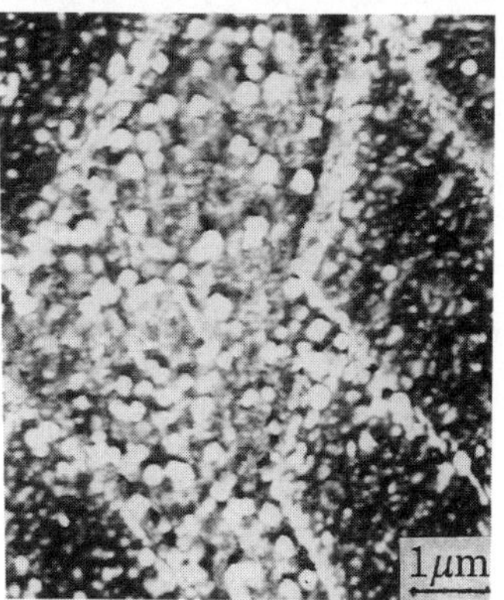

Figure 12 − Scanning electron micrograph of the bronze treated by TMT at 980℃ and tempered at 400℃ for 2 hrs

Figure 13 − Scanning electron micrograph of the bronze treated by TMT at 980°C and tempered at 450°C for 2 hrs.

temperature of samples lowers obviously while they are forged on the crank press, the actual quenching temperature after forging determined in situ by a thermocouple decreases by $70-80°C$ from 980°C. Therefore in the figure the volume fractious of α—phase precipitated at quenching temperature is more than that in the Figure 5. The scanning electron micrographs of the bronze treated by TMT at 980°C then tempered at 400°C and 450°C for 2 hours are shown in Figure 12 and Figure 13 respectively. In comparison with Figure 11(b). the volume fractious of κ—phase precipitates in Figure 12 increase, and in Figure 13 the κ—phase precipitates are further coarsened during tempering at 450°C.

Effect on Tensile property

The tensile properties of the bronze both treated by quenching and tempering and treated by TMT and tempering are determined, The tcnsile properties listed in Table III are the average of 5 tests.

TableIII — Tensile properties of the Bronze Hardened by Heat Treatment

Heat Treatment	UTS (MN/m²)	Elongation (%)
Quenched at 950°C and Tempered at 400°C for 2 hrs.	732	1.84
TMT at 980°C and Tempering at 400°C for 2 hrs.	743	1.96

DISCUSSION

The Quenching Microstructure

In Figure 5(b), the fine particles of κ—phase still percipitate in the martensite after quenching. This is in accord with the report of reference (9) and the CCT curve of a bronze with similar composition. According to the CCT curve shown in Figure 14, the κ phase and α' phase will still precipitate from the β phase during water quenching. And with the descent of temperature the remainingβ turns into the ordered β_1 phase and then into martensite β_1'. This mixed   structure of the martensite, bainite and fine κ phase precipitates may cause an apparent increase of hardness of the bronze. The hardness of the bronze quenched at 950°C will rise to HRC31.

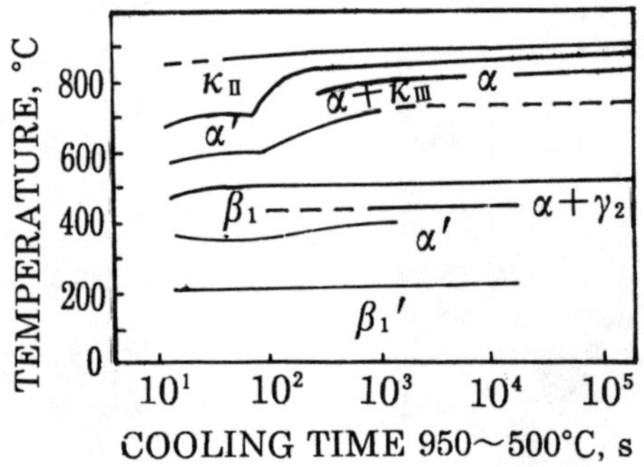

Figure 14—CCT curre of Cu$-$10.5Al$-$4.6Fe$-$5Ni alloy (cooled after heating at 950°C for 1 hr.) after Brezina (9).

Effect of Tempering Temperature

In Figure 6, there is a considerable volume fraction of fine dispersed κ−phase precipitates in the microstucture of the bronze quenched at 950°C and tempered at 400°C for 2 hours. Meanwhile the hardness of the bronze rises remarkably. The dispersed precipitation of κ−phase may be the cause of hardening. With   rise of the tempering temperature, especially over 450°C the precipitates of κ−phase are further coarsened, and the hardening effect also reduces.

Effect of Quenching Temperature

In Figure 8, the hardening effect of the bronze quenched at 850°C and tempered is lower than that quenched at 950°C and tempered. The possible causes are the coarse granular and lamellar κ−phases in the original microstructure at as−cast and forged state can not be dissolved sufficiently into β matrix and the volume fractions of massive α−phases may increase while the temperature for quenching decreases. The κ−phase particles precipitate less in the α−phase during tempering. These may cause the reduction of dispersed precipitates of κ−phase and their non−uniform   distribution in the microstructure. So the hardening effect is weakened.

Effect of TMT

In Figure 9, the hardening effect of the bronze treated in the range of 350−450°C is remarkable. But the hardening effect reduces with the decrease of the TMT temperature (see Figure 10). The reason is similar to the effect of quenching temperature.

However, in Table III the tensile property of the specimen treated by TMT and tempering is some what higher than that of the specimen quenched and tempered.

Prospect of Application

The present work showed that the hardness of the bronze treated by quenching and tempering or by TMT and tempering can be raised remarkably. The ultimate tensile strength of the bronze hardened is somewhat higher than that of the forged state, and the elongation decreases correspondingly. These types of heat treatment may be adopted for the components required surface hardness and wear resistance. By using the method of case hardening or partial quenching, the excessive reduction of the elongation can be avoided. By using the TMT technique, the energy charge may be saved.

CONCLUSIONS

A complex aluminum bronze with nominal composition Cu−11 wt pct Al−5 wt pct Ni−5 wt pct Fe may be treated by water quenching at 950°C and tempering in the range of 350−450°C for 2−6 hours. Its hardness can be raised to HRC34−39. Meanwhile the original forged microstructure with the massive α−phases and coarse granular and lamellar κ−phases transforms into the microstructure with a dispersed distribution of κ−phase precipitates in the martensite and bainite matrix. It may be the cause of hardening.

This bronze may be also treated by thermo−mechanical treatment at 980°C and tempering in the range of 350−450°C for 1−4 hours. Its hardness may rise to HRC32−38. The variation of microstructure is similar to that above−mentioned.

Decreasing the temperature for quenching or TMT will increase the quantity of the massive α−phases and reduce the volume fractions of dispersed precipitates of κ−phase. The excessive high temperature and prolonged time of tempering may lead to the coarsening of the precipitated κ−phases. All of These will reduce the hardening effect.

The bronze treated by these treatments may be applied for some components required surface hardness and wear resistance. They may be also used as the partial substitutes of the beryllium bronze in some wear resistant use, so as to reduce the costs. The case hardening or partial quenching method may be adopted to improve the elongation and be suitable for the larger componcnts.

ACKNOWLEDGEMENTS

The authors are grateful to the Shijiazhung Explosion−Proof Tool Factory, Hebei Province, China, for arranging the supply of test materials, and also wish to express thanks to Mr. Chen Yifei, Ms. Feng Yu, Mr. Tang Xiaohong, Mr. Xu Mingang, Mr. Wang Zhengde, Ms. Yin Liandi, Ms. Shen Qiong, Ms. Zu Mingming and Mr. Zu Ren for their participation of the investigation.

REFERENCES

1.   A. Jahanafrooz, F.Hasan, G.W. Lorimer and N. Ridley, "Microstructural Development in Complex Nickel−Aluminum Bronzes", Metallurgical Transactions A, Vol. 14A, 1983, 1951−1956.

2.   D.M.Lloyd, G.W. Lorimer and N.Ridley, "Characterization of Phases in a Nickel−Aluminum Bronze", Metals Technology, 1980, No.3, 114−119.

3.   R. Thomson and J.O.Edwards, "Effect of Compositional and Process Variables on the Tensile Properties of Cast Nickel−Aluminum Bronze", Transactions AFS, Vol.85, 1977, 13−18

4.   K.P.Lebedev, G. Ya.Yaroslavskij and L. S. Raines, "Optimum Composition for the Cast Aluminum Bronze", Metal Science and Heat Treatment, 1973, No.3, 25−29

5.   P. Brezina, "Microstructure Transformation and Mechanical Property of Multicomponent Aluminum Bronze type CuAl10Fe5Ni5", Giesserei−Forschung, Vol.25, 1973, 125−144

6.   F.Hasan, A.Jahanafrooz, G.W.Lorimer and N.Ridley, "The Morphology, Crystallography and Chemistry of Phases in As−Cast Nickel−Aluminum Bronze", Metallurgical Transactions A, Vol. 13A, 1982, 1337−1345.

7.   Zhengchun Fang and Hualong Zhang, "A Study on the Microstructure of the Aluminum Bronze type ZQAl9−4−4−2 for Propellers", Development and Application of Materials, 1982, No.2, 21−27.

8.   M.Cook, W.P.Fentiman and E.Davis, "Observations on Structure and Properties of Wrought Copper−Aluminum−Nickel−Iron Alloys", Journal of the Institute of Metals, Vol.80, 1951−52, 419−429.

9.   P.Brezina, "Heat Treatment of Complex Aluminum Bronzes", International Metals Reviews, Vol.27, 1982, No.2, 77−120

10.   Mingzhou Li, Weibing Zhang and Xuezhong Zhao, Production Handbook of Heavy Non−Ferrous Metal Materials, Vol.I, Metallurgical Industrial Publishing House, Beijing, P.R. of China, 1979, 197.

11.   D.E.Tyler and G.W.Lockington, "The Thermo−Mechanical Working of Aluminum Bronzes", Journal of the Institute of Metals, Vol.99, 1971, 215−222.

Significant improvement of copper-beryllium alloy's properties through a small amount of magnesium addition and multi-step aging

Y. Wang, Y. Ling
Department of Mechanical Engineering, Hangzhou Institute of Electronic Engineering, Hangzhou, People's Republic of China

S. Zhou
Department of Materials Science, Shanghai Jiao-tong University, Shanghai, People's Republic of China

ABSTRACT

By means of a small amount of magnesium addition and the application of a two-step aging instead of the conventional one-step aging for the solution treated alloy, the mechanical properties such as hardness, tensile strength, yield strength and fatigue life of the $Cu-2\%Be-0.35\%Ni$ alloy were improved significantly. Light and transmission electron microscopic studies revealed that the improvement of mechanical properties mentioned above was mainly due to suppression of discontinuous precipitation at grain boundaries and significant improvement in size and distribution of the precipitates within the matrix. The Mg-added copper-beryllium alloy and the two-step aging treatment have been successfully used for fabricating important elastic components.

INTRODUCTION

By virtue of excellent age hardenability and high mechanical properties,
Cu-Be alloys are extensively used in a variety of industrial applications. In
P. R. China, several commercial Cu-Be alloys have been produced and each of
them offers a different combination of useful properties. Among these alloys,
QBe2 is an important one. Its nominal composition can be denoted as
Cu-2%Be-0.35%Ni (in weight per cent). In this alloy, a small amount of Ni is
helpful in retarding grain growth and also in stabilizing age hardening
response[1] . This alloy has considerable tendency for discontinuous
precipitation at grain boundaries during aging after solution treatment. When
this reaction takes place, some losses in mechanical properties occur. In view
of the facts that addition of a small amount of magnesium to Cu-2%Be-0. 3%Co
alloy effectively suppressed discontinuous precipitation[2] and multi-step aging
was successfully used for improving the mechanical properties of some
precipitate-hardened alloys[3], we investigated the possibility of improving
QBe2 alloy's mechanical properties through a small amount of magnesium
addition in combination with the application of multi-step aging during aging
treatment. The results are reported in this paper.

EXPERIMENTAL

The chemical compositions of the alloys used are listed in Table I.
Strips with thickness of 0.3 and 0.8 mm and in tempers of solution treated,
solution treated and cold rolled to 1/4 hard, solution treated and cold rolled
to 1/2 hard, and solution treated to full hard were manufactured and supplied
by Shanghai Nonferrous Rolling Works. Solution treatment of the strips was

Table I-Chemical Composition (wt%)

Alloy	Be	Ni	Mg	Al	Si	Fe	Sn	Zn	Cu
QBe2	1.95	0.28	0.002	0.04	0.13	0.08	0.018	0.011	Balance
QBeMg2-0.1	1.82	0.24	0.1	0.06	0.14	0.11	0.012	0.0047	Balance

performed in an atmosphere furnace using dissociated ammonia as protective
atmosphere and cold water with agitation was used as the quenchant.
Specimens for tensile test and bending fatigue test were cut parallel to
the rolling direction from the strips with a thickness of 0.8mm and 0. 3mm
respectively. Aging of the specimens were performed in a vacuum furnace.
Tension specimens were prepared according to GB6397-86 national standard, and
tensile properties were determined on a Shimadzu Autograph DCS-25T testing
machine according to the procedure of GB228-87 national standard. In order to
determine the fatigue life of the materials, reverse bending fatigue tests
were carried out on a ZS-20D electromagnetic resonance forced-vibration device.
The plate specimen for cantilever reverse bending fatigue test is shown in
Figure 1. The frequency of reverse bending during testing was 13 Hz.
The average grain size of the alloys after solution treatment and the
volume fraction of discontinuous precipitation after aging were determined by
the chart comparision method recommended in Chinese SJ 3197-89 Standard. Thin
foils for transmission electron microscopy were prepared by mechanical
thinning. chemical thinning (the thinning solution was composed of 200 ml H_2O
+ 80 ml H_2SO_4 +20 ml HNO_3 +1 ml HC1+60 g Cr_2O_3) and ion milling. The
foils were examined in a STEM-100CX electron microscope operated at 100 KV.

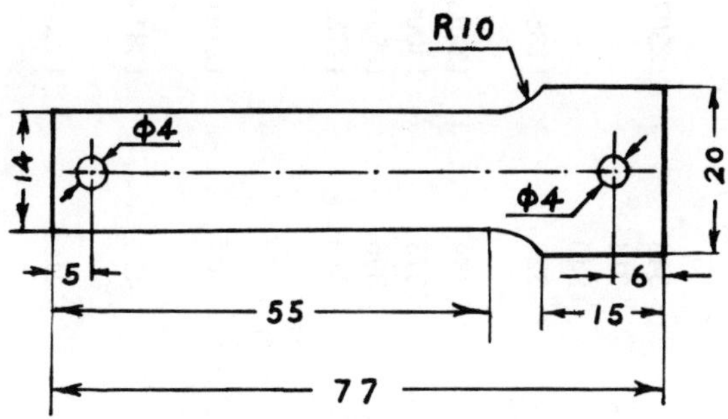

Figure 1-The plate specimen for catilever
reverse bending fatigue test.

RESULTS AND DISCUSSION

Mechanical Properties

Table II shows the mechanical properties of QBe2 and QBeMg2-0. 1 heat
treated to various tempers. The data listed in this table show that after
traditional one-step aging, the tensile strength, yield strength and hardness
of the Mg-contained QBeMg2-0.1 alloy are higher than that of the Mg-free QBe2
alloy.   Besides, when traditional one-step aging of QBeMg2-0. 1 alloy is
substituted by optimum two-step aging, higher tensile strength, yield strength
and hardness can be obtained.
Table III shows the bending fatigue life of QBe2 and QBeMg2-0. 1 heat
treated to HT temper. Obviously, the fatigue life of the two-step aged
QBeMg2-0. 1 alloy is longer than that of the traditional one-step aged QBe2
alloy.

Table III-Bending Fatigue Life of QBe2 and
QBeMg2-0.1 Heat Treated to H
Temper-Specimen With 0.3mm Thickness

| Alloy | Temper | Aging Treatment | | Stress | Fatigue Life |
		Temperature °C	Time h	Amplitude MPa	Cycles to Failure
QBe2	HT	320	2	588.4	0.765×10^5
QBeMg2-0.1	HT	240	2		
		then320	2	588.4	1.811×10^5
QBeMg2-0.1	HT	240	2		
		then320	2.5	588.4	1.235×10^5

Table II-Mechanical Properties of QBe2 and QBeMg2-0.1 Heat
Treated to Various Tempers-Strip with 0.8mm Thickness

| Temper(a) | Aging Treatment | | Tensile Strength | Yield Strength (0.2%Offset) | Elongation | Hardness | Modulus of Elasticity |
| | Temperture | Time | | | | | |
	°C	h	MPa	MPa	%	HV	MPa
Alloy QBe2(Traditional One-Step Aged)							
AT	320	2	1241	1067	8.8	346	123073
1/4HT	320	2	1189	988	8.8	314	—
1/2HT	320	2	1324	1145	4.8	340	—
HT	320	2	1299	1155	4.5	320	118661
Alloy QBeMg2-0.1(Traditional One-Step Aged)							
AT	320	3	1276	1128	6.1	379	134039
1/4HT	320	2.5	1323	1196	5.3	373	140067
1/2HT	320	2	1340	1219	5.1	396	129266
HT	320	2	1335	1188	0.0	410	127885
Alloy QBeMg2-0.1(Optimum Two-Step Aged)							
AT	240	2					
	then340	2.5	1319	1203	3.0	401	121814
1/4HT	220	2					
	then330	2.5	1372	1258	4.9	414	138140
1/2HT	200	1.5					
	then330	2	1438	1312	4.2	416	135891
HT	180	1.5					
	then320	2	1452	1324	1.9	427	131702

(a)The temper symbols are: A=Solution heat treated; 1/4H=Cold rolled to 1/4 hard; 1/2H=cold rolled to 1/2 hard;
H=Cold rolled to hard; T=Age hardened.

Metallographic Observations

Figure 2 shows the effect of Mg-addition on the suppression of grain boundary discontinuous precipitation. The dark areas beside grain boundaries in Figure 2 (a) and (c) are discontinuous precipitates. The volume fractions of discontinuous precipitation in Figure 2 (a) and (c) are rather large, whereas there is substantially no discontinuous precipitate in Figure 2 (b) and (d). It follows, then, that a small amount of Mg-addition in QBe2 alloy can strongly suppress discontinuous precipitation during aging. In view of the facts that discontinuous precipitation generally has a deleterious effect on mechanical and physical properties, whereas continuous or general precipitation usually associates with high age-hardening effect[4][5], we consider that after traditional one-step aging, the strength and hardness of the Mg-contained QBeMg2-0.1 alloy are higher than that of the Mg-free QBe2 alloy must be due to strong suppression of discontinuous precipitation thus benefiting continuous precipitation.

Table IV shows the average grain size and the volume fraction of discontinuous precipitation in QBeMg2-0.1 alloy after solution treatment and aging. Wikle[6] has pointed out, for solution and aging treated high strength Cu-Be alloys, a 0.020-0.030mm average grain size and a volume fraction of

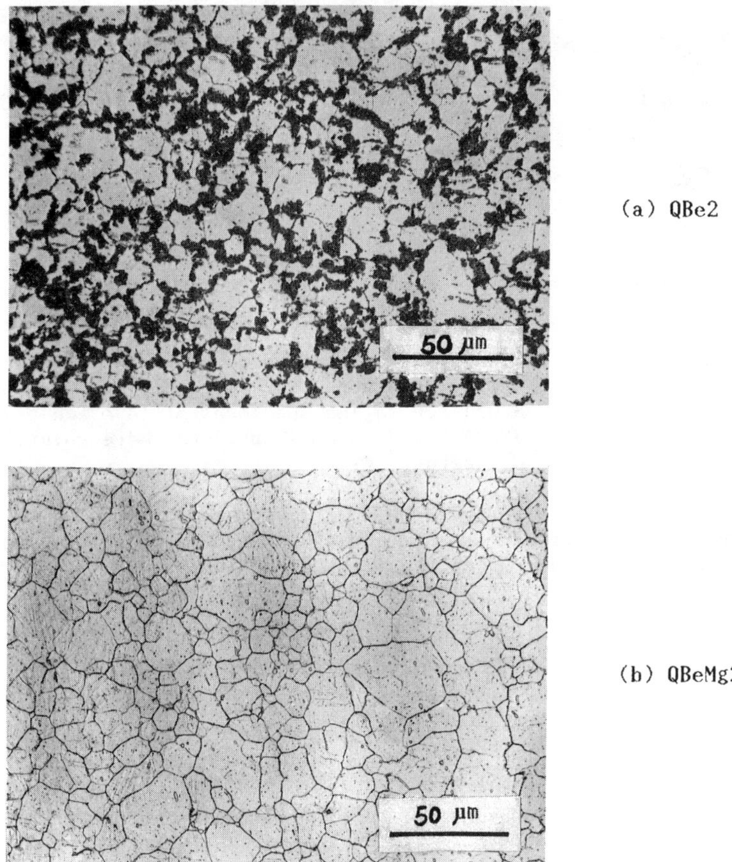

(a) QBe2

(b) QBeMg2-0.1

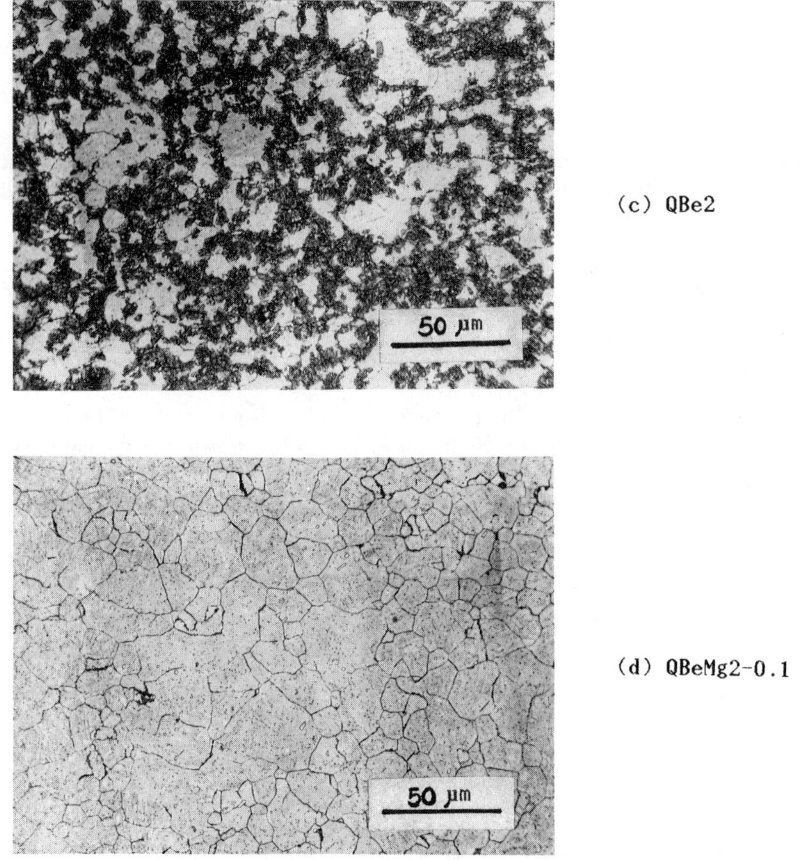

(c) QBe2

(d) QBeMg2-0.1

Figure2-Effect of Mg-addition on suppression of grain boundary
          discontinuous precipitation. Specimens of (a) and (b)
          were aged at 320°C for 2h and specimens of (c) and (d)
          were aged at 320°C for 2.5h. All specimens were solution
          treated at 790°C before aging.

discontinuous precipitation of 〈 2% are considered as desirable results. The
data shown in Table IV fully meet these requirements.

Table IV-Averag Grain Size and Volume Fraction of
          Discontinuous Precipitation in QBeMg2-0.1
          Alloy after Solution Treatment and Aging.
          Strip with 0.8mm Thickness

Average Grain Size after Solution-Treated at 790°C for 20 min (μm)	Volume Fraction of Discontinuous Precipitation after Aged at the Following Temperature for 3h		
	300°C	320°C	340°C
22	〈0.02	〈0.02	〈0.02

Figure 3 shows the micrographs of the specimens age hardened by optimum two-step aging after solution treated at 790°C and cold rolled to hard temper. In Figure 3 (a), a large volume fraction of discontinuous precipitation still can be seen, although optimum two-step aging has been used to substitute traditional one-step aging. This means that two-step aging is useless for

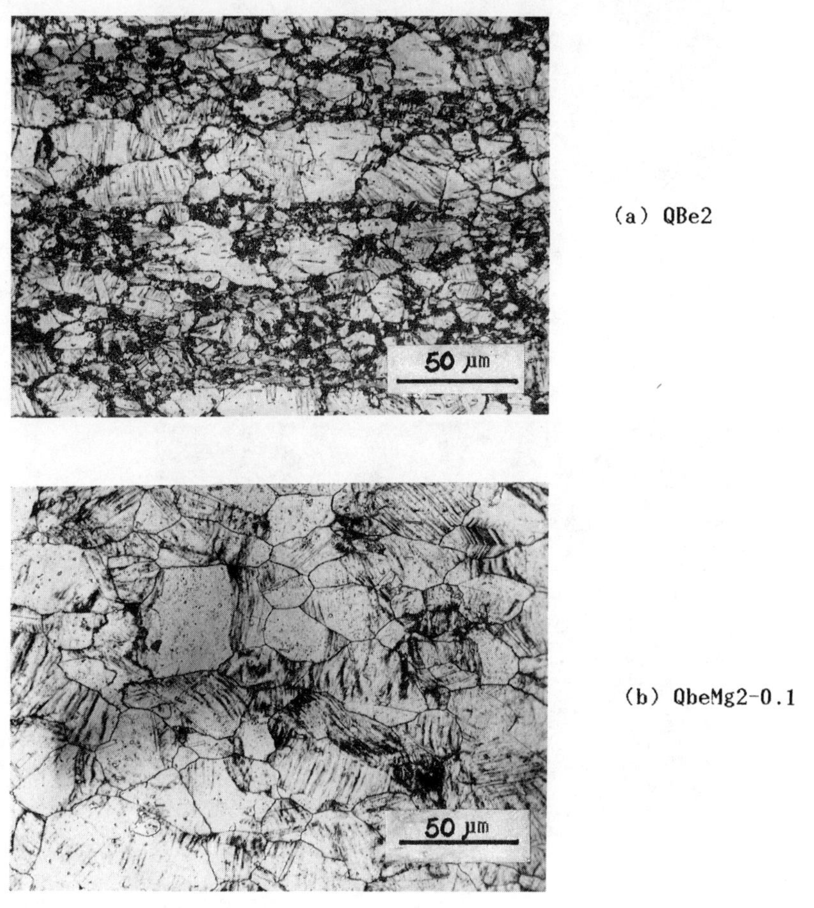

(a) QBe2

(b) QbeMg2-0.1

Figure 3-Micrographs of the specimens age hardened by optimum two-step aging after solution-treated at 790°C and cold rolled to hard temper.

significantly suppressing discontinuous precipitation, therefore it is also useless for significantly improving the mechanical properties of QBe2 alloy. In figure 3 (b), there is substancially no discontinuous precipitate. Figure 3 also shows that a small amount of Mg-addition can effectively suppress discontinuous precipitation in QBe2 alloy.

Figure 4 are TEM micrographs of the specimens solution treated at 790°C and then aged at 320°C for 2h. In figure 4 (a), discontinuous precipitation has taken place and the coarse equilibrium phase precipitates have formed. On the contrary, in Figure 4 (b), no discontinuous precipitation has taken place

on the grain boundary observed and general precipitation has well taken place
within the grains. Figure 4 clearly shows that a small amount of Mg-addition
in QBe2 alloy can strongly suppress discontinuous precipitation.

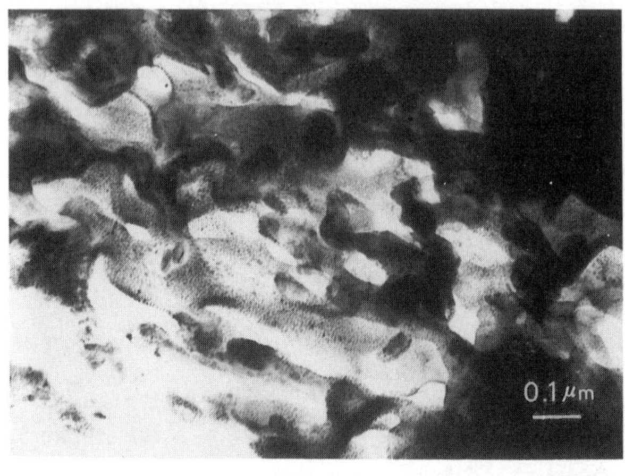

(a) QBe2

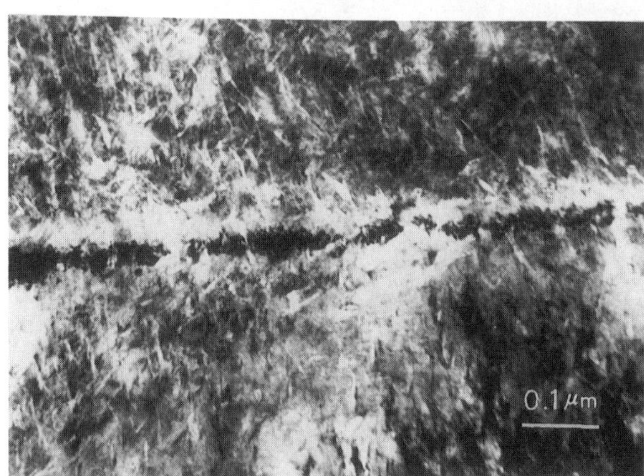

(b) QBeMg2-0.1

Figure 4-TEM Micrographs of the specimens solution treated at
        790°C and then aged at 320°C for 2h.

     Figure 5 is the TEM micrograph of QBeMg2-0.1 alloy aged at 240°C for 2h
after solution treated at 790°C . The 240°C /2h aging is the first step aging of
a 240°C /2h+340°C /2.5h two-step aging. Figure 5 shows, after the first step
aging, a great many of fine disclike G.P.(II) zones (or γ'' precipitates)
have formed and dispersed uniformly throughout the matrix. During the second
step aging, most of the G.P.(II) zones formed during the first step aging
could serve as nucleation sites for the γ' precipitate, therefore the
nucleation rate of γ' phase was significantly increased and the obtained γ'
precipitates after the two-step aging (Figure 6) are more uniform and disperse
than that of the traditional one-step aged same alloy (Figure 4 (b)) .
     We consider that strong suppression of discontinuous precipitation by a
small amount of Mg-addition and the much finer and more evenly distributed γ'

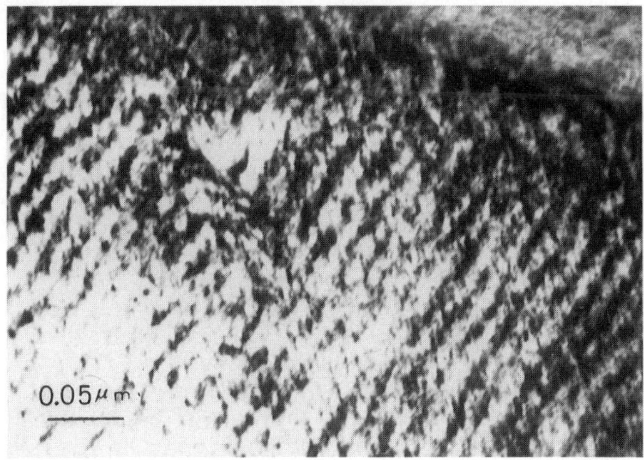

Figure 5-TEM micrograph of QBeMg2-0.1 alloy aged at 240°C for 2h
   after solution-treated at 790°C .

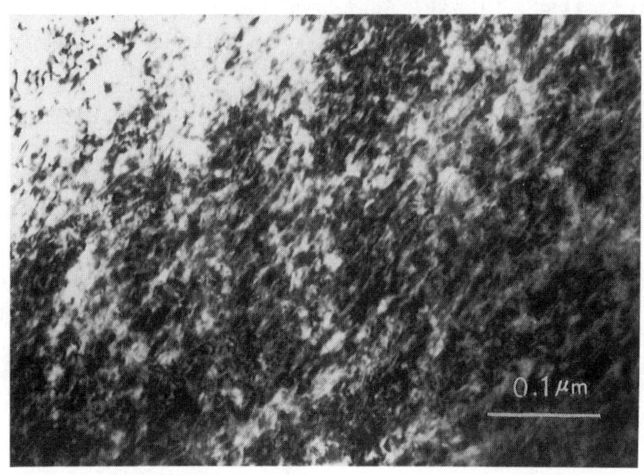

Figure 6-TEM micrograph of QBeMg2-0.1 alloy aged at 240°C for 2h
   and at 340°C for 2.5h after solution treated at 790°C .

precipitates on the matrix are the vary reasons that the two-step aged
QBeMg2-0.1 alloy having far better mechanical properties than that of the QBe2
alloy.
     Because of excellent mechanical properties, the two-step aged QBeMg2-0. 1
alloy have been successfully used to substitute foreign C17200 (Cu-1.9
Be-0. 4% Co) alloy for fabricating important elastic components such as
electrical connectors, electrical and electronic springs, switch jaws and
blades, instrument springs, etc. As a result of the substitution, the
production cost of these components has been well reduced.

CONCLUSIONS

1.   A small amount of Mg-addition can strongly suppress discontinuous precipitation and benefit general precipitation in QBe2 alloy during aging.
2.   For QBeMg2-0.1 alloy, two-step aging results in a significant improvement in the size and distribution of precipitates over that achieved by traditional one-step aging, as a result, the mechanical properties of the alloy are significantly improved.
3.   The two-step aged QBeMg2-0.1 alloy can be used to substitute C17200 alloy for fabricating important elastic components.

ACKNOWLEDGEMENTS

The experimental materials were kindly supplied by Shanghai Nonferrous Rolling Works. Financial support was received from the Ministry of Machinery and Electronics of the People's Republic of China.

REFERENCES

1.   W. D. Robertson and S. Bray, "Precipitation Hardening of Copper Alloys", <u>Precipitation From Solid Solution</u>, R. Doughton et al, Eds. American Society for Metals, Cleveland, Ohio, USA, 1959, 329-390

2.   Hiroshi Jitsu, Takashi Agatsuma and Kimio Hashizume, " Control of Grain Boundary Reaction in Beryllium Bronze by The Addition of Magnesium" , <u>Mitsubishi Denki Giho</u>, Vol.40, No.7, 1966, 1075-1079, (in Japanese) .

3.   R. F. Hehemann, "Multistage Heat Treatment" , <u>Metallurgical Transactions</u>, Vol.2, No.1, 1971, 39-44.

4.   D. B. Williams and E. P. Butler, " Grain Boundary Discontinuous Precipitation Reactions" , <u>International Metals Reviews</u>, Vol. 26, No.3, 1981, 153-183.

5.   J. b. Newkirk, "Structures Resulting From Aging and Precipitation" <u>Metals Handbook</u>, 8th Edition, Vol.8. Metallography, Structures and Phase Diagrams, T. Lyman et al, Eds., American society for Metals, Metals Park, Ohio, USA, 1973,175-183.

6.   K. G. Wikle, "Heat Treatment of Beryllium Copper Alloys, Part III-Standard Solution and Aging Treatments" , <u>Industrial Heating</u>, Vol.39, No.12, 1972, 2214-2220.

The influence of copper-deoxidation by CaB_6, P and Li on the electrical conductivity and mechanical properties of copper

H. Razavizadeh
Iran University of Science and Technology, Teheran, Iran

Mechanical properties (especially ductility) and electrical conductivity of copper are strongly influenced by quality of the products.

The Electrolytic tough pitch copper type (ECDA110) has a minimum of 99.9%Cu and a nominal 0.04% 0 contents. The normal limits of oxygen in ETP copper are between 200 and 500 PPm.

The oxygen would have two opposing effects in copper as follows:

a) Advantages if oxygen presence:

1) Oxygen dissolves in copper up to 20 ppm at room temperature and   forms Cu_2O at higher, in the presence of hydrogen, it causes both macro-and microporosity. Porosity occurs due to the change in hydrogen solubility from liquid to solid state, which happens when the hydrogen molecules are trapped by a solidification phenomenen. Hydrogen entering the melt from the atmosphare, would react with the oxygen to form secondary water, which also causes porosity. The equilibrium between the hydrogen and oxygen reaction, depends on the temprature, pressure and the rate of activity of different phases. The equilibrium between H_2 and O_2 (Fig. 1) shows that reduction of oxygen content to a certain level, would increase the hydrogen level, which causes porosity as discribed.

2)   The impurity elements are transfered by oxygen to their oxide state, which, in turn, would increase the electrical conductivity of E.T.P. copper by removal of the impurities.

b) Disadvantages of the oxygen presence

1) At temperatures around 400°C, solid Electrolytic tough-pitch copper is deoxizied by reducing gases, especially hydrogen-containing ones. Since the hydrogen atoms are so small, they are able to diffus into the solid copper and react with internally dispersed Cu_2O to form steam according to the reaction:

$$Cu_2O + H_2 \quad = \quad 2Cu + H_2O \text{ (steam)}$$

Since the steam formed by such a reaction is insoluble in copper, high
pressures are built up so that the grain boundaries of the copper would
rupture (hydrogen embriltlement). Therefore in welding or soldering
process. The oxygen free ETP copper should be used.

2) The highest ductility and toughness would be obtained when the oxygen
content of copper is reduced to its minimum level.

In the above stated conditions free oxygen copper should be selected.
To achive such a condition the deoxidizing agent, such as P,Li,Al,Mg or
Ca should be added. They would remove the hydrogen from melt too.
In industry for deoxidizing treatments the above mentioned elements are
used in the minimum amount which is required to remove the oxygen.
However, they are partially dissolved in copper and change its physical
and mechanical properties. These elements have a significant effect on
lowering the electrical conductivity which is very important in industry.
There are only a limited number of publication in the field of deoxidizing
of copper melt with Ca B_6 and its effect on the physical and mechanical
properties of copper. In this paper an attempt was made to compare the
deoxidizing effect of Ca B_6, P, Li and their influence upon the electrical
conductivity and mechanical properties of copper.

1) Effect of deoxidizing

1-1 Under argon protection
The cathodic copper containing 100 to 400 PPm oxygen was melted in a graphite
crucible in a vaccuum furnace under argon protection. Due to the re-melting
process the amount of oxygen reduced from the initial percentage to 30 to 60
PPm. The reason for this reduction is that some reactions occur between the
graphite crucible and the dissolved oxygen and hydrogen producing CO_2, C O
and H_2O. Normally the partial pressure of the gases in Argon is zero, but due
to the movement of the melt caused by electromagnetic induction, some gases
can leave the melt, therefore leaving no porosity in the solid copper.
The addition of CaB_6, P and Li was then made separately to the molten bath.
After this treatment a complete deoxidizing has taken place (Fig. 2).
In these tests P and Li were added up to 0.1%, more than this repidly causes a
reduction in electric conductivity. The only way to reduce the oxygen level to
less than 10 ppm is to use CaB_6.
Oxygen appears in copper as Cu_2O with eutectic percentage. Most of these
precipitates at grain boundaries. Electron microscopic (S.E.M) examinations
showed that after deoxidizing treatment with CaB_6, there was some Boron distributed throughout the matrix and at grain boundaries (Fig. 3). It seems

that the removed oxygen has been replaced by boron at grain boundaries. In
determination of the amount of calcium, it has been found that most of the
calcium and oxygen appeared in slag probably in the form of CaO. Small
quantities of calcium and boron have also been observed in the copper. These
may be CaB_6 particles which did not act as deoxidizing agents.

1-2 Under soot black protection

The melting operations under soot black protection, in which the surface of
the melt was coverd by a layer 30 mm of soot black, were carried out in open
atmosphere. The melt was held in on open atmosphare for 7 minutes at $1150^{\circ}C$
to provide enough oxygen. Then after covering the surface of the melt with
soot black, the deoxidizing treatments were carried out. It is possible to
completely deoxidize the melt using CaB_6 and soot black. In Figure 4, the
deoxidizing effect of CaB_6, P and Li in copper has been shown. It can be concluded that deoxidizing of copper can be carried out using CaB_6 or P but in
the case of CaB_6 the quantity required is greater than when using phosphorus.
As can be seen in Fig. 4, if after deoxidizing the residual amount of Bor is
more than 0.1%, then the amount of oxygen will be reduced to 10 ppm. In the
case of deoxidizing copper with phosphorus, if the amount of P remaining in
copper is more than 0.08% , then the amount of oxygen will be reduced under
10ppm. By deoxidizing treatment with Li is not possible to reduced the amount
of oxygen under 10 ppm.

2) Electrical conductivity

The effect of impurities on the electrical conductivity of copper containing
oxygen, differs from that on deoxidized copper because in the case of copper
which contains oxygen, impurities can easily react with oxygen and form
oxide.

The formation of oxides reduces the effect of impurities on electrical conductivity,
but since the existence of oxygen reduces the ductility, especially when the
copper may be welded or soldered, it must be deoxidised. Copper should have
a minimum electrical conductivity of 56.058 $m/mm^2\Omega$ for use in electronic and
electro-technical industries.

After deoxidizing, if the amount of remaining Li and P is more than 0.05% , the
electrical conductivity will fall below 45 $m/mm^2\Omega$ (Figur 5), on the other
hand for deoxidizing treatment with CaB_6 under soot black protection, electrical
conductivity will remain unchanged, the maximum remaining boron in these
specimens is 0.1% and their electrical conductivities are more than 56 $m/mm^2\Omega$

Figure (6).

If the deoxidizing treatment is carried out with CaB_6 under argon protection,
since the oxygen reduces to less than 10 ppm, electrical conductivity will be in
the range of 50-56 m/mm$^2\Omega$ with variation + 3 m/mm$^2\Omega$ (Fig. 6)

The reduction of electrical conductivity arises from:

1)   Solid solution of Cu-B

2)   Heterongenous particles containing boron which are distributed in matrix
        and at grain boundaries.

3)   Remaining CaB_6 (unreacted)

It is concluded from the comparison between deoxidizing effect of CaB_6, P and
Li that residual amounts of phosphorus and lithium greater than 0.01% can
rapidly reduce the electrical conductivity.
On the other hand, the remaining CaB_6 does not have such a great effect on the
electrical conductivity (Fig. 6).
The reduction in electrical conductivity of copper by the remaining phosphorus
may be due to the formation of solid solution of Cu-P because copper can dis-
sove 1.75% P at room temperature. For the reduction in electrical conductivity
by the remaining lithium, since copper does not dissolve lithium, the LiH and
LiOH compounds may reduce the electrical conductivity.

3) Age Hardening
It is found that during copper deoxidation process, with the aid of CaB_6 compound, some boron remains in the copper as an alloying element. Figer   7
shows the effect of aging and hardness with 0.39 wt%B alloy which, prior to
aging, was held at 980 C for 30 minutes. After quenching in iced water, it
was cold worked $\varepsilon = 70\%$. The other specimen was subjected to the same
amount of cold working without undergoing the solution process. The hardness of the cold-worked specimen, which did not receive the solution treat -
ment, was nearly reduced to half after 30 minutes at 200 $^\circ$C, where as the
specimen which was subjected to thermomechanical treatment, had constant
hardness up to 300 $^\circ$C. This hardness reduced only by an amount equal to 30
HV at 350 $^\circ$C. The latter specimen, at 350 $^\circ$C still had a hardness which nearly
twice that of the other.

It can be concluded that in the case of the specimen subjected to thermomechanical treatment, an increase of hardness due to the age hardening effect of precipitate particles compensated for the softening effect of recrystallization (which
is responsible for the hardness decrease in the other specimen). The two curves
gradually begin to converge to a point above 400 $^{\circ}$C.

Figures (8,9) shows the effect of aging and hardness of 0.2%B,0.835% P and
0,512% Li alloys which, prior to solution heat treatment, were held at Cu-P:
680 $^{\circ}$C,Cu-Li:150 $^{\circ}$C and Cu-B:880 $^{\circ}$C for 1 houre. After quenching in iced-
water, the hardness was determinated after aging treatment at 200 and 400 $^{\circ}$C .
It can be concluded that only in the case of the Cu-P and Cu-B subjected to
thermotreatment, an increase in hardness, due to the age hardening   effect
precipitate particles, compensated for the softing effect of recrystallization.
But Alloys of Cu-P Alloy are not commen in industry becouse the electrical conductivity is very low.
A test to ascertain the effect of aging time on tensile properties used two spe -
cimens containing 0.02% boron and a third specimen which was boron free.
The two specimens containing boron were held at 250 $^{\circ}$C and 300 $^{\circ}$C respec -
tively, while several boron free specimens were held at 300 $^{\circ}$C for different
periods of time. The tensile strength was measured and plotted against time
(Fig .10). As can be seen (Fig.10), tensile strength of the (specimen containing boron) held at 250 $^{\circ}$C, rises to a maximum value after 5 hours and then
decreases very slowly, so that after 190 hours the tensile strength showed a
fall which only amounts to 50 Nmm^{-2}. The other specimen containing boron
held at 300 $^{\circ}$C, showed a maximum tensile strength after about 2 hours, after
which its strength began to fall, first rapidly (so as to reach to a value of 330
N mm^{-2} after 5 hours) then more slowly. The tensile strength of this specimen
eventually reaches a constant value of 250 Nmm^{-2} after 50 hours. The boron
free specimen, shows a constant valued of tensile strength (220 Nmm^{-2})
throughout the experiment.

4) Effect of recrystallization temperature
To determine the deoxidizing effects of the recrystallization temperature, the
specimens were cold worbed 70% at room temperature and recrystallized   at
different temperature for 1 houre. Fig.11 shows if copper containts boron
the recovery of copper won't change too much but when copper containts O_2,
P or Li ther recovery of copper will be change considerbly. Also it is concluded from comparison between Cu-B, Cu-P, Cu-Li and Cu-O that when   the
recrystallization temperature increased more than 500 $^{\circ}$C the hardnesse of
Cu-B and Cu-O will be constant, but it will be change considerbly by Cu-P and

Cu-Li (Fig . 12) . The reason for this effect is found that increasing   tem -
perature will cause a grain growth by Cu-P and Cu-Li (Figures 13 and 14) .

Conclusion:

1) Deoxidizing effect of copper by CaB_6 is comparable with P and Li.

2) After deoxidizing of copper the remaining CaB_6 does not have a great effect
on the electric conductivity of copper .

3) Increasing mechanical properties of copper is possible by using CaB_6
deoxidizing .

4) Bor-residual is not influenced practicaly the recrestallization temperature
of copper .

5) Copper-Bor can be used as heat resistant alloy.

References:

1) N.P. Allen U.T.Hewitt: J.Inst. Metals 51 (1933), P.257/308.

2) Kh.G. Schmitt - Thomas, H.Meisel, H.J.Dorn U.H.Razavizadeh :Metall
     12 (1975).P. 1198-1204.

3) H.Razavizadeh, Dr.Ing. Dissertation, T.U.Muenchen 1977.

4) K.h.G. Schmitt - Thomas, H.Meisel, S.Mirdamadi- Tehran; Metall 32 (1978),
     P. 1103 - 1108.

5) S.Mirdamadi-Tehrani;Dr.Ing. Dissertation, T.U.Muenchen (1979)

6) Hekmat Razavi-zadeh and S.Mirdamadi.T. Journal of Metals, 39, (2),1987,
     P.42/47.

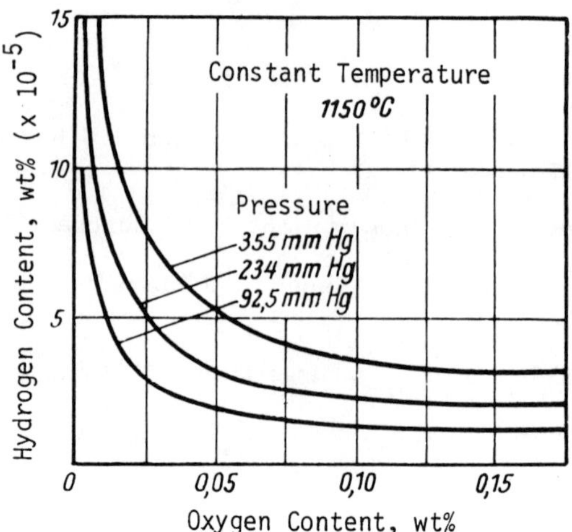

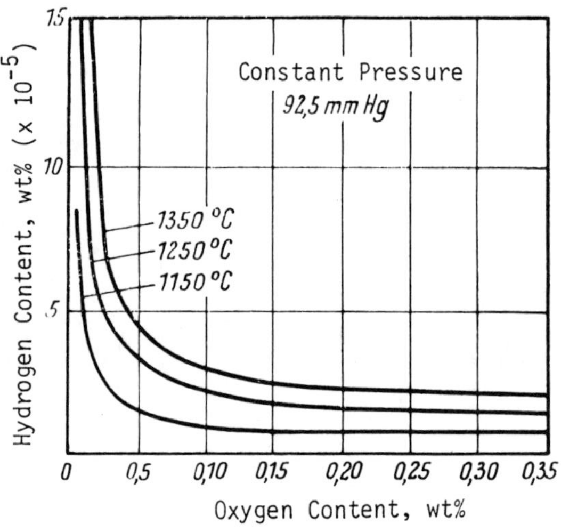

Figure 1. Equilibrium between hydrogen and oxygen content of molten copper at (a) 1150 °C with variation of pressure, and (b) 92.5 mmHg with variation of temperature. [1]

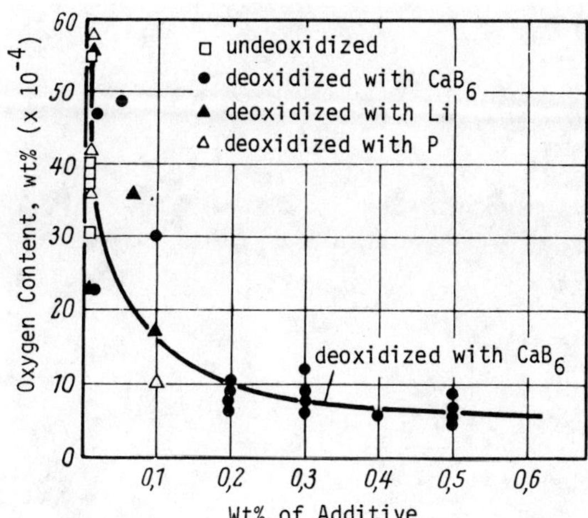

Figure 2. The deoxidizing effect of calcium boride, lithium and phasphorus in copper under argon protection.

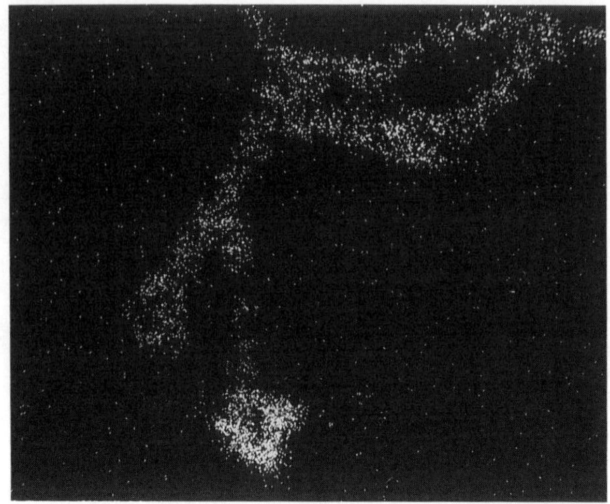

a

b

Figure 3 . Electron micrograph analysis of deoxidized copper with CaB_6 X 1340
(a) Boron distribution (b) calcium distribution.

O –Content: 0.0008   Wt%
B –Content: 0.2   Wt%
Ca–Content: 0.0145 Wt%

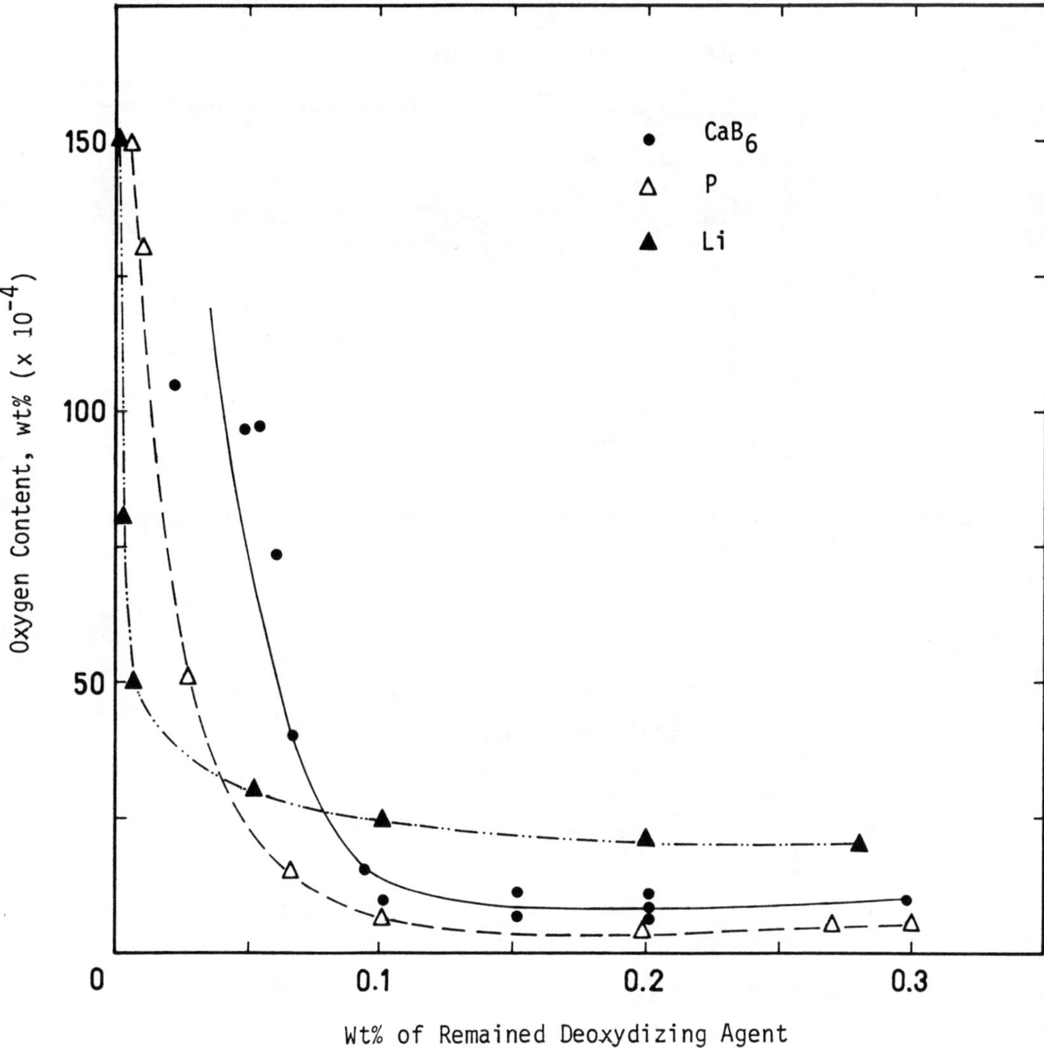

Figure 4:   The deoxidizing effect of Calcium boride, Lithium
and Phosphorus in Copper under soot black protection

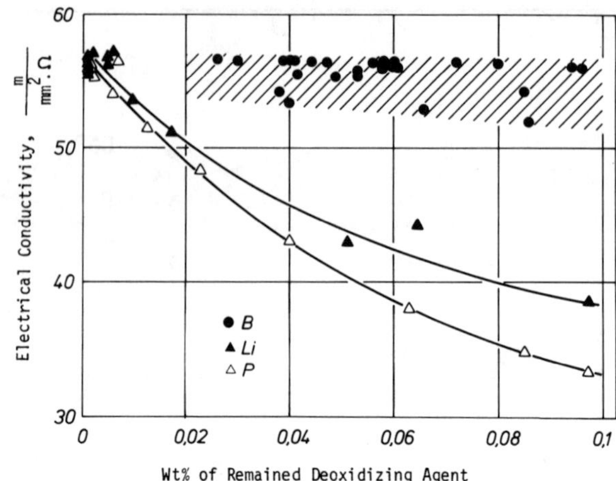

Figures 5. The effect of remaining deoxidizing agent on the electrical conductivity of copper.

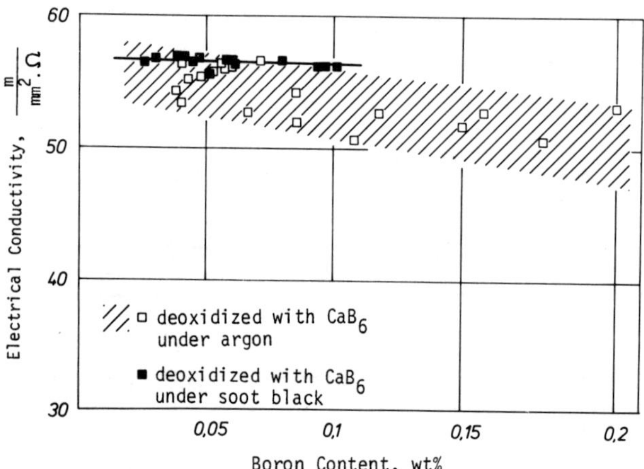

Figure 6. The effect of remaining boron on the electrical conductivity of copper.

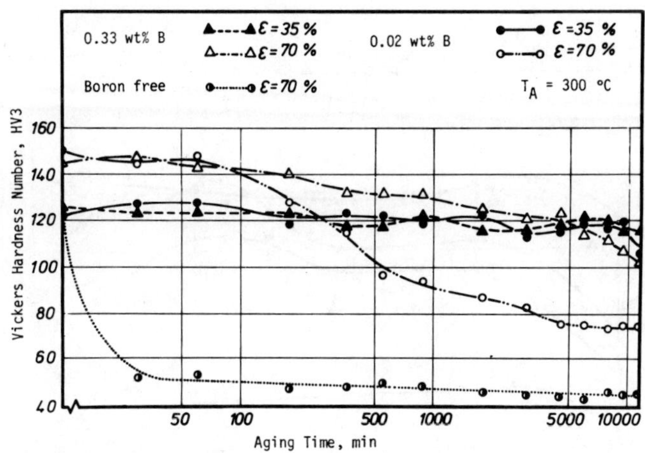

Figure 7. Hardness variations with temperature of the alloy Cu 0.39 Wt%B for thermomechanically and work-hardned conditions 980 C,0,5 hw [5]

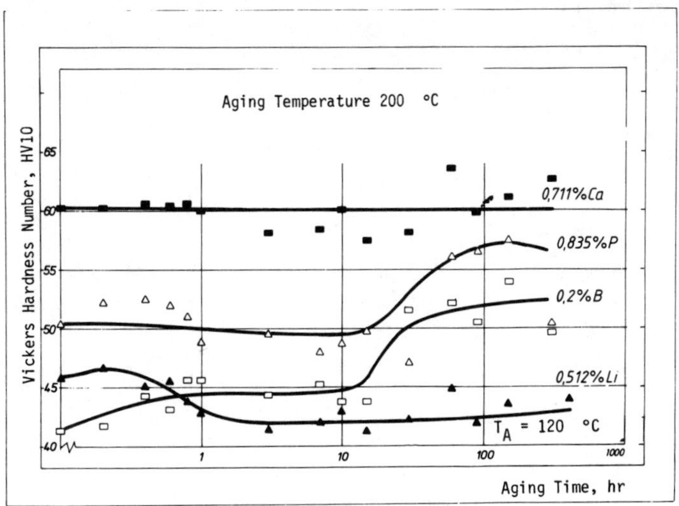

Figure 8. Hardness variations with temperature of the alloy Cu-Li, Cu-Ca, Cu-P and Cu-B.

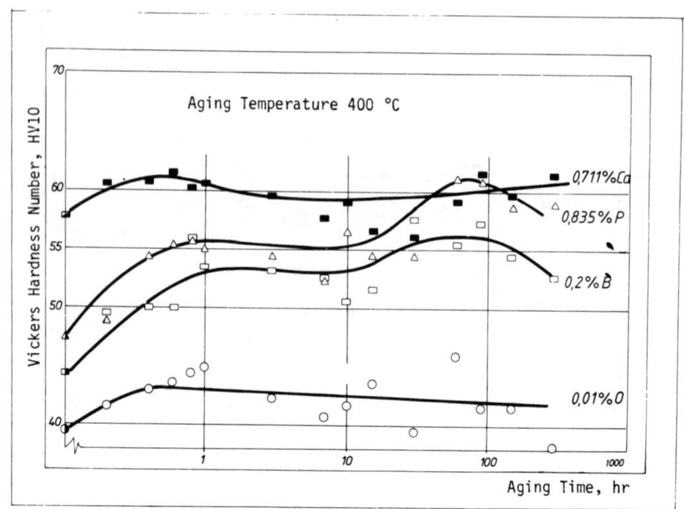

Figure 9. Hardness variations with temperature of the alloy Cu-B , Cu-Ca , Cu-O and Cu-B.

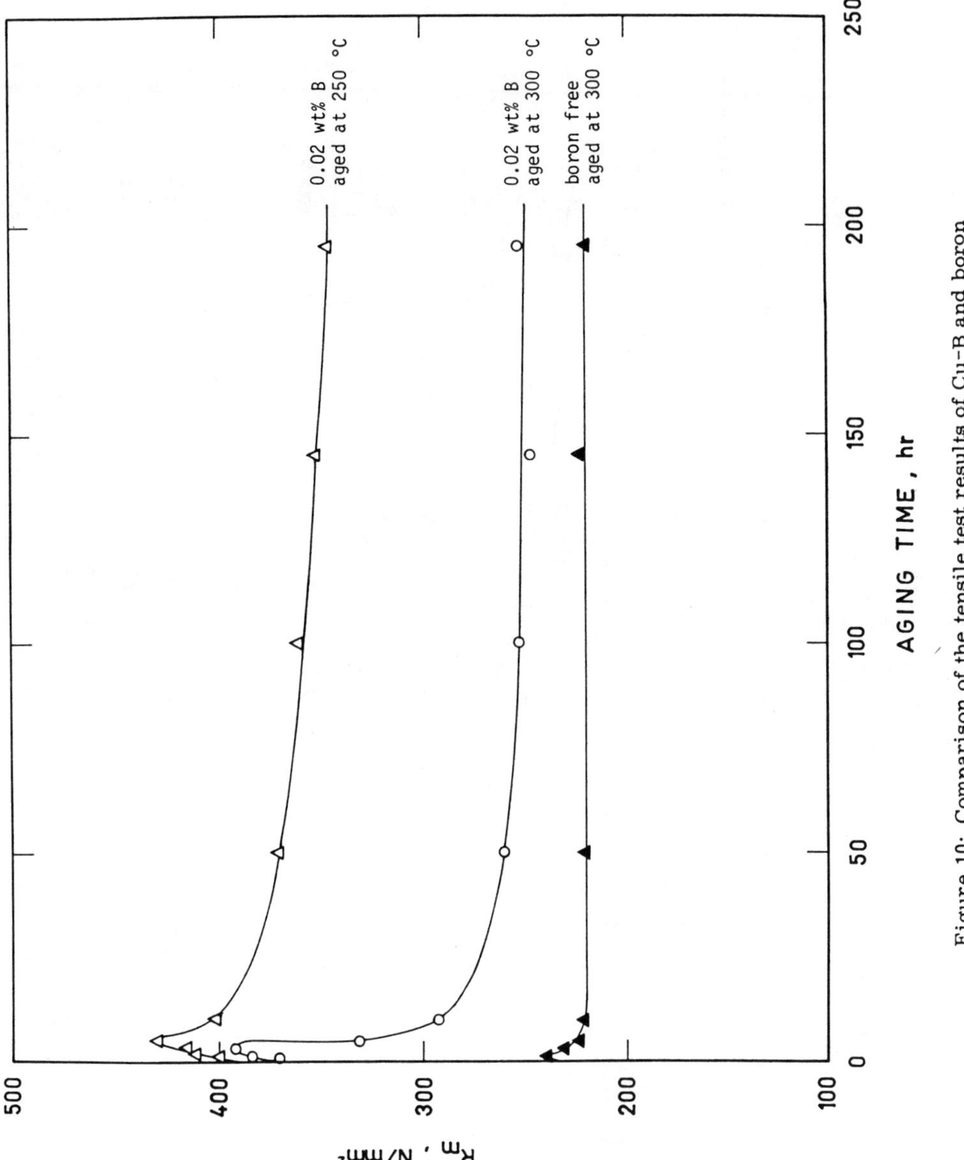

Figure 10: Comparison of the tensile test results of Cu-B and boron free Samples: $\varepsilon = 70\%$, 980 C, 0.5h, W.5.

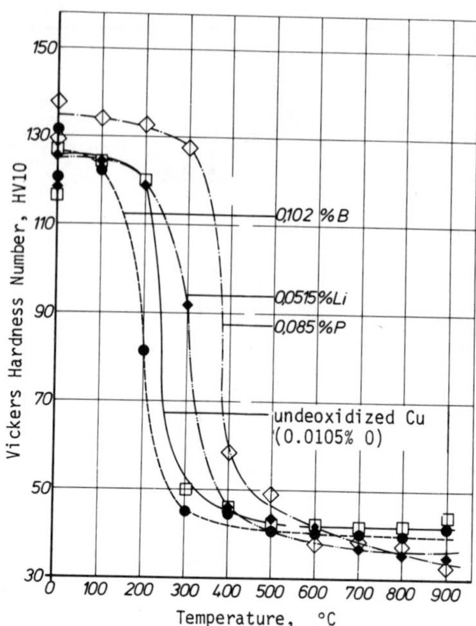

Figure 11. Hardness as a function of annealing temperature for 1 houre annealing time the alloys were originally rolled at 25 °C to a 70% reduction in thickness.

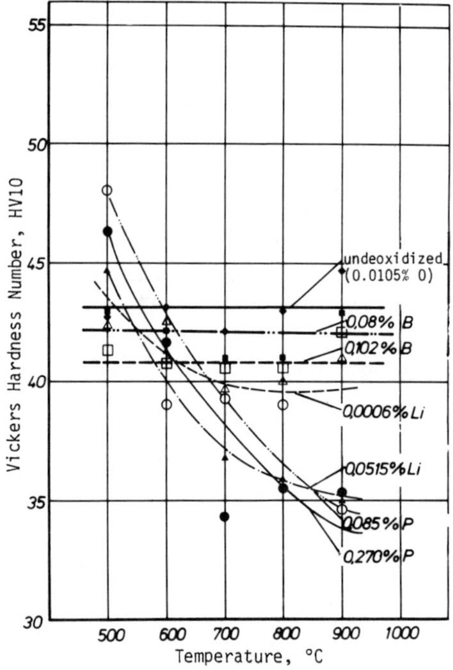

Figure 12. Hardness as a function of annealing temperature T>500 °C for 1 houre annealing time ε =70% .

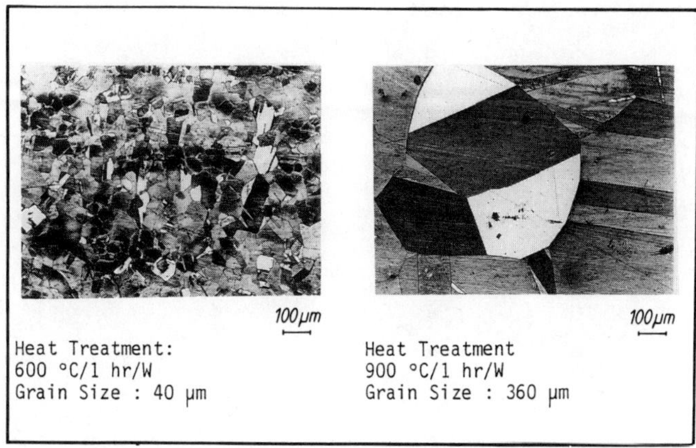

Figure 13. Microstructure of a Cu-P alloy cold rolled to 70%, then annealed for
different temperature for 1 houre annealing time .
O-content: 0.001 wt% ; P-content: 0.085 wt%

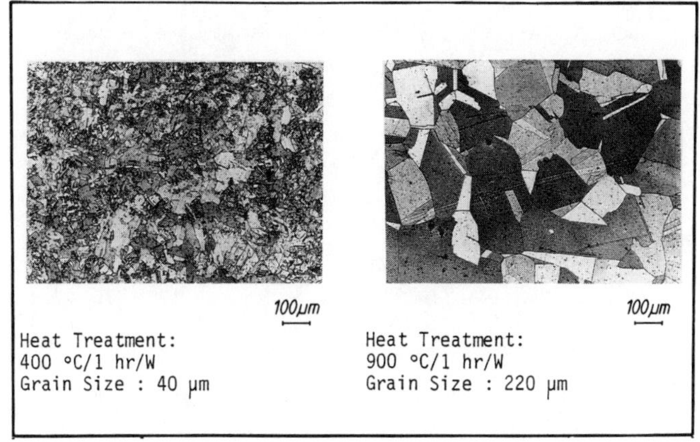

Figure 14. Microstructure of a Cu-Li alloy cold rolled to 70% , then annealed for
different temperature for 1 houre annealing time.
O- content: 0.003 wt %   ; Li- content: 0.05 wt%

Fatigue propagation behaviour of Mn-Al-Ni bronzes

J.I. Dickson, Li Shiqiong, L. Handfield and J.-P. Baïlon
Département de Génie métallurgique, École Polytechnique, Montréal, Québec, Canada

ABSTRACT

The fatigue propagation behaviour in air and in 3.5% NaCl solution of Mn-Al-Ni bronzes containing 8-11% Mn was studied for different material conditions and compared to that of 12-14% Mn-Al-Ni bronzes. Accelerated corrosion-fatigue propagation for the tests at 1 Hz in the salt solution was observed in some of the alloys. Accelerated propagation at high da/dN was associated with increased crack tip plasticity, resulting in easier propagation across grain boundaries and interfaces. A corrosion-fatigue effect at lower da/dN was associated with interfacial cracking and preferential dissolution of the b.c.c. ß phase along α-ß boundaries. Heat treatments improved the resistance to fatigue propagation of the higher Mn alloys and decreased the corrosion-fatigue effects.

INTRODUCTION

The corrosion-fatigue behaviour in salt water of ship propeller bronzes is of practical interest, since propellers are subject to cyclic stresses at frequencies for which corrosion-fatigue effects can be expected. The fatigue propagation behaviour in air and in aqueous 3.5 % NaCl solution of 12-14% Mn-8% Al-2% Ni bronzes, Cu alloy C95700, ASTM B-148, has previously been reported (1,2). Accelerated propagation was observed at higher crack growth rates in the NaCl solution at a cycling frequency of 1 Hz and was associated with facilitated propagation across or along α-β interfaces. A subsequent study (3-5) showed the beneficial influence on the fracture toughness of Mn compositions lower than those of C95700 alloy specification, albeit accompanied by a loss in yield strength and, for laboratory-cast alloys, in tensile strength. A beneficial effect of annealing at 720°C was also noted on the fracture toughness/tensile strength ratio.

The objectives of the present study were to compare the fatigue and corrosion-fatigue propagation behaviour of Mn-Al-Ni bronzes containing 8-11% Mn with those of 12-14 Mn-Al-Ni bronzes previously studied corresponding to C95700 compositions and to identify the influence of microstructure on this behaviour.

EXPERIMENTAL PROCEDURES

The experimental procedures employed were essentially identical to those described previously (1,2). Plates 295 mm x 148 mm x 25 mm were cast at the Metals Technology Laboratories, CANMET. Two compact tension specimens of thickness B=22.3 mm and width W=101 mm were machined from a plate. The heats with 8% and 10% Mn nominal compositions analyzed 8.4-8.8% and 10.5-10.9% Mn, respectively. The analysis of the other major elements were 6.7-7.1% Al, 2.0-2.1% Ni and 2.9-3.3% Fe. Material in the laboratory as-cast condition will be referred to as AC material. To obtain a microstructure more typical of a large casting, the AC plates were first annealed two hours at 900°C followed by cooling at 10°C per hour to 230°C. Material in this slowly-cooled (NSC) condition subsequently annealed for two hours at 690°C and then water quenched will be referred to as in the HT1 condition. NSC material annealed for 24 hours at 720°C and air cooled will be referred to as in the HT2 condition. The microstructure of these alloys has been reported elsewhere (3-5) and consists principally of f.c.c. α grains, with small ligaments of b.c.c β phase at some of the α grain boundaries. In addition, fatigue propagation tests were performed on specimens taken from the blade of a large 13.5 % Mn bronze ship propeller, referred to as propeller #2 by Couture and others (6).

The tests were performed at room temperature, employing a sinusoidal waveform, a cycling frequency of 20 or 1 Hz and an R-ratio (K_{min}/K_{max}) usually of 0.1. Crack lengths were calculated from compliance measurements and compared with crack front markings on the fracture surfaces associated with identifiable stages of the tests. Good agreement (generally < 0.3 mm) was obtained. The tests in well-aerated 3.5% NaCl solution were performed under freely-corroding conditions, employing a cell attached to the specimen. The fracture surfaces tested in the NaCl solution were ultrasonically cleaned in a 0.3% solution of Micron commercial cleaner and then in a solution of a few drops of Teepol 610 wetting agent in methanol, prior to observations by scanning electron microscopy (SEM). Metallographic observations of fracture surface profiles near the specimen mid-thickness were also carried out.

RESULTS AND DISCUSSION

Fatigue Propagation

Typical log-log curves of crack propagation rate da/dN as a function of ΔK, the range of stress intensity factor, or as a function of ΔK_{eff}, the effective range, (which is often lower than ΔK as a result of crack closure reducing the stress or strain amplitude at the crack tip) are presented in Figures 1-12. The values of $\Delta K_{eff} = K_{max} - K_{op}$, where K_{op} corresponds to the stress intensity value at which the crack tip starts to open, were obtained by regression analysis of the load and crack mouth opening displacement data points for the hysteresis loops recorded, employing the same type of analysis as in the previous study (1,2).

For the 8 and 10 % Mn AC bronze, slightly accelerated corrosion fatigue propagation was noted on a log da/dN-log ΔK basis for the tests at 1 Hz in the NaCl solution for da/dN values above approximately 2×10^{-5} mm/cycle, as shown in Figure 1 for the 8% Mn alloy. No corrosion-fatigue effect was noted for tests at 20 Hz. Duplicate tests in air at 20 Hz and in the NaCl solution at 1 Hz from different lots of 8 and 10 % Mn AC bronzes gave very similar curves (Figures 1 and 2), except for differences at low da/dN associated with differences in crack closure effects. At low da/dN, the curves in the NaCl solution, especially those for the tests at 20 Hz, lay to the right of those in air (Figure 1). These results indicate important corrosion-product induced crack closure in the NaCl solution, especially for tests at 20 Hz. For these AC bronzes, comparison of the curves of log da/dN vs log ΔK_{eff} for the tests in air and at 1 Hz in the NaCl solution generally indicated some accelerated cracking in the NaCl solution for da/dN $> 10^{-5}$ mm/cycle (Figure 3). This effect was not evident in comparing results on a ΔK basis, since it was masked by the important corrosion-product induced crack closure effect for lower da/dN values in the NaCl solution. Accelerated cracking on a ΔK_{eff} basis was also not evident at 20 Hz in the solution (Figure 3).

Comparison of the crack growth curves for the 8 and 10% Mn AC bronzes with those previously obtained for the 12 and 14% Mn AC alloys (1,2) did not show a large influence of the Mn composition (Figure 4), in spite of an important influence of Mn content on the amount of β phase present and of an important difference in the strength properties of the α and β phases (3-5).

Figure 5 compares the crack growth curves obtained in the NaCl solution at 1 Hz and R=0.1 for the different AC materials. In contrast to the tests in air, it is the two AC alloys with the lower Mn contents which give slower crack propagation for da/dN $> 10^{-4}$ mm/cycle, as a result of less important corrosion-fatigue. Below da/dN $\approx 5 \times 10^{-5}$ mm/cycle, there is little difference in these log da/dN - log ΔK curves for the 8-14% Mn AC bronzes.

Figures 4 and 5 also compare the crack propagation curves in air (R=0.1, f=20 Hz) and in the NaCl solution (R=0.1, f=1 Hz) for specimens from the as-received (AR) propeller with those for the AC condition. For da/dN > 2 x 10^{-4} mm/cycle, it is the propeller specimen which gives the fastest crack growth for a given ΔK value. In the 3.5% NaCl solution, the propeller specimen showed a very similar log da/dN-log ΔK curve as those for the 12 and 14 % Mn AC materials. In this solution, the propeller specimen showed faster propagation on a ΔK basis than in air for da/dN > 10^{-4} mm/cycle. On a ΔK_{eff} basis, the crack propagation was faster in the NaCl solution for da/dN > 3 x 10^{-6} mm/cycle. For the test in the NaCl solution, the low da/dN portion of the curve obtained with a ΔK decreasing procedure agreed quite well with the portion of curve obtained in air.   The low da/dN portion subsequently obtained in the NaCl solution employing an increasing ΔK procedure indicated slower crack growth as a result of greater crack closure (Fig. 5).

The log da/dN-log ΔK crack growth curve for the AR propeller specimen in the NaCl solution at R=0.1 and f=1 Hz was compared on a ΔK and on a ΔK_{eff} basis with those for the 12 and 14 % Mn NSC bronzes previously tested (1) in this solution at R=0.5 and f=1 Hz. For these high propagation rates, the ΔK_{eff} measurements should be particularly accurate, especially for tests at R=0.5, for which less crack closure is expected. The log da/dN-log ΔK curve for the propeller specimen was slightly to the right of those for the NSC specimens (Figure 6). On a ΔK_{eff} basis (Figure 7), the curve for the propeller specimen lay to the left of those for the two NSC alloys. These results indicate that the accelerated corrosion-fatigue propagation at high da/dN is particularly important for the propeller specimen tested in the NaCl solution at 1 Hz and that significant closure occurred in this R=0.1 test at high da/dN.

The fatigue propagation results for the 8 and 10 % Mn HT1 or HT2 alloys (comparison of Figures 2 and 8) did not show accelerated crack propagation in the NaCl solution. Comparison of these curves on a ΔK_{eff} basis also did not indicate a clear corrosion-fatigue effect. Figures 2 and 8 also show that the differences in crack propagation curves in air for the 8 and 10 % Mn alloy conditions studied are quite small at high da/dN but are somewhat larger for the tests in the NaCl solution at 1 Hz. For the latter tests, the 8% Mn HT1 and HT2 and 10% Mn HT1 materials gave very similar curves for da/dN > 2 x 10^{-4} mm/cycle. The curve for the 10% Mn HT2 was somewhat to the right, indicative of slower propagation. The curves for the AC materials were to the left, consistent with the corrosion-fatigue accelerated cracking effect noted for these materials.   This comparison indicates that the HT1 and HT2 heat treatments reduce the magnitude of the corrosion-fatigue propagation effect. The slower crack growth observed at low ΔK for the HT1 and HT2 alloys suggest that these two heat treatments, for the lower da/dN values in the tests in the NaCl solution, result in more important crack closure compared to the AC material condition.

The tests on the 12 and 14% Mn HT2 alloys were performed in the NaCl solution at 1 Hz and R=0.5 to compare their behaviour under reduced crack closure effects with results previously obtained on the 12 and 14% Mn bronzes in the AC, NSC and HT1 conditions (1). These tests resulted in log da/dN-log ΔK and log ΔK_{eff} curves quite similar to those for the 8 and 10% Mn HT2 material at R=0.1 (Figure 9), especially at higher da/dN. This similarity shows that the increased resistance to crack propagation which would be expected for the two higher Mn HT2 alloys at R=0.1 would be largely associated with an increase in crack closure.

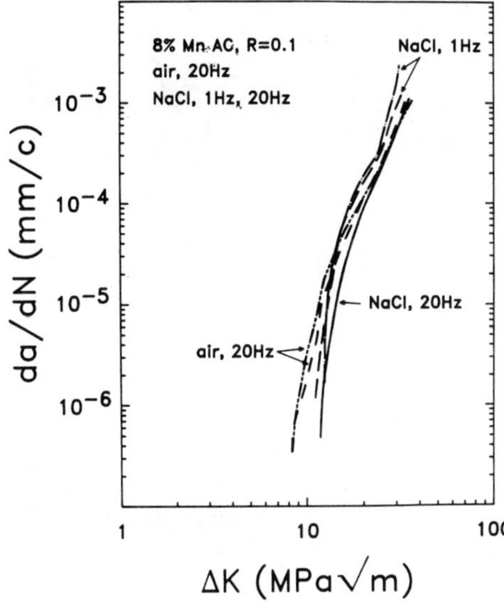

Fig. 1. Crack growth curves for the 8% Mn AC alloys.

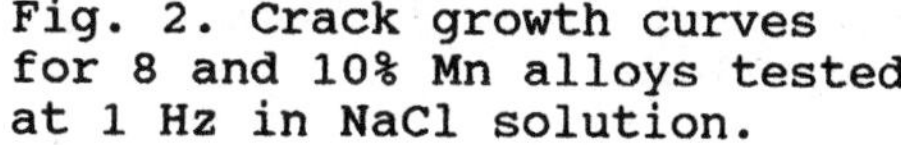

Fig. 2. Crack growth curves for 8 and 10% Mn alloys tested at 1 Hz in NaCl solution.

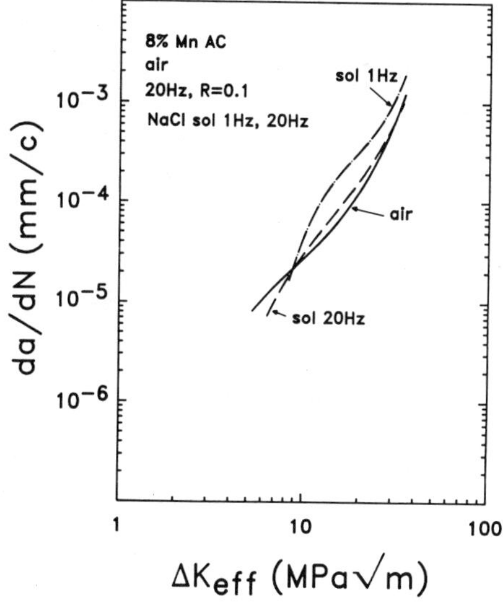

Fig. 3. Log da/dN-log ΔK_{eff} curve for the 8% Mn AC alloys.

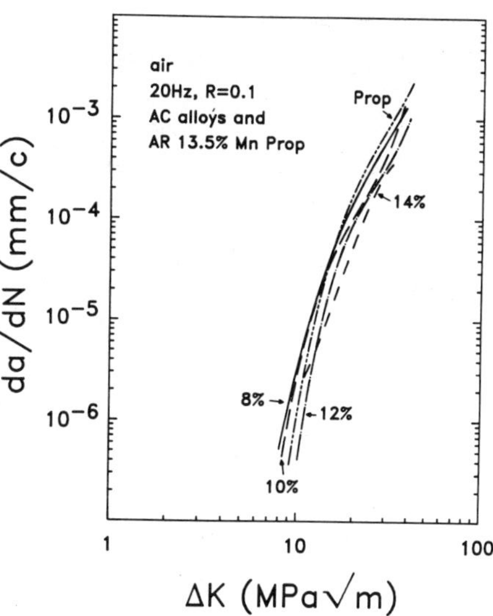

Fig. 4. Crack growth curves in air for the AC alloys an as-received (AR) propeller.

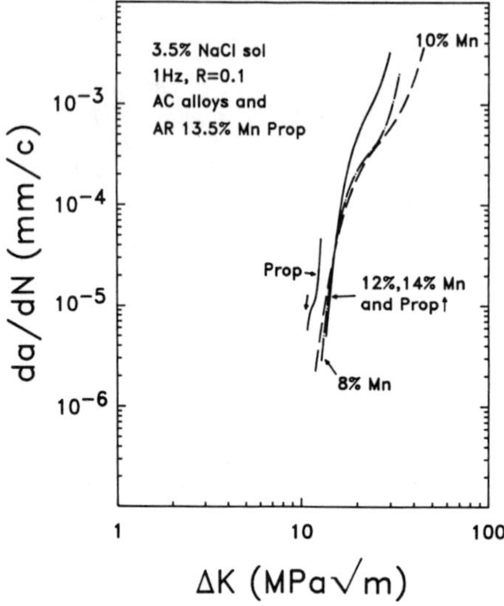

Fig. 5. Crack growth curves at 1 Hz in the NaCl solution for the AC alloys and the AR propeller.

Fig. 6. Crack growth curves at 1 Hz in the NaCl solution for for the NSC alloys and AR propeller.

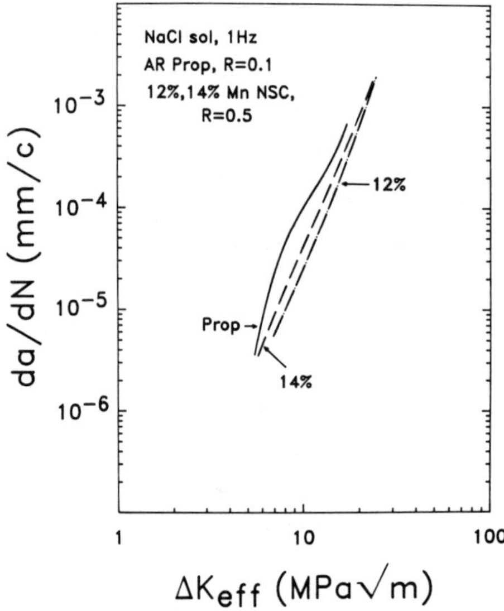

Fig. 7. Log da/dN-log ΔK_{eff} for the 12 and 14% Mn NSC alloys and AR propeller in NaCl solution.

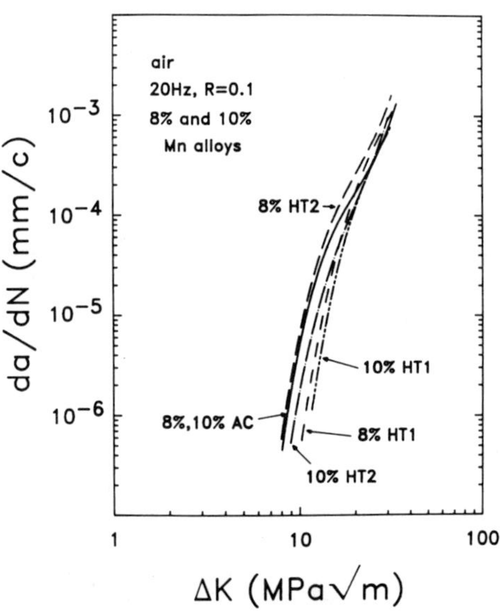

Fig. 8. Crack growth curves in air for the 8% Mn alloys.

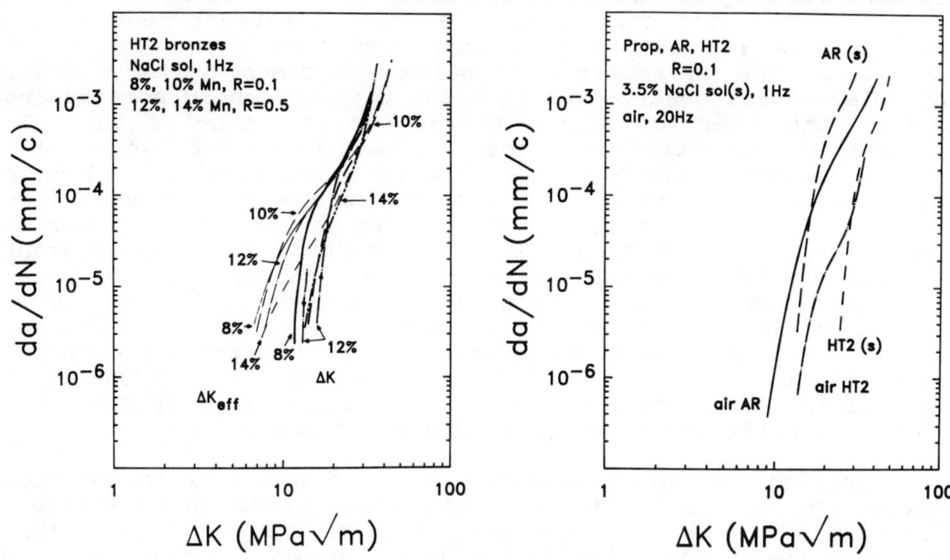

Fig. 9. Crack growth curves at 1 Hz in NaCl solution for the HT2 alloys.

Fig. 10. Crack growth curves for the AR and HT2 propeller.

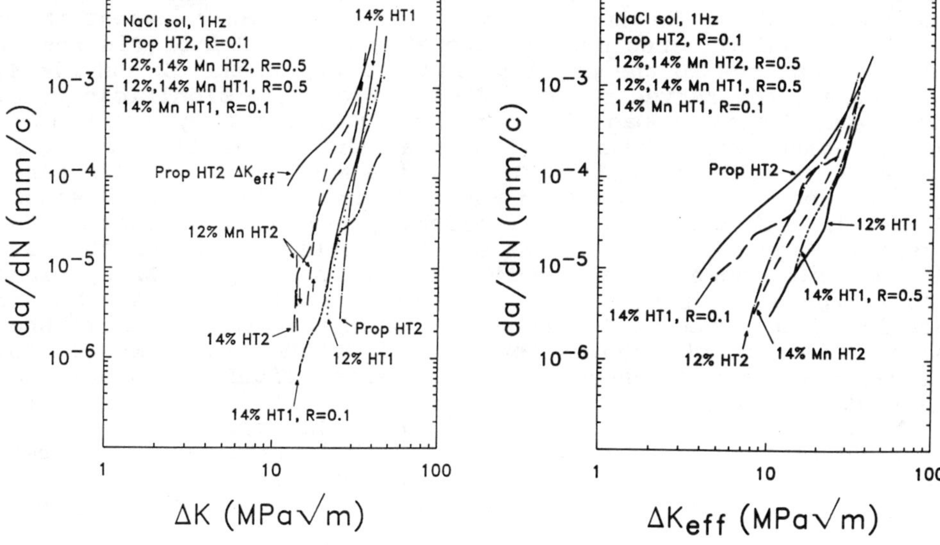

Fig. 11. Log da/dN-log ΔK curves for the HT2 materials.

Fig. 12. Log da/dN-log ΔK_{eff} curves for the HT2 materials.

The test in air on the propeller specimen given the HT2 heat treatment was only carried out to a da/dN value of approximately 2.6 x 10^{-4} mm/cycle; nevertheless, this covered the region of primary practical interest (Figure 10) and showed that the HT2 heat treatment was very effective in improving the resistance to fatigue propagation for long cracks. The HT2 heat treatment caused this curve (Figure 10) to be displaced considerably to the right compared to the as-received propeller specimen, with a clear beneficial effect observed on both the near-threshold portion and the high da/dN portion of the curve. The influence of this HT2 treatment was expected since the HT1 heat treatment, which also strongly increases the β volume fraction, had resulted (1) in a similar effect. The HT1 and HT2 heat treatments have much less influence on the fatigue propagation rates in air for the lower Mn alloys, (Figure 8) for which the volume fraction of β phase present is relatively low. For the test in the NaCl solution on the 13.5% Mn HT2 propeller specimen, (Figure 10), the propagation on a ΔK basis was slower in the NaCl solution than in air at low da/dN and similar to that obtained in air at higher growth rates. Compared to that for the AR propeller specimen, the entire log da/dN-log ΔK curve for the HT2 propeller specimen in the 3.5% NaCl solution has been displaced to ΔK values approximately twice as high.

Comparing the curves obtained at R=0.5 and 1 Hz in the NaCl solution for the 12 and 14% Mn HT2 bronzes to those obtained at R=0.1 on the 13.5% Mn HT2 propeller specimens, Figure 11 shows that the former lie considerably to the left, suggesting that the good resistance to fatigue propagation observed for the HT2 propeller specimens is at least in part associated with a sizeable crack closure effect even at high da/dN. This important crack closure effect is confirmed by the considerable difference between the ΔK and the ΔK_{eff} values obtained even for da/dN values above 10^{-4} mm/cycle for the propeller HT2 specimen tested in the NaCl solution as well as by the good agreement for da/dN > 3 x 10^{-4} between the curve of log da/dN as a function of log ΔK_{eff} for this propeller specimen and those for the 12% Mn HT2 and to a lesser extent those for the 14% Mn HT2 specimens tested at 1 Hz and R=0.5 (Figures 11 and 12). Compared to the previous results (1) on the 12 and 14% Mn HT1 bronzes, the curves for the HT2 materials lie significantly to the side of faster propagation (Figure 11), indicating that the HT2 heat treatment was not as beneficial as the HT1 heat treatment on the crack growth behaviour at 1 Hz in the NaCl solution. The curve for the propeller HT2 specimen was obtained at R=0.1 and, because of more important crack closure effects, lies to the right of the curves in Figure 11 obtained at R=0.5. For da/dN greater than 3 x 10^{-5} mm/cycle, however, the curve for the propeller HT2 specimen lies to the left of that previously obtained for the 14 % Mn HT1 material also tested at R=0.1 and f=1 Hz in the NaCl solution. When the different crack propagation curves for the higher Mn HT2 and HT1 materials are compared on a ΔK_{eff} basis, Figure 12 shows that the tests at R=0.5 agree better with each other, especially for higher da/dN values, although those for the HT1 alloys still lie to the right of those for the HT2 material. THis last aspect suggests that the better resistance to propagation provided by the HT1 treatment is only partly associated with greater crack closure.

Macrofractographic Observations

The fracture surfaces produced in the 3.5% NaCl solution in general were smoother than in air. In particular, the propeller specimens tested in air presented very large fractographic facets, associated with crystallographic cracking in the matrix ß phase, while such facets were absent for the test in the NaCl solution.

For the tests in air, the fracture surface produced at da/dN values approximately less than 3×10^{-5} mm/cycle were covered with dark brown corrosion product, which was particularly dark for the lowest growth rates and for the 8-10% Mn AC alloys. This rate agreed well with that at which log da/dN-log ΔK curves changed from a very steep slope below this da/dN value to a more gradual slope above it. Similar observations were previously reported for the 12 and 14% Mn alloys (1,2). For the tests in the NaCl solution, corrosion product covered the entire fracture surfaces; however, after ultrasonic cleaning, the fracture surfaces presented a similar sharp transition in appearance, which corresponded well to the transition in slope of the crack growth curves. For the 8 and 10% Mn alloys, the fracture surface corrosion product produced in the salt solution appeared considerably thicker and was more difficult to clean for tests at 20 Hz than at 1 Hz. These different observations indicate the occurrence of important corrosion product crack closure especially in the NaCl solution and at 20 Hz. This important crack closure also allows to explain the differences in the low da/dN portions of log da/dN-log ΔK crack growth curves employing an increasing ΔK procedure and those obtained with a decreasing ΔK procedure. At low da/dN, the corrosion product on the fracture surface increases in thickness as da/dN decreases. As a result, the crack growth curve obtained employing an increasing ΔK procedure results in a thicker amount of corrosion product in the wake of the crack and therefore in greater crack closure and slower crack growth for a given nominal ΔK value.

Microfractographic Observations

Intergranular cracking

For tests in air on the 8 and 10% Mn AC, HT1 and HT2 alloys, the transgranular cracking at low crack growth rates was highly crystallographic. For da/dN between 3×10^{-5} to 10^{-4} mm/cycle , the fractographic features was primarily that of ductile, crystallographic striations, while above 10^{-4} mm/cycle, the striations were less crystallographic. The 8% and to a lesser extent the 10% Mn AC bronzes tested in air showed some intergranular cracking at low da/dN (Figure 13). This intergranular cracking was a maximum ($\approx$40% for the 8% Mn AC alloy) for a da/dN range of approximately 10^{-6} to 5×10^{-6} mm/cycle. Similar intergranular cracking for tests in air has been noted in a number of previous studies (7-11), including on copper-based alloys. The evidence indicates that it is an environmental effect, which is associated with the presence of water vapour (8-10) and/or of hydrogen (11).

The fracture surfaces produced in the NaCl solution presented a greater amount of intergranular cracking than for tests in air. A low cycling frequency, a low Mn content and the AC material condition favoured intergranular cracking at low da/dN values. For the 8% Mn AC specimens in the NaCl solution, intergranular cracking was observed up to da/dN $\approx 3 \times 10^{-4}$ mm/cycle for the tests at 1 Hz and up to da/dN $\approx 2 \times 10^{-5}$ mm/cycle at 20 Hz. These results suggest that the maximum crack growth rate at which intergranular cracking occurs is approximately inversely proportional to the cycling frequency, in general agreement with results on Ni-Al bronzes (12).

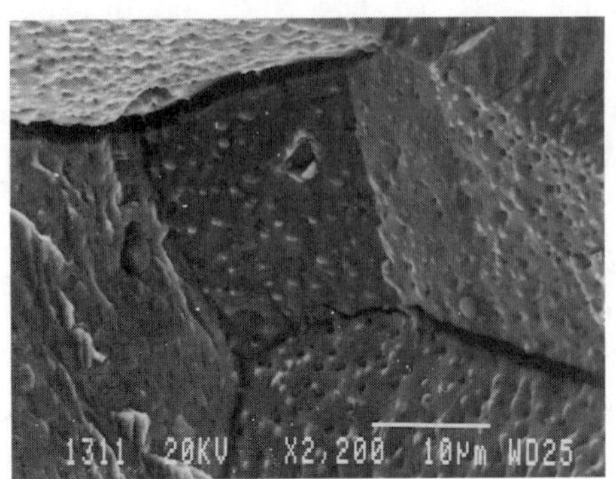

Fig.13. Intergranular facets in 8% Mn AC alloy tested in air, $da/dN \approx 10^{-6}$ mm/cycle.

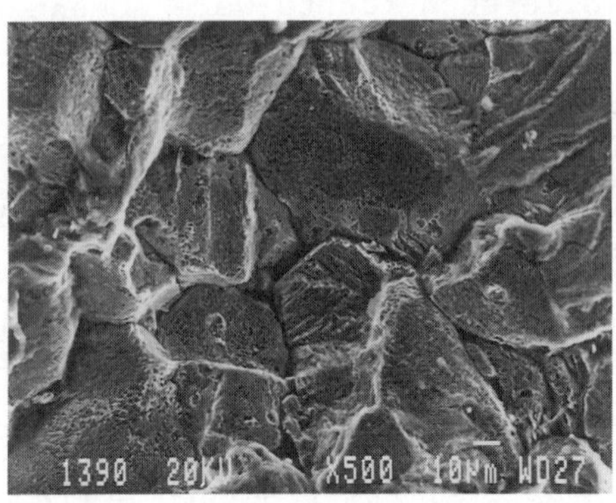

Fig.14. Intergranular facets in 8% Mn AC alloy tested in NaCl, solution, 1 Hz, $\approx 10^{-6}$ mm/cycle.

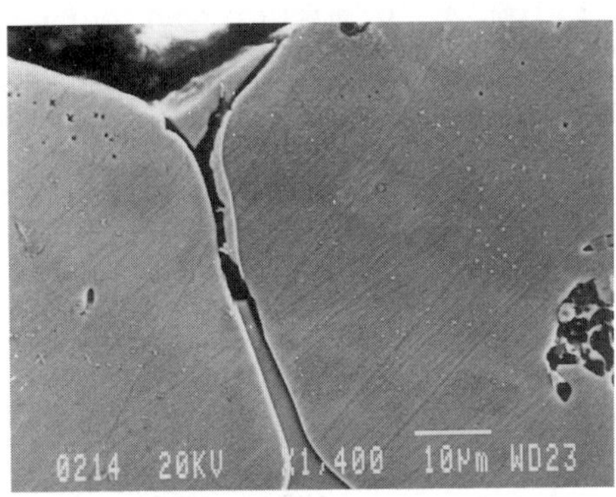

Fig. 15. Dissolution of ß phase in 10% Mn HT1 alloy, tested in NaCl solution, 1 Hz.

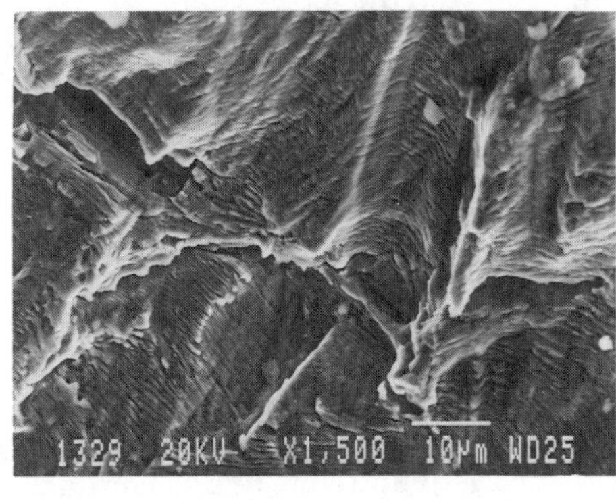

Fig.16. Aspect of striations in 8% Mn AC alloy, tested in air, $da/dN \approx 10^{-3}$ mm/cycle.

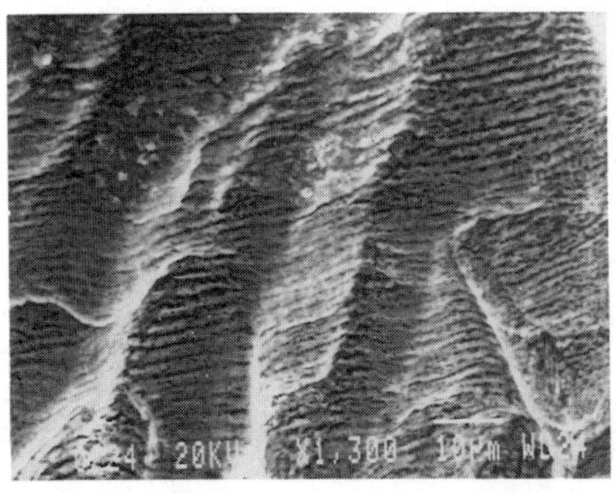

Fig.17. Aspect of striations in 8% Mn AC alloy, NaCl solution, 1 Hz, $da/dN \approx 10^{-3}$ mm/cycle.

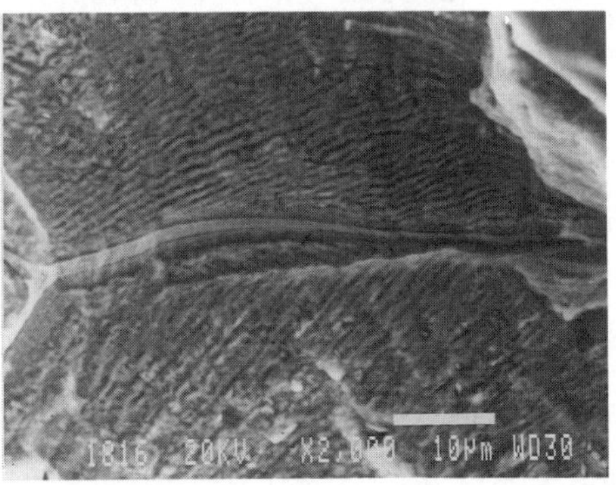

Fig.18. Aspect of striations in 10% Mn HT2 alloy, NaCl solution 1 Hz, $da/dN \approx 10^{-3}$ mm/cycle.

High magnification observations of intergranular facets showed that those produced in the NaCl solution were considerably rougher than in air and, at lower propagation rates, wide secondary cracks (Figure 14) were often present between such facets, strongly suggesting the occurrence of dissolution along these interfaces or grain boundaries. Observations on metallographic profile sections often indicated noticeable preferential dissolution of the ß phase (Figure 15) in the immediate vicinity of an interface with an α grain at secondary cracks. Preferential crack tip dissolution of the ß phase is consistent with the interfacial cracking observed at lower growth rates in the NaCl solution of these 8-10% Mn alloys.

For the da/dN range in which this intergranular cracking was obtained, accelerated propagation at 1 Hz in the NaCl solution for the 8 and 10% Mn AC was generally only observed on the log da/dN-log ΔK_{eff} curves (Figure 3). These results indicate that the increased corrosion product-induced crack closure at low da/dN values in the NaCl solution masked, on a ΔK basis, the accelerated crack growth. Accelerated propagation was not observed for the 8 and 10% Mn HT1 and HT2 bronzes tested in the NaCl solution, for which much less intergranular cracking was noted.

Influence of Environment on Striations

For the high da/dN values for which accelerated propagation in the NaCl solution was indicated on the crack growth curves for the 8 and 10% Mn AC bronzes, the fracture surfaces were covered with ductile striations. A relationship was found between the manner in which striations interacted with grain boundaries and the occurrence of corrosion-fatigue propagation (13). For tests in air, the striation aspect indicated that the crack frequently had difficulty to cross grain boundaries. The striations showed some tendency to change orientation from grain to grain (Figure 16), the interstriation spacing quite often decreased as the crack front approached grain boundaries and, immediately after crossing a boundary, striations at times had a semi-elliptical shape (Figure 16), indicating crack initiation in the new grain along a small segment of the boundary. In contrast, for tests at 1 Hz in the NaCl solution, the observations at high da/dN generally indicated that the crack crossed grain boundaries easily (Figure 17) along a wide segment and with the interstriation spacing generally not decreasing on approaching these boundaries. The ductile striations observed at high da/dN were generally almost perpendicular to the macroscopic propagation direction for the tests at 1 Hz in the NaCl solution, while, for tests in air, the striations were often more crystallographic. The striations observed at high da/dN for tests at 20 Hz in the NaCl solution indicated that the crack crossed grain boundaries only somewhat more easily than for tests in air.

The aspects of striations observed for the 8 and 10% Mn HT1 and HT2 alloys tested at 1 Hz in the NaCl solution indicated that the fatigue crack experienced considerable difficulty to cross boundaries containing a sizeable ß ligament (Fig. 18). The curved aspect of the striations at times indicated that the crack crossed the boundary by causing initiation in the new α grain along a small segment of the ß-α interface. The striation aspect indicated that the crack experienced less difficulty in crossing boundaries containing only a very thin ß ligament and little difficulty in crossing boundaries in which no ß ligament was observed. Since many of the boundaries between α grains in the 8 and 10% Mn HT2 bronzes

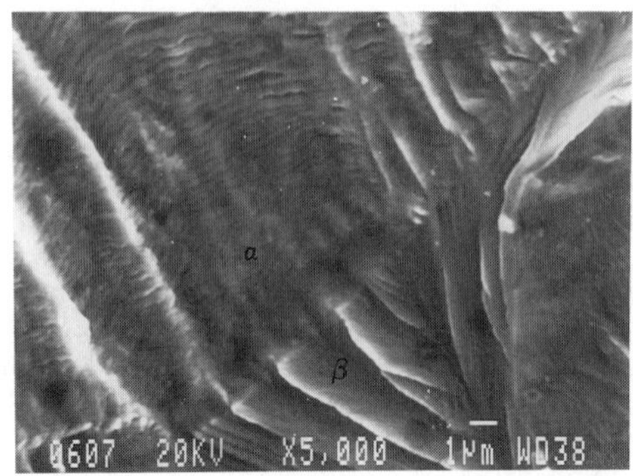

Fig.19. Aspect of striations in α phase, AR propeller, tested in air, da/dN ≈ 7 x 10^{-4} mm/cycle.

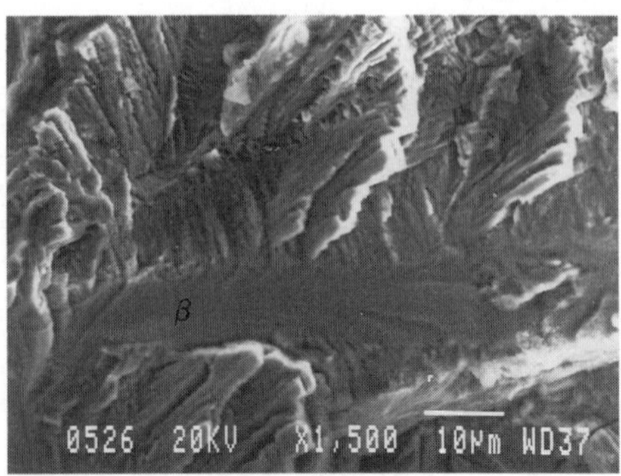

Fig.20. Fracture facets showing fracture leading in ß phase, AR propeller, air, ≈10^{-5} mm/cycle.

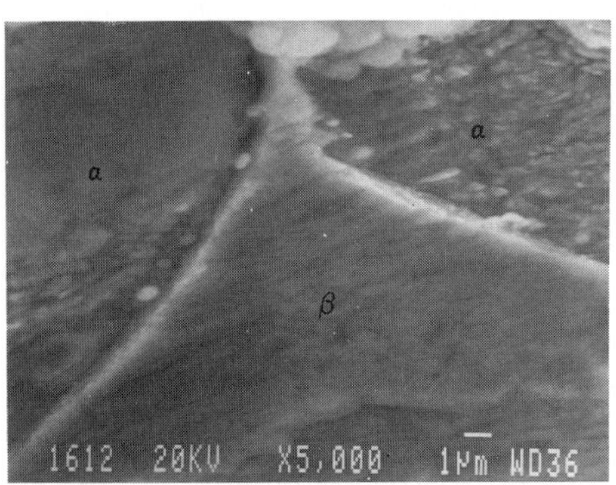

Fig.21. Striations which continue across α and ß phases, AR propeller, da/dN ≈ 10^{-3} mm/cycle.

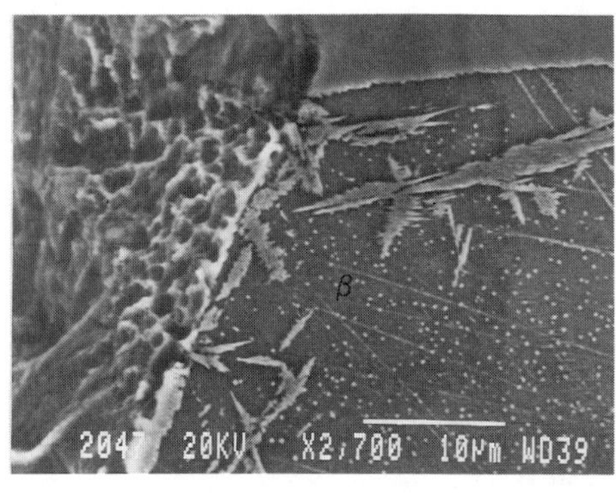

Fig.22. Dimples in ß phase of HT2 propeller, tested in air, da/dN ≈ 2 x 10^{-4} mm/cycle.

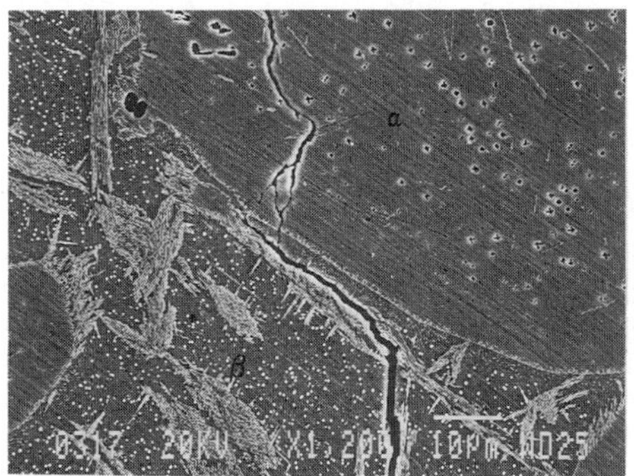

Fig.23. Crack path in α and ß phases, HT2 propeller tested in air, da/dN ≈ 10^{-5} mm/cycle.

Fig. 24. Path of a secondary crack following ß slip traces, HT2 propeller, NaCl sol., 1 Hz.

contained ß ligaments, these observations were consistent with the absence of accelerated cracking at 1 Hz in the NaCl solution for these bronzes with respect to the tests at 20 Hz in air or in the NaCl solution. The aspect of striations on the 8 and 10% Mn HT1 bronzes suggested the occurrence of a corrosion-fatigue effect intermediate between that for the AC and HT2 alloys, with testing at 1 Hz in the NaCl solution facilitating propagation across some α grain boundaries at high da/dN. The corresponding propagation results for the 8 and 10% Mn HT1 alloys did not show clear accelerated propagation compared to tests in air but did show faster propagation than for the tests at 20 Hz in the NaCl solution.

Microfractography of the AR Propeller Specimens and HT2 Bronzes

For the AR propeller specimen tested in air, metallographic sections of the profile of the primary crack and of secondary cracks which branched off the primary crack showed that the crack path was primarily in the α phase, with some portions following α-ß interfaces or crossing ß ligaments. For da/dN < 3 x 10^{-5} mm/cycle, the amount of ß phase cracking appeared to be close to the minimum required to assure macroscopic crack growth. For higher da/dN values, the fractographic observations indicated that, while more ß phase cracking occurred, the presence of this phase and of α-ß interfaces still acted as obstacles to propagation. Fractographic observations which indicated that the crack had difficulty crossing ß-α interfaces, included the frequent presence of a finer inter-striation spacing near the site at which initiation of fatigue cracking occurred in an α grain and the frequent presence of curved striations near ß-α interfaces (Figure 19), indicating that the crack front was held back at the interface. Striations were generally not visible on the ß facets. The fractographic observations, however, also frequently showed that the crystallographic cracking in the ß phase generally proceeded locally ahead of the macroscopic crack front, with the cracking then spreading laterally into the adjacent α grains (Figure 20). Therefore, while initiation of cracking in the ß phase was an obstacle to crack propagation, once initiated, the crack propagated rapidly in this phase.

Both the macroscopic and microscopic crack path for the test on the AR propeller specimen in the NaCl solution tended to be significantly less irregular than in air at all da/dN values and crossed α-ß interfaces with little deviation, even for lower da/dN. The observations also indicated that the microscopic fracture facets in both phases were also flatter than for the test in air. The microfractographic observations for da/dN > 10^{-4} mm/cycle were similar to those observed in air for high crack growth rates, except in the vicinity of α-ß interfaces, where the aspect of striations suggested that, in the NaCl solution, the crack crossed α-ß interfaces with little difficulty (Figure 21). Some of the ß ligaments presented similar striations (Figure 21), as previously observed on 12 and 14% Mn HT1 bronzes tested at 1 Hz in NaCl solution. These observations indicate that the corrosion-fatigue effect in the NaCl solution is largely associated with easier propagation across α-ß interfaces. The metallographic observations also often showed some dissolution at interfaces of large $κ_I$ particles, situated very near the fracture surface.

The observations on the 12 and 14% Mn HT2 bronzes as those on the propeller specimen indicated that testing in the NaCl solution at 1 Hz favoured the presence of observable fatigue striations on ß phase ligaments, which were rare or absent for the tests in air. This result indicates a corrosion-fatigue effect of increased crack tip plasticity in the ß phase. This result is consistent with the often observed effect, reviewed recently (14), of corrosion fatigue to facilitate crack initiation by resulting in more important localized slip at the surface in contact with the aggressive environment. A similar effect of greater crack tip plasticity resulting from the crack tip being in contact with the aggressive environment also permits to explain the less crystallographic striations observed in the α phase, which, at a given value of ΔK, results in the crack crossing grain boundaries or interfaces more easily. For the 8 and 10% Mn AC materials, the observations indicate that this type of effect occurred for the striations produced in the α phase at high da/dN. For the propeller specimen as well as for the 12 and 14% Mn HT1 bronzes (1,2), the observations indicate such an effect for both phases.

For the 13.5% Mn propeller specimen given the HT2 heat treatment and tested in air, an approximately equal amount of cracking was observed in both phases. At higher propagation rates, the crystallographic ß facets (Figure 22) presented some dimples, formed at fine κ_{IV} particles, similar to those observed on crystallographic ß facets in fracture toughness tests (3-5). The number of dimples observed increased with increasing da/dN. For this test in air, observations of secondary cracks (Figure 23) showed that the cracking in the α phase was quite devious, while the crack path in the ß phase was crystallographic and cleavage-like, with crack segments often roughly parallel to elongated Widmanstaetten secondary α grains within the ß matrix. The cracks appeared to have trouble to propagate across such α grains, resulting in frequent propagation along or near their interfaces or occasionally in secondary cracks terminating at such interfaces.

For the HT2 propeller specimen tested in the NaCl solution, the profile of the fracture surface was rougher than for the AR propeller specimen tested in this solution. The crack path was crystallographic in the ß phase and deviated on crossing α-ß interfaces. The crack path within the α grains quite often was close to but not along an α-ß interface. On metallographic sections some dissolution at interfaces between α grains and large propeller-shape κ_I precipitates was observed in the immediate vicinity of the fracture surfaces. For da/dN < 10^{-4} mm/cycle, secondary α grains within the ß phase were observed to influence the local crack propagation in a manner similar to that for the test in air. One of the secondary cracks observed contained segments in the ß phase along three different sets of slip traces (Figure 24). This observation indicates that the fatigue cracking in the ß phase involves slip deformation and that testing in the NaCl solution increases the crack tip plasticity. The profile of this fracture surface was also rougher, especially for the low and intermediate crack propagation rates than that of the as-received propeller specimen tested in the NaCl solution but not nearly as rough as the profile of the HT2 propeller specimen tested in air.

The good resistance to fatigue propagation in both test environments of the 13.5% Mn propeller after the HT2 heat treatment can be associated in part with this microstructure requiring an important amount of crack propagation in the harder ß phase. The crack often has difficulty to cross from the ß into the primary or into the secondary α phase indicating that the microstructure contains many obstacles tending to retard, deviate or arrest the crack. The devious crack path and rough fracture surface which resulted for the tests produced on the HT2 propeller specimen in both the air and NaCl solution environments also contributes to the good resistance to crack propagation, by causing ΔK_{eff} to be signicantly lower than the nominal ΔK value.

The 12 and 14% Mn HT2 specimens tested in the NaCl solution at R=0.5 presented microfractographic features at high da/dN which were similar to those for the HT2 propeller specimen tested in this solution. The 14% Mn HT2 specimen was adequately cleaned to permit microfractographic observations for lower da/dN values. These indicated crystallographic cracking in both phases as well as, for da/dN $\approx$ 10^{-6} mm/cycle, some cracking along α-β interfaces.

Influence of the Microstructure

The small influence of Mn composition, within the range studied, on the fatigue crack propagation curves of the AC materials can be associated with the majority of the propagation occurring in the softer α phase. While only the fatigue crack propagation of the 12 and 14% Mn NSC alloys was studied (1,2), it is these higher Mn alloys which contain more β phase and the NSC condition was that which gave the smallest volume fraction of primary β phase of the material conditions tested (3-5). Moreover, the log da/dN-log ΔK curves for these NSC bronzes were similar to those for the AC materials of similar Mn composition (1,2). Little influence of Mn content is therefore expected on the crack propagation curves of the 8-14% Mn NSC bronzes.

There is a considerably more important effect of the Mn composition on the propagation curves of the HT1 and HT2 alloys, with the higher Mn alloys showing considerably improved resistance to fatigue propagation, with respect to the AC and NSC materials. The improved resistance to crack propagation displayed by the 12-14% Mn HT1 and HT2 alloys can be related to their volume fraction and distribution of β phase, which results in a larger amount of the fatigue crack propagation occurring in this hard phase. As well, a strong tendency is obtained in the tests in air for large fracture facets because of a tendency for crystallographic cracking in the β phase and because of the large β grain size. The resulting macroscopically rough fracture surface gives important roughness-induced crack closure effects as well as important crack deviation effects, both of which favour a large difference between the nominal or applied and the effective ΔK values. For the tests in the NaCl solution, the fracture surfaces of the higher Mn HT1 and HT2 alloys were covered with thick corrosion product, favouring corrosion product-induced crack closure.

The present study combined with that of the fracture toughness of these bronzes (3-5) suggests that the HT2 heat treatment results in an interesting combination of mechanical properties for the C95700 bronzes. For these higher Mn alloys, the HT2 heat treatment had a beneficial effect on the yield strength (3-5), dynamic fracture toughness (3-5) and resistance to fatigue propagation of long cracks in both air and in 3.5 % NaCl solution. A difficulty in applying this heat treatment to a large casting however, is the possibility of creep distortion during the anneal at 720°C.

The corrosion-fatigue crack propagation effect observed at

high da/dN in all the Mn-Ni-Al bronzes studied in this and in the previous investigation (1,2) was associated with propagation across or along α-ß interfaces or across α-α grain boundaries being facilitated in some manner during testing at low frequency in the NaCl solution. The manner in which this effect of easier propagation across such boundaries occurs depends on the microstructure and therefore on the alloy composition and condition. The general tendency observed is that the corrosion-fatigue crack propagation effect at high da/dN is associated with a decrease in the influence of simultaneous presence of the α and ß phases to retard crack propagation during the tests in air.

Relevance of the Present Results to In-Service Fatigue Behaviour

The present study focused on the crack propagation behaviour of long cracks in Mn-Ni-Al bronzes employed or of possible interest for ship propeller applications. Since in propellers, the fatigue crack initiation and short crack propagation behaviour can be of similar or even greater importance, the fatigue properties should not be evaluated solely based on the propagation behaviour of long cracks. A study (15) is in progress of the fatigue initiation and short crack propagation behaviour of these materials. With the principally duplex microstructure of these Mn-Ni-Al bronzes, for tests in air, crack initiation occurs in the softer α phase, and the resulting microcracks can then encounter considerable difficulty in crossing the first α-ß interface encountered (15). The presence of the ß phase therefore acts as an obstacle to the propagation of these microcracks, although when propagation does occur in this phase, it occurs relatively rapidly.

In a relatively non-aggressive environment, such as laboratory air, the mainly duplex microstructure of the bronzes studied can be expected to have some retardation effect on the propagation of both long and short cracks. Such a microstructure should therefore favour good overall resistance to fatigue propagation. This is also suggested from previous studies of fatigue of austenitic-ferritic stainless steels (16-18).

For tests at 1 Hz in the NaCl solution, the dissolution effects observed at α-ß interfaces and at precipitates as well as the tendency for enhanced cyclic plasticity at the crack tip suggest that some of these Mn-Ni-Al bronzes should also be susceptible to accelerated crack initiation in this and similar aqueous environments, which has been confirmed in longer duration tests on cylindrical specimens (15). These tests also suggest that the high Mn alloys in the HT2 condition resist relatively well to dissolution facilitated crack initiation in 3.5% NaCl solution.

The present results as well as that of the previous study (1-3, 19) of the corrosion-fatigue behaviour of Mn-Ni-Al bronzes also indicated that part of the good resistance to fatigue propagation of long cracks for the higher Mn alloys in the HT1 and HT2 conditions was associated with the large ß grain sizes obtained and the very crystallographic cracking for tests in air. This resulted in important roughness-induced crack closure and crack deviation effects, both of which favoured a true or effective value of ΔK which was substantially lower than the nominal value. For the tests in the NaCl solution, very important corrosion product-induced crack closure effects were observed especially for low crack growth rates, with these effects often being more important in the portion of the crack growth curve obtained with a ΔK increasing procedure. It appears probable that the difference between the nominal and effective values of ΔK would often be considerably lower for the propagation of short or small cracks. For this reason, the high threshold values for crack propagation and the good resistance to crack propagation in the low da/dN range compared to the behaviour in air indicated from the tests carried out in the NaCl solution cannot be assumed to represent the behaviour which would be obtained for small cracks at low nominal ΔK values in propellers under actual service conditions.

Since testing in an aggressive aqueous environment tends to increase the fatigue propagation threshold for long cracks but to reduce the endurance limit, a Kitagawa and Takahashi type analysis (20) predicts that this increases the propagation distance over which accelerated short crack propagation behaviour is expected. The exact propagation behaviour for short cracks in the aggressive environment should be influenced considerably by the rapidity with which the corrosion product-induced crack closure builds up as the crack develops. It appears doubtful that the improved propagation resistance observed at low da/dN for these tests on long cracks in the NaCl solution with respect to those in air will occur to a similar extent in service, where low da/dN values generally correspond to short cracks for which crack closure effects should be smaller and for which other effects accelerating propagation are possible (21). On the other hand, the present tests were carried out at 22-24°C, while the service temperature for propellers will often be considerably lower. This difference in temperature (17,18) as well as that in the saline environment (seawater compared to 3.5% NaCl solution) can significantly influence the magnitude of both the corrosion-fatigue effects and of the corrosion product-induced crack closure effects.

CONCLUSIONS

The alloy conditions studied (AC, HT1 and HT2) had only a slight influence on the fatigue propagation behaviour in air of the 8-11% Mn-Ni-Al bronzes. These bronzes were also less susceptible to corrosion-fatigue accelerated propagation in 3.5% NaCl solution than 12-14% Mn alloys. The 8 and 10% Mn AC bronzes showed some corrosion-fatigue accelerated propagation associated, at high da/dN, with increased crack tip plasticity and facilitated crossing of grain boundaries and, at lower da/dN, with intergranular cracking and dissolution in the region of α-β interfaces.

The 13.5% Mn propeller specimens showed a stronger susceptibility to corrosion-fatigue accelerated propagation than the AC alloys. The propeller specimens given the HT2 heat treatment showed substantially improved resistance to propagation in both air and in the NaCl solution, with an important portion of this increased resistance associated with increased crack closure and more important crack deviation effects. The 12 and 14% Mn HT2 bronzes tested at 1 Hz in the NaCl solution also showed increased propagation resistance compared to the AC and NSC alloys. Improved resistance to crack propagation at 1 Hz in the NaCl solution with increasing Mn content was also noted for the HT2 alloys and was partly associated with increased crack closure effects.

ACKNOWLEDGMENTS

The present study was performed under DSS (Canada) contract 24ST.23440-5-9054. Financial support from NSERC (Canada) and FCAR (Quebec) research support programs is gratefully acknowledged. The authors thank Dr. Mahi Sahoo (CANMET) for useful discussions.

REFERENCES

1. J.I. Dickson, L. Handfield, S. Lalonde, M. Sahoo and J.P. Bailon, J. Materials Engineering, Vol. 10, 1988, pp. 45-56.
2. J.I. Dickson, L. Handfield, J.P. Bailon and M. Sahoo, in "Fatigue 84", C.J. Beevers (ed.), EMAS Ltd, Cradley Heath, 1984, Vol I, pp. 191-200.
3. J.I. Dickson, J. Hallen-Lopez, L. Handfield, Y. Blanchette and S. Lalonde, "A Study of the Fracture Toughness of Bronze Propeller Alloys", D.S.S Report 64SS.23440-5-9054, Part I, 1989.
4. J. Hallen-Lopez, J.I. Dickson and M. Sahoo, Transactions of the American Foundrymen's Society, Vol. 99, 1989, pp. 489-500.
5. J.I. Dickson, L. Handfield, J. Hallen-Lopez, Y. Blanchette and M. Sahoo, in "Proceedings of the International Symposium on Fracture Mechanics", W.R. Tyson and B. Mukherjee (eds.), Pergamon Press, Oxford, 1988, pp. 327-336.
6. A. Couture, M. Sahoo, B. Dogan and J.D. Boyd, Transactions of the American Foundrymen's Society, Vol. 95, 1987, pp. 87-92.
7. Y. Higo, A.C. Pickard and J.F. Knott, Metal Science, Vol. 15 (1981) pp. 233-240.
8. J.P. Bailon, J.I. Dickson and Li Shiqiong, in "Basic Mechanisms in Fatigue in Metals", P. Lukas and J. Polak (eds.) Academia Press, Prague, 1988, pp. 361-371.
9. J.P. Bailon, M. Elboujdaini and J.I. Dickson, in "Fatigue Crack Growth Threshold Concepts", D.L. Davidson and S. Suresh (eds.), AIME, New York, pp. 63-82.
10. N. Marchand, J.P. Bailon and J.I. Dickson, Metallurgical Transactions, Vol. 19A, 1988, pp. 2575-2587.
11. T.S. Sudarshan, M.R. Louthan, Jr, T.A. Place and H.H. Mabie, in "Fatigue Life Analysis and Prediction", V.S. Goel, ed., ASM, Metals Park, OH, 1986, pp. 47-52.
12. R.N. Parkins and Y. Suzuki, Corrosion Science, Vol. 23, 1983, pp. 577-599.
13. J.I. Dickson, Li Shiqiong, L. Handfield and J.-P. Bailon, Microstructural Science, Vol. 18, 1990, pp. 475-484.
14. J.I. Dickson, Li Shiqiong and J.P. Bailon, "Microstructural and Fractographic Aspects of Corrosion Fatigue", submitted to Materials Characterization.
15. P. Leblanc and J.I. Dickson, research in progress.
16. J.A. Moskowitz and R.M. Pelloux, in "Corrosion Fatigue Technology", ASTM STP 642, 1976, pp.133-154
17. M. Ait Bassidi, J. Masounave, J.I. Dickson and J.P. Bailon, Canadian Metallurgical Quarterly, 23, 1984, pp. 17-24.
18. M. Ait Bassidi, J. Masounave and J.-P. Bailon, in "Duplex Stainless Steels", R.A. Lula (ed.), 1983, ASM, pp. 445-463.
19. J.I. Dickson, S. Lalonde, L. Handfield and J.-P. Bailon, in "Strength of Metals and Alloys: Proceedings of ICSMA-7", Pergamon Press, Oxford, 1986, Vol. III, pp. 2129-2136.
20. H. Kitagawa and S. Takahashi, in "Proceedings ICM II", ASM, 1976, pp. 627-631.
21. R.P. Gangloff, Metallurgical Transactions, Vol. 16A, 1985, pp. 953-969.

Variation of the short-range order parameter in αCu-Al alloys during DSC experiments

A. Varschavsky and E. Donoso
University of Chile, Faculty of Physics and Mathematics,
Institute for Materials Research and Testing, IDIEM, Santiago, Chile

ABSTRACT

Differential scanning calorimetry (DSC), performed in quenched α Cu-Al alloys reveals that ordering takes place in two stages. Stage 1 ordering is associated with the migration of excess vacancies and stage 2 ordering occurs by the migration of equilibrium vacancies. The magnitude of both stages varies with the alloy composition. At higher temperatures a marked surge of energy absorption occurs, stage 3, being attributed to the destruction of order. For furnace-cooled alloys the foregoing stage is the only one appearing. The transformed fraction for stages 1 and 2 along a DSC run was evaluated by means of an overall kinetic expression in terms of the first short-range order parameter. A flow diagram of the factors controlling the fractional increase of short-range order during anisothermal experiments was proposed.

INTRODUCTION

There has been continuing interest in the local-order structure of the α phase in Cu-Al alloys since the electrical resistivity studies on neutron-irradiated material by Wechsler and Kernohan (1) indicated that a composition-dependent, diffusion-controlled, solid-state reaction occurs under certain conditions. Perhaps the most comprehensive study of such reaction in this system is that of Matsuo and Clarebrough (2) who identified the presence of short-range order (SRO) by calorimetry and resistivity measurements suggesting the role of vacancies on the kinetics of ordering. Further evidence of SRO in α Cu-Al alloys has been reported by Borie and Sparks (3) using diffuse X-ray scattering and by Panin et al. (4) using density and resistivity measurements. In the foregoing literature, it is clear that SRO in αCu-Al has been well identified.

The different features of ordering have been investigated for instance, by diffuse scattering of X-ray (3, 5-9), small angle X-ray scattering (10), electron microscopy and electron diffraction (11-15), by determination of elastic and plastic properties (16-18), of strengthening and fatigue properties (19-24) as well as by electrical resistivity (25-28) and thermal analysis (16, 17, 21, 29-34). Two of those studies (14, 16) inferred that after isothermal anneals at a temperature of 523 K (which is around the estimated critical value for domain dissolution (34)), during 1800 s in quenched Cu-19 at % Al, a disperse order (DO) state could be detected. These findings were fairly consistent with the measured diffusion time (2400 s) at 518 K for furnace cooled alloys pre-annealed during 80 h, where no excess vacancies were available (34). On the other hand, for the same alloy, analysis of data obtained from non-isothermal experiments at different heating rates employing differential scanning calorimetry (DSC) (32), confirmed that the relaxation times at peak temperatures were about three orders of magnitude less han the calculated diffusion times required for an equilibrium DO state to be developed at those temperatures (14, 16). At lower aluminum concentrations, such differences become even larger (34). Furthermore, the critical temperature for domain dissolution, which occurs by a first order transition (34), is lower than the temperatures at which the ordering process goes to completion for the range of heating rates usually employed in DSC scans (2, 16, 32). Thus, it seems unlikely that ordered regions would develop in quenched alloys during rising temperature experiments where ordering reactions have to take place before disordering commences. Besides, although the study of short-range ordering phenomena in α Cu-Al alloys is being pursued (35, 36), there is comparatively little research concerning the kinetics of the SRO process from a quantitative treatment of DSC traces. This is particularly appropiate if a statistical SRO model is suitable, since for these alloys, the changes in internal energy can be represented effectively by the change in the order parameter for the first coordination shell as it was demonstrated by Kuwano and co-workers (37).

The principal objectives of the present work are: a) to evaluate the kinetics of SRO reactions in quenched and furnace-cooled alloys, b) to compute boundary values for the first SRO-parameter from the features displayed by the DSC traces (thus enabling to enhance the capabilities of the technique), c) to investigate the relative dominance of the stages involved in the ordering process after different quenching conditions in an attempt to clarifie the effect of concentration of excess vacancies and, d) to determine the vacancy behavior during the return of SRO to equilibrium.

MATERIAL AND EXPERIMENTAL METHOD

The three αCu-Al alloys studied contained, respectively, 3.00 ± 0.04 ,

6.00 ± 0.07, and 9.0 ± 0.10 wt pct aluminum (99.97 wt pct). They were prepared in a Baltzer VSG 10 vacuum induction furnace from electrolytic copper (99.95 wt pct) in a graphite crucible. The ingots were subsequently forged at 923 K to a thickness of 10 mm pickled with a solution of nitric acid (15 pct) in destilled water to remove surface oxide, annealed in a vacuum furnace at 1123 K for 36 hour to achieve complete homogeneity, and cooled in the furnace to room temperature. They were cold rolled to 1.5 mm thickness with intermediate annealing periods at 923 K for one hour. After the last anneal the material was finally rolled to 0.75 thickness (50 pct reduction). A subsequent heat treatment was performed at 873 K for one hour, followed by quenching. Other batches of alloys were furnace-cooled at a rate of 15 K h^{-1}. Microcalorimetric analysis of the samples was performed in a Dupont 2000 Thermal Analyser. Specimen discs of 0.75 mm thickness x 6 mm diameter were prepared and examined for each material condition. Differential scanning calorimetric measurements of the heat flow were made by operating the calorimeter in the constant heating mode (heating rates of 0.83, 0.33, 0.17, 0.083, and 0.033 K s^{-1}).

THEORETICAL DEVELOPMENT

The First Short-Range Order Parameter Approach to DSC Traces

In this section a method is presented to determine the kinetics of the thermal events associated with the different stages observed during non-isothermal heating of the alloys under study through a common parameter representative of the state of ordering. The following considerations are contained in the proposed method: (a) The equilibrium short-range order can be described by a Warren-Cowley parameter $\alpha_e = 1-P(Cu-Al).\bar{c}^{-1}$, $P(Cu-Al)$ is the conditional probability to find an Al atom next to a given Cu atom and $\bar{c}$ is the atomic aluminum concentration. For simplicity the high temperature approximation will be used (38)

$$\alpha = 2(1 - \bar{c})\ W/RT \qquad (1)$$

where $W = V_{AB} -(1/2)(V_{AA} + V_{BB})$ is the ordering energy, V_{ij} is the bonding energy of $i - j$ atomic pair, T the absolute temperature and R the gas constant. (b) The kinetics of SRO can be described for all stages by a single relaxation time

$$(d\alpha/dt) = -[(\alpha - \alpha_i)/\tau] \qquad (2)$$

being α_i the value of α at the end of the stage considered, $\tau = 1/k$ where $k = k_o \exp(-E/RT)$, k_o the pre-exponential factor, E the effective activation energy describing the overall process of the thermal event and t the time. (c) As it is well known in DSC, $dy/dt = (1/A)(da_t/dt)$, where y is the degree of transformation (fraction transformed), da_t/dt is the rate of heat flow and monitors the course of the reaction. a_t is the area under the peak to time t, and A is the total area under the peak. For a particular transformation considered, y can be expressed in terms of α as:

$$y = (\alpha - \alpha_{io})/(\alpha_i - \alpha_{io}) \qquad (3)$$

where α_{io} refers to the start of the transformation. Replacing α from eqn. 3 in 2 and considering that $k = 1/\tau$, a first order kinetic law is obtained as expected from the consideration made in (b)

$$(dy/dt) = k(1 - y) \qquad (4)$$

Under non-isothermal conditions, integration of eqn. 5 yields:

$$\ln[(1 - y)^{-1}] = k_o \theta \tag{5}$$

where θ is identical to the reduced time and can be computed from (39)

$$\theta = (T^2 R/\phi E) \exp(-E/RT) \tag{6}$$

where $\phi = dT/dt$ is a constant heating rate. Knowledge of boundary values for the short-range order parameters α_{oi} and α_i corresponding to a specific thermal event allows to evaluate the kinetics of α in the range of temperatures scanned where such event takes place. Therefore, from eqns. 3 and 5 it turns out that

$$\alpha = \alpha_i - (\alpha_i - \alpha_{oi}) \exp(-k_o \theta) \tag{7}$$

(d) Boundary values of α for each thermal event displayed in the quenched and furnace-cooled alloys can be estimated from equivalent equilibrium short-range order temperatures (T_E^i s) considered below. Fig. 1a shows schematically a typical DSC thermogram for an α Cu-Al alloy after being quenched. It is characterized by two exothermic reactions. Stages 1 and 2, and an endothermic reaction stage 3. These stages have been reported in the literature in connection with the following processes: stage 1 with short-range order development associated with th migration of excess vacancies (12, 40); stage 2 with short-range order formation produced by the migration of equilibrium vacancies (2, 13, 16) and stage 3 with disordering (2, 13, 16).

Determination of Boundary Values

Since energy evolutions are due to the return of ordering and the energy absorption to the destruction of ordering, then the degrees of order present at room temperature after quenching may be specified in terms of an equivalent temperature T_{E1} at which these degrees of order would be in equilibrium. For a given quenching temperature, this temperature can be calculated form a knowledge of the net absorption or evolution of energy up to some temperature at which a constant rate of destruction of order has been attained and the rate of absorption of energy on continuous heating above such temperature. From fig. 1a, if $\Delta H_1(T_1, T_2)$, $\Delta H_2(T_2, T_3)$, $\Delta H_3(T_3, T_{eq})$ are the energies associated to the respective peaks, and $\Delta H_{E1} = \Delta C_{pe}(T_{E1} - T_{eq})$, where ΔC_{pe} and T_{eq} represent the constant diferential specific heat and equilibrium temperature respectively, at which energy commences to be absorbed at a constant rate, one has $\Delta H_{E1} = \Delta H_1(T_1, T_2) + \Delta H_2(T_2, T_3) - \Delta H_3(T_3, T_{eq})$. Therefore T_{E1} can be determined from the expression

$$T_{E1} = T_{eq} + [\Delta H_1(T_1, T_2) + \Delta H_2(T_2, T_3) - \Delta H_3(T_3, T_{eq})]/\Delta C_{pe} \tag{8}$$

hence

$$\alpha_{E1} = 2(1 - \bar{c}) \bar{c} W/RT_{E1} \tag{9}$$

Similarly at the beginning of stage 2

$$T_{E2} = T_{eq} + [\Delta H_2(T_2, T_3) - \Delta H_3(T_3, T_{eq})]/\Delta C_{pe} \tag{10}$$

and

$$\alpha_{E2} = 2(1 - \bar{c}) \bar{c} W/RT_E \tag{11}$$

Finally $T_{E3} = T_3$, because T_3 is the temperature at which the transition occurs from an evolution to an absorption of energy at the end of stage

2, which indicates the attainement of the equilibrium degree of order at that temperature. Hence,

$$\alpha_{E3} = 2(1 - \bar{c}) \; \bar{c} W/RT_3 \tag{12}$$

Therefore, between T_1 and T_2 the short-range order parameter varies according to:

$$\alpha = \alpha_{E2} - (\alpha_{E2} - \alpha_{E1}) \exp (-k_{01} \; \theta) \tag{13}$$

where k_{01} is the pre-exponential factor corresponding to stage 1. Consequently between T_2 and T_3 , α can be expressed by

$$\alpha = \alpha_{T3} - (\alpha_{T3} - \alpha_{E2}) \exp (-k_{02}\theta) \tag{14}$$

where k_{02} is the pre-exponential factor for to stage 2, and finally

$$\alpha = \alpha_{eq} - (\alpha_{eq} - \alpha_{T3}) \exp (-k_{03}\theta) \tag{15}$$

between T_3 and T_{eq}, where k_{03} is the corresponding pre-exponential factor for stage 3. For the furnace – cooled alloys a single disordering stage 3 is observed (21, 27, 28) which is shown schematically in fig. 1b. In this situation

$$T_E = T_{eq} -[\Delta H_3(T_0, \; T_{eq})]/\Delta C_{pe} \tag{16}$$

where T_0 is the starting temperature in abscense of quenched in vacancies, as it is the case (cooling rate: 15 K h^{-1}). Therefore,

$$\alpha_E = [2(1 - \bar{c})]\bar{c}W]/RT_E \tag{17}$$

which yields

$$\alpha = \alpha_{eq} - (\alpha_{eq} - \alpha_E) \exp (-k_0 \; \theta) \tag{18}$$

between T_0 and T_{eq} , where k_0 is the corresponding pre-exponential factor.

The return of SRO to equilibrium via two processes, each one obeying a first order kinetic law, can be decpicted by a single overall kinetic relationship. This expression can be written as a convex linear combination:

$$y = \psi \, y_1 + (1 - \psi) \, y_2 \tag{19}$$

where $y_1 = 1 - \exp (-k_{01} \; \theta_1)$, $y_2 = 1 - \exp (-k_{02} \; \theta_2)$, being θ_1 and θ_2 the θ function when E_1 or E_2 are respectively used and $\psi = \Delta H /(\Delta H_1 + \Delta H_2)$. ΔH_1 and ΔH_2 are the enthalpies associated to stages 1 and 2 respectively. It should be noticed that ψ is a measure of the excess vacancy mechanism contribution to the return of SRO. Thus, if $\psi = 1$ equilibrium SRO order is attained via and excess vacancy mechanism only, while $\psi = 0$ indicates that ordering thakes place exclusively assisted by equilibrium vacancies. Certainly, under the quenching conditions imposed to the samples used in the present work, both mechanisms are contributing to complete the ordering process. Hence, the transformed fraction for both stages can be expressed as $y = (\alpha - \alpha_{E1})/(\alpha_{T3} - \alpha_{E1})$ and therefore eqn. 19 yields:

$$\alpha = \alpha_{E1}+(\alpha_{T3}-\alpha_{E1})\{\psi[1-\exp(-k_{01}\theta_1)] + (1-\psi)[1-\exp(-k_{02}\theta_2)]\} \tag{20}$$

where α_{T3} is the equilibrium first SRO-parameter when stage 2 goes to completion and α_{E1} is the value retained after quenching.

On the basis of the above considerations, it is now possible to evaluate the variation of the short-range order parameter for quenched and annealed alloys during a non-isothermal scanning calorimetric run. The influence of the aluminum atomic concentration is also determined.

RESULTS AND DISCUSSION

SRO Kinetics

Before the aims of the preceding section could be accomplished, it is necessary to evaluate for all stages activation energies, pre-exponential factors and the corresponding values of T_{E1} and T_{E2} for the alloys in the quenched condition. Values for T_3 and T_{eq} are measured directly in the DSC traces. Similarly for the furnace-cooled alloys, temperatures T_E need to be computed as well as activation energies and pre-exponential factors, while T_0 and T_{eq} are measured on the thermogrmas as in the former case.

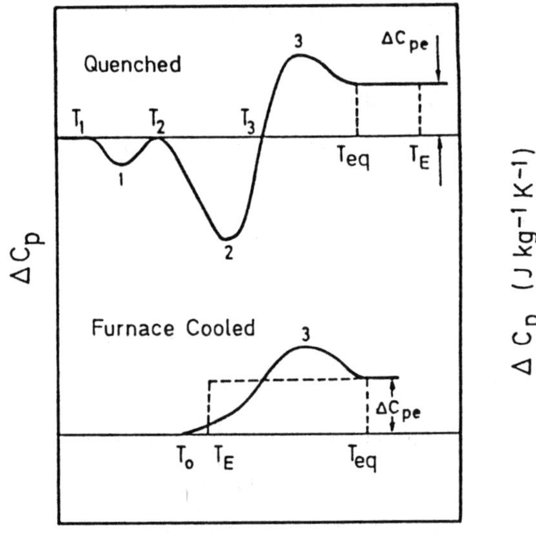

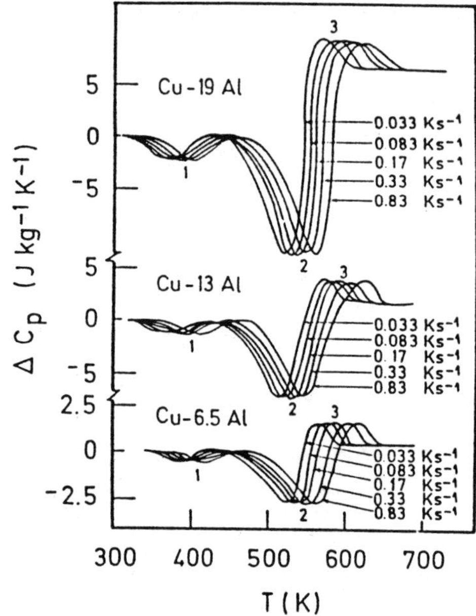

Figure 1. a) DSC thermogram showing a schematic representation of a two stage ordering process followed by a disordering stage in a queched alloy. T_{E1} may be simply specified as the temperature at which the equilibrium degree of order is retained at room temperature; b) DSC trace of a disordering process in a furnace-cooled alloy.

Figure 2. DSC thermograms for αCu-Al alloys quenched from 723 K. Each curve is labeled with the heating rate (K s^{-1}).

Fig. 2 shows the DSC traces for the quenched alloys obtained at different heating rates. It can be observed that the three stages are present and also that the corresponding peak temperatures T_p, become larger as the heating rate is increased. It implies that the transformations are kinetically controlled. The activation energy for each stage was determined from plots of $\ln(\phi/T_p)$ vs $1/T_p$ for the 19, 13, and 6.5 at % Al alloys. These plots of slope $(-E/R)$ resulted in straigth lines. They are not shown in the present work for the sake of brevity.

Table 1 —Activation Energies for the Three Stages Observed in
Quenched αCu–Al Alloys

Material at % Al	E_1 kJ mol^{-1}	E_2 kJ mol^{-1}	E_3 kJ mol^{-1}
19	87.7	148.6	150.1
13	89.6	161.9	159.8
6.5	93.5	180.3	179.1

It can be seen from Table I that activation energies for stage 1, although somewhat larger , are consistent with those for vacancy migration (40), while the corresponding ones to stage 2 agree reasonable well with those for self-diffusion reported in the literature (25, 28). These values decrease with aluminum content, consistently with the decrease in atomic mobility usually found in solid solutions of copper as the solute concentration decreases (31). The above tendency for both activation energies is also consistent with the observed increase of peak temperatures of stages 1 and 2, as the alloys become less concentrated. It can be also noticed that activation energies for disordering are practically the same than those computed for stage 2.   This finding is inconsistent with a diffusion controlled nucleation and growth of ordered particles in a disordered matrix. If it would be so, ordering rates slower than disordering rates would be expected, which is not the case in the present work.

Pre-exponential factors were obtained from plots $\ln[1/1-y)]$ vs θ. These plots,which resulted in straigth lines of slope k_0,are shown in fig. 3.   Such straigth lines confirm that a first order kinetics law can be associated to each stage and consequently the variations of α can be described by a single relaxation time as previously assumed. The results are shown in Table II.

Table II —Values of Pre – Exponential Factors for the Three Stages
Observed in Quenched αCu–Al Alloys

Material at % Al	k_{01} x 10^{10} s^{-1}	k_{02} x 10^{14} s^{-1}	k_{03} x 10^{14} s^{-1}
19	4.6	2.6	2.0
13	6.7	6.0	3.8
6.5	17.0	35	26

By computing the energies involved in the different peaks, values for T_{E1} and T_{E2} can be obtained. Temperatures T_1, T_2, T_3 and T_{eq} are measured on the DSC traces as said above. Boundary values for the short-range order parameters were calculated employing $W = -367$ x 10^3 J mol^{-1}(17, 36).

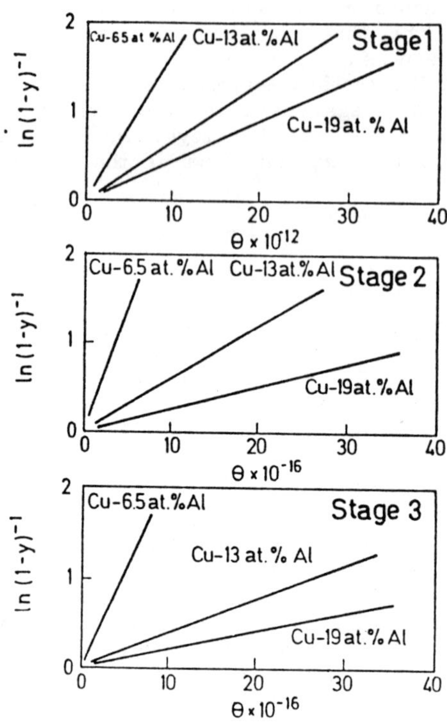

Figure 3.   Plots of $\ln[(1-y)^{-1}]$vs . θ.
For the quenched alloys (a),
(b) and (c) refer to stage
1, 2 and 3 respectively.

With the above data and making
use of eqns .20 and 15, the variation of
the short-range order parameter was
plotted in fig. 4 for the three alloys
assuming that W is independent of the
aluminum content. As stages 1 and 2
were treated simultaneously, the α vs.T
curves have to be continuous at T_2
and T_3 as a physical requirement. It
can be noticed the shift of the curves
to higher temperatures as the aluminum
concentration decreases due to
the decrease in atomic mobility. The
hiperbolas drawn with long dashed
lines represent the equilibrium short-
range order parameter α_e vs T curves,
which decrease drastically as the
alloy become more diluted. It should
be remarked that T_3 and T_{eq}lie on that
curves in all cases. If one confront
the equilibrium short-range order
parameter for the first coordination
sphere of annealed Cu-15 Al(21)with the
value at T_3(= 600 K) for Cu-13 Al one
obtains from fig. 4, -0.17 for the
later, while the reported value for the
first is approximately -0.18 at
the same temperature. For Cu-13 Al at
503 K an equilibrium $\alpha_e$value of -0.142
is reported (41) while for Cu-17 Al at
the same temperature, -0.217 was
estimated (41). This estimate is in
in reasonable agreement with the
value of -0.238 calculated at
T_3 (= 590 K) for Cu-19 Al. Such
comparisons suggest that the value
of W employed was a good choice,
since otherwise the curves would shift
vertically, giving values of α_e at T_3 and T_{eq} in noticeable disagreement with
those already reported. Considerable data at different alloy concentrations
and equilibrium temperatures (36, 41) can be   confronted with the present
results being all together in quite reasonable agreement.

For the furnace-cooled alloys the corresponding thermograms are shown
in fig. 5 at the indicated heating rates. As pointed out before, the single
stage 3 appearing is that corresponding to a disordering process. There
are considerable shifts in the maxima of the rate of transformation curves
to higher temperatures with increasing in heating rate, impliying again that
this process is kinetically controlled. From such traces, boundary values
for the short-range order parameter can   be evaluated employing eqns. 16
and 17, similarly as for the quenched alloys. Activation energies and also
k_0 values were computed as before. All that information is listed in
Table III.

With the above data, the variation of α with temperature is shown in
fig. 6.   That the path of kinetic curves for α are higher than those
corresponding to equilibrium, stems from the fact that during the destruction
of DO on continuous heating, the alloys fail to attain a permanent equilibrium
degree of order until T = T_{eq} .   This feature means that the alloys are in

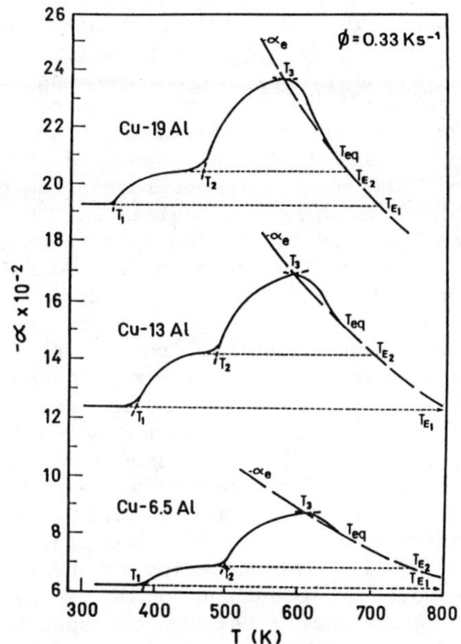

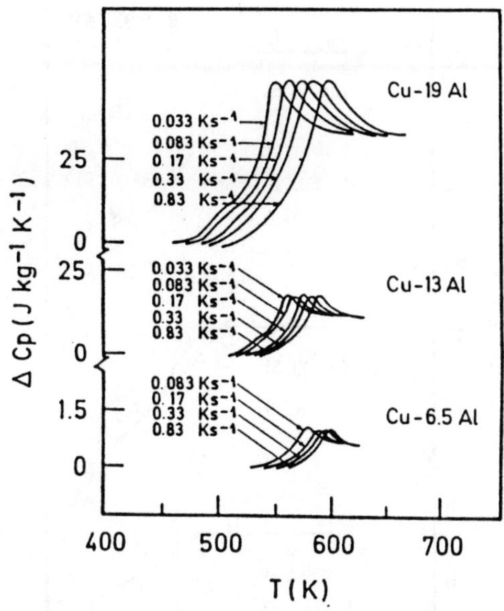

Figure 4. Variation of the short-range order parameter in αCu-Al alloys quenched from 723 K, at a heating rate of 0.33 (K s⁻¹). The hyperbolas represent equilibrium short-range order parameters as a function of temperature.

Figure 5. DSC thermograms for αCu-Al alloys furnace-cooled at 15 (K h⁻¹). Each curve is labeled with the heating rate ϕ(K s⁻¹).

Table III -Values of Activation Energies and Pre-Exponential Factors
for Furnace Cooled Alloys at 15 K h⁻¹

	Units	Cu-19 Al	Cu-13 Al	Cu-6.5 Al
E	J mol⁻¹	156×10^3	151×10^3	171×10^3
k_0	s⁻¹	1.4×10^{14}	4.2×10^{14}	25.0×10^{14}

a state of order higher than that of equilibrium. Furthermore, the kinetic and equilibrium curves intercept each other at temperatures slightly higher than T_E in all cases, which can be naturally attributed to the expected behaviour of eqn. 18 between T_0 and T_{eq}.

Effect of Quenching Conditions

The transformed fraction for the SRO reaction via both non overlaping processes can be also expressed as

$$y = \psi[1 - \exp(-k_{01}\theta_1)] + (1 - \psi)[1 - \exp(-k_{02}\theta_2)] \qquad (21)$$

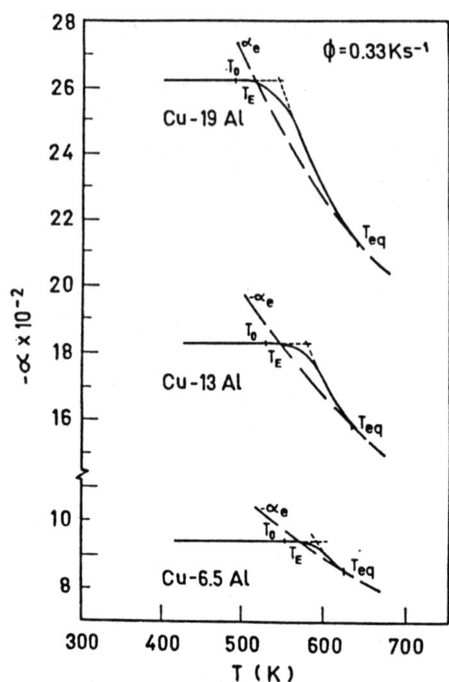

Figure 6. Variation of the SRO – parameter in furnace-cooled αCu-Al alloys (cooling rate 15 (K h^{-1}) at a heating rate of 0.33 (K s^{-1}). The hyperbolas represent equilibrium values of the short–range order parameter as a func – tion of temperature.

As it was pointed out before, ψ is a measure of the dominance of stage 1, which is in turn determined by the quenching conditions, namely quenching temperature, quenching rate and specimen shape. Disc shaped samples of Cu–19 Al quenched from two different temperatures using a high quenching-rate-device($\approx$2x10^{4}Ks^{-1}) were supplied to us. The upper part of fig. 7 shows DSC traces at ϕ=0.33 Ks^{-1} under different quenching conditions . Curve (a) is the result of such a disc sample quenched from 573 K. It can be seen that ψ = 0. In this case equilibrium attainement is reached via an equilibrium vacancy mechanism only. Curve (b) corresponds to one of these disc specimens quenched from 873 K. There is only a single peak corresponding to stage 1 SRO ordering, and ψ = 1. This behavior is observed because the quenching rate is high and additionally because the specimen surface-to-volume ratio (disc) is large. Curve (c) corresponds to samples quenched from 873 K under normal conditions having a plate shape before quenching, which is the case of the present experiments. Although for this sample the quenching temperature is the same as for the disc, the quenching rate is slower and less vacancies are frozen in. As a result, insufficient excess vacancies remain to complete SRO through stage 1.

The lower part of fig. 7 shows the corresponding y against T kinetic curves. It can be seen, consistently with the thermograms, that equilibrium SRO is attained at lower temperatures when only stage 1 is present. In the kinetic path for curve (c) both stages are evidenced. Curve (a) only exhibits stage 2, being the ordering process completed at a higher temperature (equilibrium first SRO-parameter is smaller). The kinetic parameters for curves (a) and (b), determined in the same manner as for curve (c) are, E = 146/88 kJ mol^{-1} and k$_0$ = 2.8 x 10^{14}/7.2 x 10^{10} s^{-1} obeying also a first order kinetic law. These results are not shown here for brevity sake.

Influence of Quenched-In Vacancies

The fractional increase in α during a DSC experiment depends on many factors, which make the above treatment somewhat simplified. As the alloy composition increases in quenched alloys, the excees vacancy concentration, the vacancy mobility, the bound to free vacancy ratio and the degree of SRO also do increase during stage 1. But the increase in SRO degree produces a vacancy concentration and a vacancy mobility decrease effects. Also the bound to free vacancy ratio is expected to become lower. Therfore, the net free vacancy concentration and the effective vacancy mobility are strongly

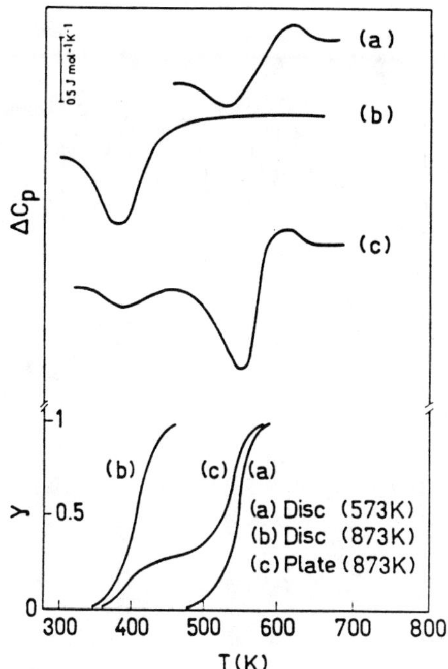

Figure 7. DSC thermograms for Cu-19 at % Al under different
quenching conditions (upper curves) and the
corresponding reacted fractions y (lower curves) .
The quenching temperatures are indicated.   Both
discs were quenched using a high-quenching-rate
device. ϕ = 0.33 K s^{-1}.

dependent on the interplay of all these factors during stage 1.   They are
shown schematically in fig. 8,   The controlling factors for the same material
variables during stage 2 are also shown there.

On the basis of the above flow chart, the fractional increase of the
first SRO-parameter $R_1 = (\alpha_{E2} - \alpha_{E1})/\alpha_{E1}$ during stage 1 is now considered.
As discussed before, there are interplaying simultaneously some effects
in favour and others in opposition to a net relative increase in the first
SRO-parameter. It is then plausible that R_1 goes through an extreme value
for intermediate Al contents.   Actually a maximum was observed for Cu-13
Al, as shown in fig. 9.   In the same figure it can be noticed that the
fractional increase of the first SRO-parameter, $R_2 = (\alpha_{T3} - \alpha_{E2})/\alpha_{E2}$,
corresponding to stage 2 decreases sistematically as the alloy becomes more
concentrated.   If the same above arguments are valid for stage 2, the
observed tendency is indicative that the factors retarding the net relative
increment in R_2 becomes much stronger than in stage 1 with increasing
aluminum content.   This behavior is not unexpected since in stage 2, as
the degree of SRO is larger than is stage 1, both the free vacancy mobility
and vacancy concentration reducing effects would be consequently more
conspicuous.   Thus, extreme values of R_2 might be not present in the whole
range of alloy compositions, as effectively observed.

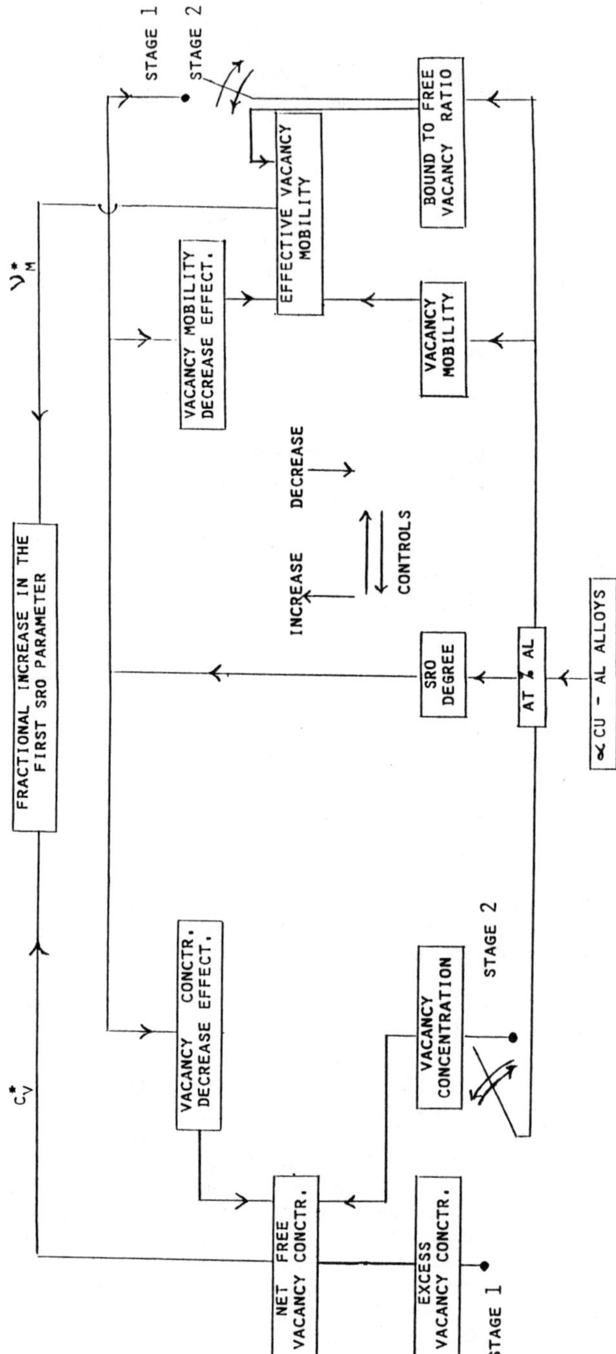

Figure 8. Flow chart illustrating the controlling factors which determine the fractional increase in the first SRO-parameter. c_V^* (effective vacancy concentration). ν_M^* (effective vacancy mobility).

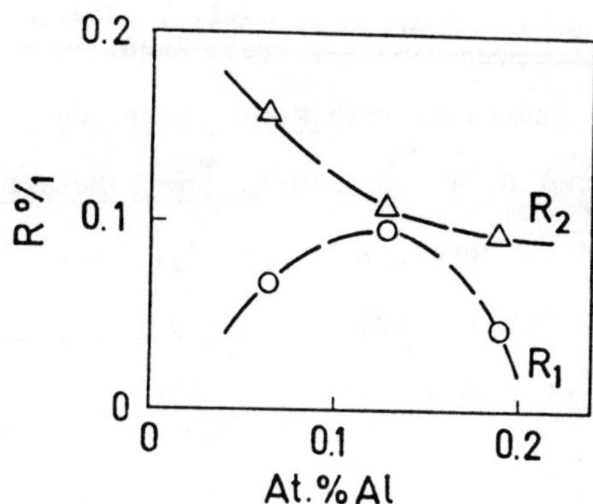

Figure 9. Fractional increase of the first SRO-parameter for stages 1 (R_1) and 2 (R_2) as a function of aluminum content.

CONCLUSIONS

The investigation by differential scanning calorimetry on the ordering process in αCu-Al alloys leads to the following conclusions:

1. Rising temperature experiments are better interpreted in terms of an homogeneous SRO rather than an heterogeneous DO model.

2. Under non-isothermal heating after the used quenching conditions, SRO takes place in two stages, stage 1 at lower temperatures and stage 2 at higher temperatures. The SRO kinetics, until equilibrium is attained, becomes faster as the aluminum content increases. Stage 1 ordering is assisted by the migration of excess vacancies, while stage 2 is assisted by the migration of equilibrium vacancies.

3. The concentration of excess vacancies is mainly controlled by the quenching temperature, the shape of the sample being quenched, the quenching rate and the number of vacancy sinks. These factors, in turn control the dominance of each stage.

4. An expression which describe the overall SRO kinetics for both stages 1 and 2 was developed. The kinetic process can be also evaluated in terms of the first SRO-parameter making use of a method suitable to obtain boundary values. In furnace-cooled alloys only stage 3 is observed.

5. The fractional increase of short-range order during stages 1 and 2 in quenched alloys is a complex interplay of vacancy concentration and vacancy mobility effects which can be rationalized by a flow diagram.

ACKNOWLEDGEMENTS

The authors wish to thank the Fondo Nacional de Desarrollo Científico y Tecnológico (FONDECYT), Project N° 90-0934, and the Departamento Técnico de Investigación (DTI) de la Universidad de Chile, Project I-3110-9012, for financial support. They are also indebted to the Instituto de Investigaciones y Ensayes de Materiales, Facultad de Ciencias Físicas y Matemáticas, Universidad de Chile, also for financial support and for the facilities provided for this research project.

REFERENCES

1. M. S. Wechsler and R. H. Kernohan, J. Phys. Chem. Solids, Vol.7, 1958,307.
2. S. Matsuo and L. M. Clarebrough, Acta Metall., Vol. 11, 1963, 1195.
3. B. Borie and C. J. Sparks Jr., Acta Cryst., Vol. 17, 1964, 827.
4. V. Y. Panin, V. P. Fadin and L. D. Kuznetsova, Phys.Metall.Metalloved., Vol. 19, 1965, 316.
5. V. I. Iveronova, A. A. Katnslelson and G. P. Revkevich, Phys. Metall. Metalloved., Vol. 26, 1968, 106.
6. R. O. Scattergood, S. C. Moss and M. B. Bever, Acta Metall., Vol. 18, 1970, 1087.
7. N. Kuwano, Y. Tomokiyo, C. Kinoshita and T. Eguchi, Trans. Jpn. Inst. Met., Vol. 15, 1974, 338.
8. Y. Kitano and Y. Kimura, J. Phys. Soc. Jpn., Vol. 32, 1972, 1430.
9. J. E. Epperson, P. Furnrohr and C. Ortiz, Acta Cristallogr., Vol. A34, 1978, 667.
10. R. W. Cahn and R. G. Davies, Philos. Mag., Vol. 5, 1960, 1119.
11. W. Gaudig and H. Warlimont, Z. Metallkd., Vol. 60, 1969, 488.
12. Y. Tomokiyo, N. Kuwano and T. Eguchi, J. Phys. Soc. Jpn., Vol. 35, 1973, 618.
13. Y. Tomokiyo, K. Kabu and T. Eguchi, Jpn. Inst. Met., Vol. 15, 1974, 39.
14. A. Varschavsky, M. I. Pérez and T. Löbel, Metall. Trans., Vol. 6A, 1975, 577.
15. W. Gaudig and H. Warlimont, Acta Metall., Vol. 26, 1978, 709.
16. J. M. Popplewell and J. Crane, Metall. Trans., Vol. 2, 1971, 341.
17. C. Kinoshita, Y. Tomokiyo, H. Matsuda and T. Eguchi, Trans. Jpn. Inst. Met., Vol. 14, 1973, 91.
18. M. Zehetbauer, L. Trieb and H. P. Aubauer, Z. Metallkd., Vol. 67, 1976, 431.
19. A. Varschavsky, Mater. Sci. Eng., Vol. 22, 1976, 141.
20. A. Varschavsky and E. Donoso, Mater. Sci. Eng., Vol. 32, 1978, 65.
21. E. Donoso and A. Varschavsky, Mater. Sci. Eng., Vol. 37, 1979, 151.
22. A. Varschavsky and E. Donoso, Mater. Sci. Eng., Vol. 40, 1979, 119.
23. A. Varschavsky and E. Donoso, Mater Sci. Eng. A, Vol. 10, 1988, 231.
24. A. Varschavsky and E. Donoso, Mater. Sci. Eng. A, Vol. 104, 1988, 141.
25. G. Veith, L. Trib, W. Puschl and H. P. Aubauer, Phys. Status Solidi, Vol. 27, 1975, 59.
26. G. Veith, L. Trieb, W. Puschl and H. P. Aubauer, Scr. Metall., Vol. 9, 1975, 737.
27. M. J. S. Wechsler and R. H. Kernohan, Acta Metall., Vol. 7, 1959, 599.
28. L. Trieb and G. Veith, Acta Metall., Vol. 26, 1978, 185.
29. C. R. Brooks and E. E. Stansbury, Acta Metall., Vol. 11, 1963, 1303.
30. Y. Tomokiyo, N. Kuwano and T. Eguchi, Trans. Jpn. Inst. Met., Vol. 16, 1975, 489.
31. A. Varschavsky, Metall. Trans., Vol.13A, 1982, 801.
32. A. Varschavsky and E. Donoso, Metall. Trans., 14A, 1983, 875.
33. A. Varschavsky and E. Donoso, Metall. Trans., 15A, 1984, 1999.
34. A. Varschavsky and E. Donoso, J. Mater. Sci., Vol. 21, 1986, 3873.
35. P. V. Petrenko and A. A. Tatarov, Phys.Metall. Metalloved., Vol. 56, 1983, 507.
36. W. Pfeiler and R. Reihsner, Phys. Status Solidi (a), Vol. 97, 1986, 377.
37. N. Kuwano, I. Ogatta and T. Eguchi, Trans. Jpn. Inst. Met., Vol. 18, 1977, 87.
38. P. A. Flinn, Phys. Rev., Vol. 104, 1956, 350.
39. C. Sandu and R. Singh, Thermochim. Acta, Vol. 159, 1990. 267.
40. C. Y. Li and A. S. Novick, Phys. Rev., Vol. 103, 1956, 294.
41. V. S. Zubchenko, N. P. Kulish, P. V. Petrenko, S. P. Repetsky and A. A. Tatarov, Phys. Metall. Metalloved., Vol. 50, 1980, 94.

Hot workability and thermo-mechanical processing of copper alloys

H.J. McQueen
Mechanical Engineering, Concordia University, Montreal, Quebec, Canada

ABSTRACT

The mechanical behavior of Cu alloys in the regime above 0.5 Tm (400-1000° C) at 10^{-3} to 10^2 s^{-1} is reviewed. The evolution of the flow curve, of the strain hardening rate and of the microstructure are analysed together. The dependence of flow stress, and ductility on temperature and strain rate are fitted to constitutive equations. The relative importance of dynamic recovery and of dynamic recrystallization are examined for different levels of solute and particle distribution. With respect to multistage processing, the softening between stages is studied to see both the influence of deformation history and the effect on subsequent stages. The combined influence of preheating, straining, cooling rate and aging (TMP) on product properties are reviewed.

INTRODUCTION

The hot workability of pure copper places little constraint on hot forming since its flow stress is much reduced and its ductility augmented above 450°C ($0.6T_m$, melting K) which is not severely damaging to steel tooling. Most of the Cu production is hot rolled or extruded (bar and tube) with the objective of additional cold processing; thus its ability to statically recrystallize (SRX) during air cooling is very beneficial. Simpler Cu alloys share many of these characteristics, but more complicated alloys depart from them to varying degrees. The examination of such behavior and its explanation in terms of the atomistic mechanisms is the objective of this paper.

For the range 0.01-100 s^{-1}, the flow curves both stress-strain (σ–ε) and θ-σ ($\theta = d\sigma/d\varepsilon$) are described, the former being marked by a peak (σ_p, ε_p) followed by softening to a steady state regime. The dependence of σ_p on temperature T and strain rate $\dot{\varepsilon}$ for various Cu alloys is reviewed. The dynamic softening mechanisms recovery (DRV) and recrystallization (DRX) are explained and shown to have a strong role in defining ductility of alloys in which solute or precipitates retard their operation. The occurrence of static recovery (SRV) or recrystallization (SRX) between stages of forming provide opportunities for manipulating microstructures and working forces. The potential for preserving hot-work subgrains and grain structures in association with control of phase distribution to achieve specific product properties leads to thermomechanical processing (TMP). Many aspects of the above phenomena are common to many metals and as a consequence have been described thoroughly in many recent reviews [1-29]. Such common behavior is outlined first in a brief form following the order set out above. Since hot workability requires high strains and constant strain rates, compression and torsion testing are generally employed with the latter being outstanding for simulating multistage deformation [6,9,10,26,27].

GENERAL HOT WORKING BEHAVIOR

At elevated temperature as at low, the principal deformation mechanism is generation and displacement of dislocations. However, through cross-slip, climb and annihilation (DRV), the rate of strain hardening is much reduced so that σ - ε curves rise more slowly and, as for Al, may reach a plateau in which the dislocation density becomes dynamically stable [4-9,18-21]. Before such a steady state is reached in metals of low stacking fault energy (SFE) such as Cu, Ni and γ Fe, the substructure becomes sufficiently dense to cause nucleation of DRX (at ε_c, σ_c) [1-9,15-18,22-25]. The additional softening leads first to a peak (at ε_p, σ_p) and then to work softening which leads into a steady state regime ($\varepsilon > \varepsilon_s$) (Figures 1 and 2). There, the flow stress σ_s remains constant as a result of repeated waves of DRX which maintain the grains constant in size and equiaxed [1-9]. At low $\dot{\varepsilon}$, the flow curves may exhibit several peaks if the waves of DRX do not overlap [3,17,30,31]. As T rises and $\dot{\varepsilon}$ falls, the strain hardening decreases; yet ε_p and σ_p decrease, as well as the steady state stress σ_s (and ε_s).

When the rate of strain hardening ($\theta = d\sigma/d\varepsilon$), is plotted against σ (Figure 3), the curves decline linearly from an initial athermal rate as a result of dislocation tangling. As subgrains gradually form throughout the grains (starting at the grain boundaries GB), the line curves to a second linear segment associated with additional dislocation storage in the cell walls [8,21,22]. At σ_c, the curve deflects downwards to reach $\theta = 0$ at σ_p. The saturation stress σ_s^* due to recovery alone can be determined by extrapolation of the linear segment to $\theta = 0$. The difference in σ_s and σ_s^* represents the softening due to DRX. As T rises or $\dot{\varepsilon}$ falls, the curves descend more steeply so that θ is less and all of the stresses defined above move to lower values [8,21,22].

The T and $\dot{\varepsilon}$ dependence of σ_p (or of some other specific stress) has commonly been described by three functions [2,11,22,25,30]

$$A\,(\sinh \alpha\sigma)^n = \dot{\varepsilon}\exp(Q_{HW}/RT) = Z \tag{1}$$

$$A'\exp \beta\sigma = Z \tag{2}$$

$$A''\,\sigma^{n'} = Z \tag{3}$$

where A, A', A'', α, n, n', $\beta(=\alpha n)$, R (= 8.31 J/mol K) and Q_{HW} are constants. The sinh term is suitable over a broad range (Figure 4) since for $\alpha\sigma \ll 0.8$ it becomes approximetly the same as the power law (Equation 3) which has proven suitable for creep (n ' ≈ 4.5). The activation energy Q_{HW} is usually about 20% higher than that for creep where DRV prevails [6,7,26,28,29,38-40]. Q_{HW} usually increases with solute and precipitate levels. The θ-σ analysis was proposed to represent a DRV mechanism which applies across a broad T range by changing character. The activation enthalpy rises with rising T to become constant at a value similar to that from the sinh-Arrhenius analysis [22,32].

Microscopic examination of deformed specimens reveal that the original grains elongate and are still visible into the work softening domain. However, before the peak, new grains appear at the GB and gradually consumes the old by repeatedly forming necklaces along the periphery of the old grains [1-5,7,8,15,16,30,31]. Continuing into the steady state regime, the grains reach a stable size D_S dependent on Z or σ (Figure 5) as shown in the equations below. Transmission election microscopy (TEM) reveals the progress of substructure formation already described in the θ-σ analysis. Substructure also forms in the new grains and reaches an average stable size d_s which is larger than that present near the peak (Figure 6) [7,8,15,16,22,23,25]. The subgrains are more recovered and larger as Z decreases according to the following equations:

$$d_s = a + b \log Z \quad \text{or} \quad D_S = a' + b' \log Z \tag{4}$$

$$\sigma = e + f\, d_s^{-1} \quad \text{or} \quad \sigma = e' + f'\, D_S^{-q} \tag{5}$$

where a, b, a ', b ', e, f, e ', f ' and q(≈ 0.8) are constants. At low T, the subgrains are usually elongated [23,25]. The room temperature yield strength σ_y was also found to depend on the hot work substructure [7,9,28,29]

$$\sigma_y = \sigma_0 + k'\, d_s^{-1} \tag{6}$$

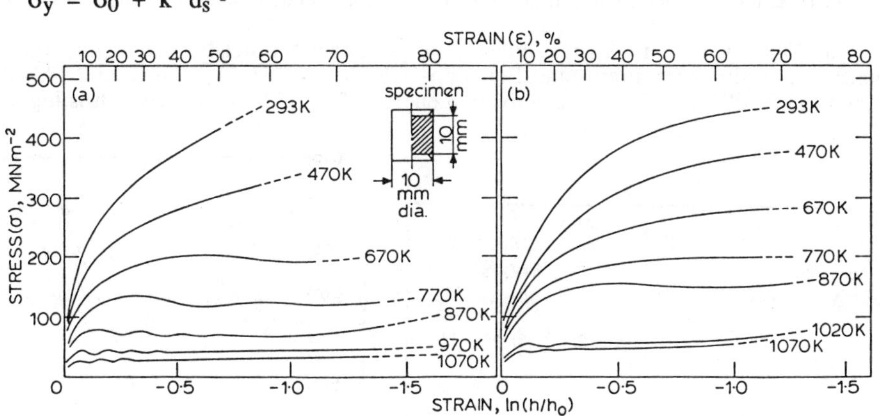

Fig. 1. Compression stress-strain curves for Cu at rates of: a) 8.4×10^{-4} s^{-1}; b) 8.4×10^{-2} s^{-1}. After Korbel et al. [40].

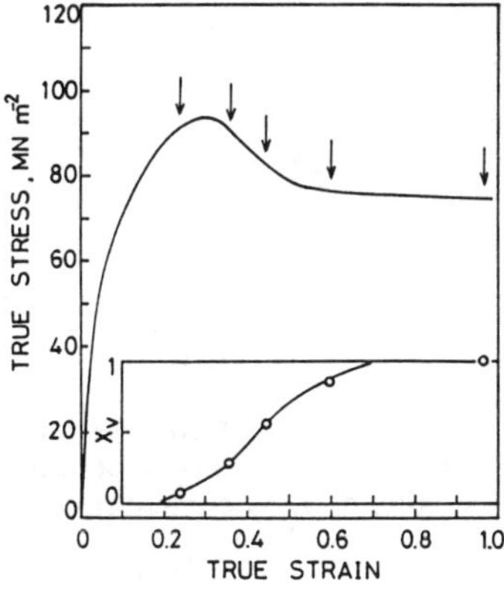

Fig. 2. Progress determined microscopically of dynamic recrystallization X_v along the flow curve is illustrated (500°C, 2×10^{-3} s^{-1}). After Blaz et al. [39].

where σ_0 is the strength without substructure and k ' is the strengthening coefficient. The exponent of -1 differs from -0.5 because the strength of the walls rises as d_s declines [7,9,29]. The dependence on DRX grain size is similar, differing from the Hall-Petch relation because the substructure in the grains becomes denser.

The ductility at high temperatures is strongly affected by the occurrence of GB sliding although at hot-work $\dot{\varepsilon}$, it contributes only 1 or 2% of the total. Differential sliding, occurring on GB at different orientations to the tensile axis, leads to stress concentration and cracking at triple junctions [1-5,7,9,13,14,18,22-25]. Such cracking at warm T (about 0.4 T_m) may result in a drop in ductility as T rises. When DRX becomes operative at higher T, the ductility rises dramatically. The migration of the GB reduces stress concentrations and leaves the cracks isolated and unable to propagate. The ductility improves as DRX speeds up with rising T and $\dot{\varepsilon}$. However, raising $\dot{\varepsilon}$ reduces the proportion of GB sliding but raises the stress which tends to increase stress concentration and delays DRX nucleation [4,9,15,24]. The result is that ductility usually passes through a maximum as $\dot{\varepsilon}$ rises; the maximum occurs at lower $\dot{\varepsilon}$ as alloying retards GB migration.

FLOW CURVES AND DYNAMIC RECRYSTALIZATION IN COPPER

Flow curves of Cu with peaks, work softening and steady state regimes (most clearly exhibited in torsion) were related to the formation of DRX grains by Tegart and colleagues [1,2,33,34] and by others more recently (Figures 1 and 2) [7,8,35-41]. The decline in rate of strain hardening as T rises and $\dot{\varepsilon}$ diminishes (Figure 3) was related to the increase in the level of recovery which made itself apparent in the formation of subgrains which change from small, rough and possibly elongated at high Z to large, neat and equiaxed at low Z [34,35,40-47]. As a function of strain, subgrains form first near the grain boundaries and develop higher misorientation [34,42,48,49]. At high $\dot{\varepsilon}$ there is clear evidence of nucleation with a new distribution of fine grains whereas at low $\dot{\varepsilon}$, there is much strain induced GB migration and no change in distribution [50].

In unalloyed Cu, the flow stress either at the peak or in the steady state regime follows the common relationships with strain rate: Equation 1 (Figure 4) [51-53], Equation 2 [42,54] and Equation 3 [35-37,39,41,55,56]. The activation energy of 261-314 kJ/mol, which is higher than

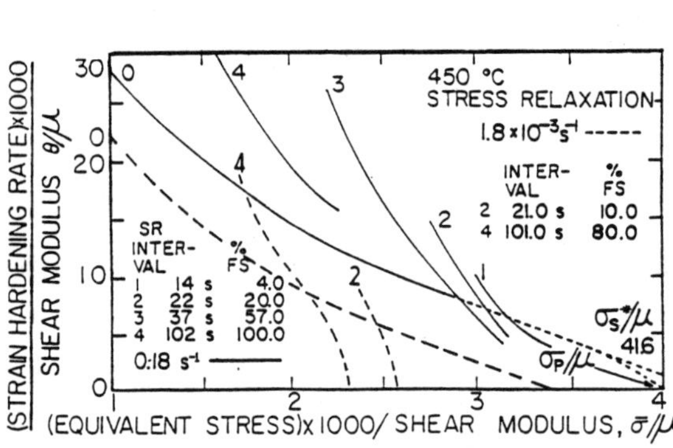

Fig. 3. Plot of normalized $\theta(= d\sigma/d\varepsilon)$ and stress showing the increase in dynamic recovery as $\dot{\varepsilon}$ decreases curve closer to origin. Additional lines are shown for reloading after $\varepsilon_1 = 0.15$ and variable delays. After McQueen and Vasquez [47].

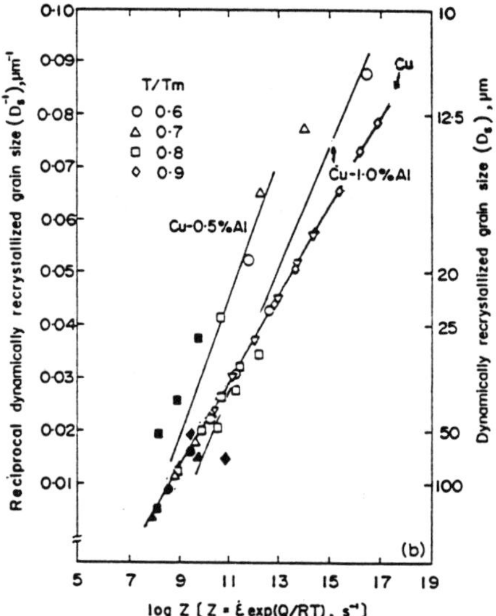

Fig. 5. The reciprocal grain size is related to log Z by Equation 4. For a given condition, the grains are smaller as alloy content rises; the lines are shifted because of changes in Q_{HW}. After Ueki et al. [14].

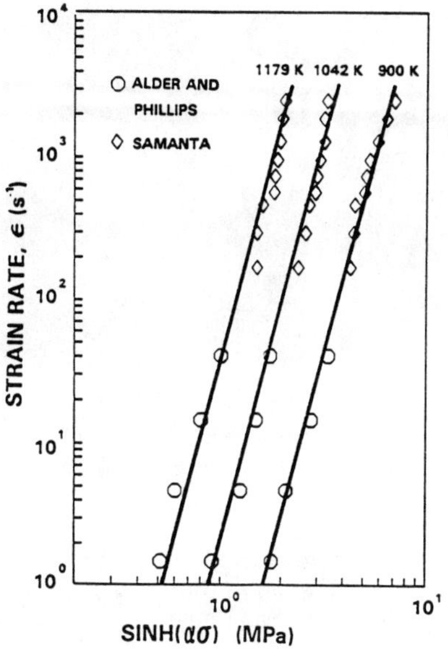

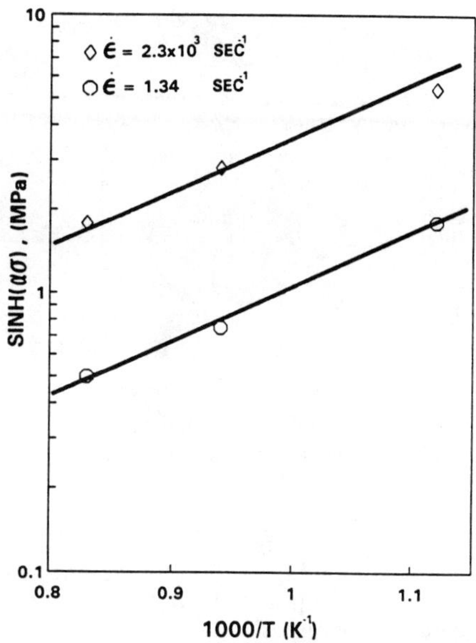

Fig. 4. The hot compression data was analyzed by means of Equation 1; the conformity to straight lines shows that it expresses the $\dot{\epsilon}$ and T dependences suitably. After Samanta [51,52].

TABLE 1. CONSTITUTIVE CONSTANTS FOR Cu ALLOYS

Alloy	Eqn.	α	n	β	n '	Q	Reference	
		MPa^{-1}		MPa^{-1}		kJ/mol		
Cu99.9	1	0.028	5.9			314	Samanta	52,53
Cu99.99	1	0.028	5.2			301	Sellars, Tegart	51
Cu99.99	1	0.016	8.2			340	Blaz	43
	(steady state)					261	Korbel	
Cu99.99	3				5-7	237	Sample	70
Cu99.99	3				6.5	227	Ohtakara	55
Cu	3	(creep)			4.8		Barrett	--
	3	(creep)			6.5		Feltham	--
Cu99.99	3				4.3	197	Ueki	41
0.5Al	3				4.3	227	Horie	--
1.0Al	3				4.3	258	Nakamura	--
Cu99.97	3				5.0		Bromley	35
1.0Al	3				4.7	200	Sellars	--
3.7Al	3				4.3			
7.6Al	3				4.1			
Cu99.99	3				6.8	261	Blaz	39
Cu30Ni	2					301	Evans	42
Cu	(diffusion)						Jones	
Cu34Ni	2			0.118		400	Livesey	54
42Ni	2			0.083		±20	Sellars	
74Ni	2			0.074				

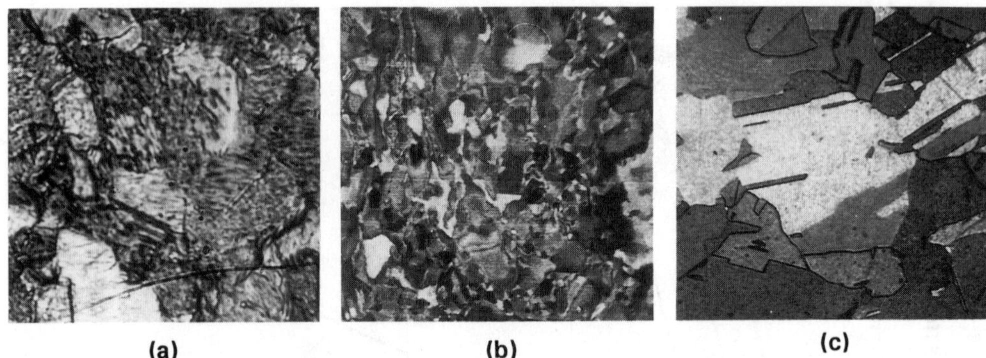

(a)   (b)   (c)

Fig. 6. Dynamically recrystallized grains in Cu deformed in hot torsion ($800°C$, 7 s^{-1}) observed by: a) optical microscopy and b) SEM in channeling mode contact (x1000); c) SRX after 5s (x400). After McQueen [46].

that for recovery, creep or grain boundary migration [1,35] is considered indicative of dynamic recrystallization [1,7,37,50,56]. The hot yield strength has been found to follow an ordinary Hall-Petch relationship with the initial grain size [57]. Similar behavior was found in Cu-Ni alloys but disappeared at high T [42].

In the steady state regime, the grain structure is almost equiaxed and remains stable (Figures 5,6) [35,41,42,45,46,49]. The dimensions are related to Z as in Equation 4 [45] and to stress by Equation 5 with a power of 1.0 for Cu [1,36,43] or 0.75 for Cu-0.8A1 [35] and 0.6 for Cu-15Zn and Cu-30Zn [40,50]. The grains contain a substructure which is not as well formed as in Al or in Ni, and which increases in size and degree of polygonization as Z decreases (Figures 7,8) [34,37,38,42,43,45-49,58]. Equation 5 suitably relates σ_p to d_s [42,47,59,60]. There is little difference in the principal features of the substructure before and after the stress maximum except for the appearance of the occasional nucleus [1,37,38,45,46]; in a contrasting report, the subgrains change from elongated to equiaxed [41].

Since D_s depends only on Z (Figure 5), the degree of grain refinement depends on D_0. As for other metals [8,16,17], refinement ($D_s < D_0/2$) is associated with a single peak flow curve and coarsening with multiple peaks [39]. This was obeyed well at small D_0 with nucleation only at GB (not at twin boundaries), but not so well at high D_0 because of nucleation at deformation bands [16,39,61] which are commonly observed as sites [61-65]. In multiple peaks, the flow softening parts of the cycles are related to the growth of new grains which consume only part of the matrix; their growth ceases because the internal dislocation density approaches that of the remaining matrix. In grain refinement, the new grains form as repeated necklaces near the interfaces of old and new grains [31]. If one considers each necklace as a cycle, they are so closely spaced in time ($t_c = \varepsilon_c /\dot{\varepsilon}$) that the flow curve is smooth; it becomes cyclic when these cycles of nucleation are separated. When the necklaces of new grains completely replace the old grains, this first wave of DRX causes work softening to the steady state plateau [16,23,39] during which new DRX grains are continually replacing deformed ones [1-4,7,8,17].

The observation of DRX in Cu single crystals is significant in that the nucleation must originate from the substructure alone and not in association with GB as in polycrystals [28,66-70]. The critical condition is σ_c which is much higher than for polycrystals under the same conditions; ε_c varies for specimens of the same batch and orientation (Figure 9). For the orientation with the most recovered uniform substructure, σ_c is the highest [15,66-69]. Holding crystals for periods just below the critical stress does not induce DRX which indicates a critical local dislocation configuration must be attained. TEM examination failed to discover any nuclei but did show an increased heterogeneity in cell size and misorientation. The stress dropped and the DRX region grew rapidly with multiple annealing twins. With continued straining, nucleation occurred at very low stresses in regions adjacent to the recrystallized grains [15,70]. In unstable crystals, deformation bands of high shear and misorientation formed and served as sites for DRX nucleation as has been observed for SRX in annealing of such crystals after cold working [63,71,72].

SOLID SOLUTION AND PRECIPITATION STRENGTHENING

Solid solution alloying with Zn, Sn, Al or Ni results in flow curves with the characteristics arising from DRX. Solutes raise the hot strength partly due to the lowering of the SFE which restricts recovery (clearly shown as Zn rises from 2.5 to 36% in Figure 8 [40]), and partly due to the reduction in mobility of the boundaries which slows dynamic recrystallization [10,42,50,71]. Brass and cupronickel quite clearly have a much less polygonized substructure than Cu when similarly deformed [10,40,42,50] and when the brass is rolled there is much evidence of deformation and shear bands as at room temperature [10,40] In Cu - 30Ni, σ_p increases inversely as D_S (Equation 5) in similarity to previous tests on Cu [42,59,60]. For this alloy, Q_{HW} was 301 kJ/mol slightly higher than pure Cu and higher than that for diffusion 243 kJ/mol. In plane strain compression at 1.0 s^{-1} and 10 s^{-1}, commercial alloys with Ni contents between 34 and 68% Ni with about 0.2Fe and 1.1Mn exhibited flow curves with peaks (≈ 0.3) and softening towards steady state at $\varepsilon = 2.7$, the degree of softening being enhanced more at high ε and low T by deformation heating [53]. The behavior was represented by Equation 2 with β decreasing from 0.118 to 0.074 MPa^{-1} but with the common Q_{HW} value of 400 ± 20 kJ/mol. This constitutive equation was employed to successfully predict extrusion pressures. The DRX grains increased from ~ 6μm at 700ºC to ~ 20 μm at 1000ºC and contained rather poorly formed subgrains [53].

In Cu-Al alloys, the flow stresses and peak strains rise as the A1 content is increased to 7.6% [35,61]. The rate of strain hardening is substantially higher in the alloys so that when compared at equal peak stress (different Z), the peak strain was less in the alloy. The primary effect of Al is to impede dislocations by lowering SFE with little effect on GB as indicated by similar strains ε_x of work softening in all alloys [35]; D_S follows Equation 4 and is smaller for higher Al content [41,61]. The mechanical parameters follow Equation 3 with stress exponent n varying from 5 to 4.1 as concentration increases but with a common activation energy of 200 kJ/mol [35]. In contrast, Q_{HW} appears to rise from 200 kJ/mol = Q for diffusion to 227 and 258 kJ/mol for 0.5 and 1% Al respectively; moreover, ε_x rises with ε_p and σ_p [41]. An activation energy that is similar to that for creep even though there is DRX and is not dependent on solute content is in strong contrast to Cu-Zn, Cu-Ni and Ni-Fe alloys, superalloys and stainless steels [2,8,11,22,23,43].

In aluminum bronze (Cu-9.9Al-3.6Fe-1.6Mn), DRX occurs when α phase is present in significant quantity; whereas when there are only $\beta + \kappa$ phases, DRV alone takes place (Figure 10) [73]. The β phase is sufficiently stable that the elongated structure is preserved even during annealing for 2 h at 900ºC. When substantial α phase is present, a high dislocation density builds up in it inducing DRX and with further ε, also in the β phase because of slip incompatibility. Moreover, when DRX occurs in both phases, the resulting spheroidized structure exhibits a great reduction in flow stress possibly due to superplastic mechanisms. In addition, SRX may take place during annealing when α phase was present during deformation [73].

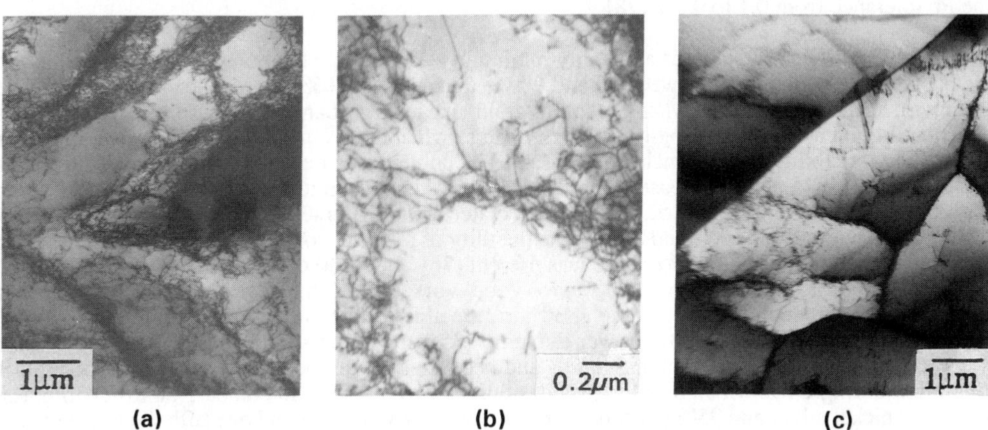

Fig. 7. The substructures observed in Cu deformed at 450°C are denser for: a) 1.8x10^{-1} s^{-1} than for b) 1.8x10^{-3} s^{-1}. c) After peak at 950°C, 4 s^{-1} [45]. After Vazquez et al. [60].

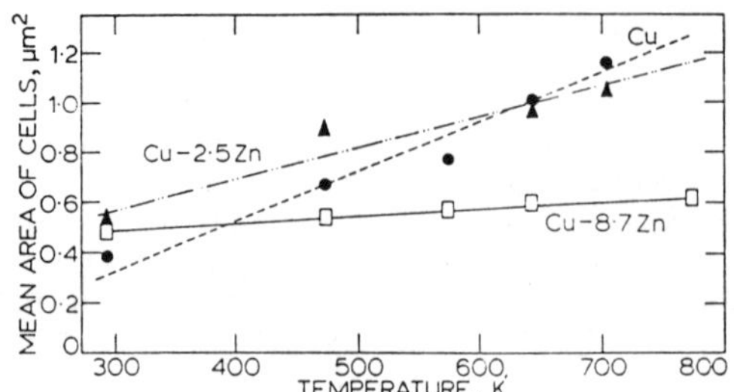

Fig. 8. Mean area of cells measured by TEM rises most rapidly with the metal of highest SFE: Cu, 72ergs/cm^{-2}; Cu-2.5Zn, 65ergs/cm^2; and Cu-8.7Zn, 45ergs/cm^{-2}. After Korbel et al. [40].

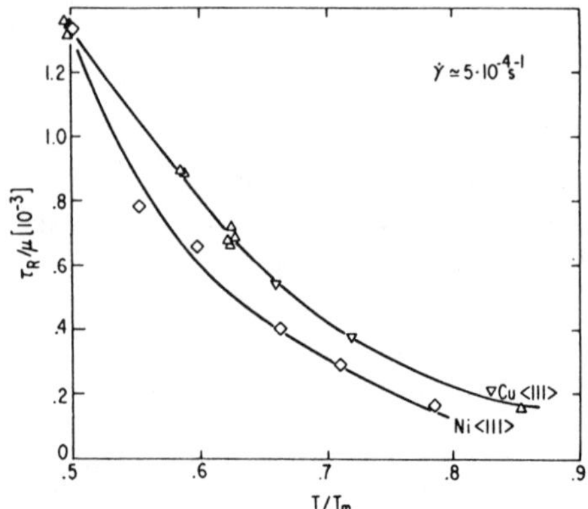

Fig. 9. In single crystals, the critical stress for DRX decreases as T rises indicating that a critical dislocation configuration becomes less dense. After Gottstein and Kocks [68].

HOT DUCTILITY

OFHC copper ruptures at ε_f through fissuration starting from grain boundary pores when T rises above 250°C; however, pore formation seems to be much slower than in nickel [2,5,7,9,13,14,74]. As T mounts above 500°C, the ductility rises rapidly as DRX occurs and becomes more rapid (Figure 11); the migration of GB isolates the pores and prevents further growth [1-4,7,9,13,14,18,20,24,33,35,41,74,75,76]. The depth and width of the ductility minimum decreases as ε rises from 0.0001 to 0.1 s^{-1} as a result of reduced GB sliding and augmented DRX; single crystals do not exhibit the ε_f minimum [74]. Tough pitch Cu and internally oxidized Cu-0.1Al ($\approx$ 50nm) have reduced ductility since the oxide particles impede recrystallization [74,75]. The ductility of Cu and brass decreases as ε is reduced and grain size increases [50]. Because of GB cracking at triple junctions the alloys usually exhibit a minimum at intermediate temperatures, which is not usually observed in Cu itself [50,75]. In coarse dual phase $\alpha + \beta$ brass, the ductility is good at low temperatures but rises very little with temperature in comparison to single phase α alloys [75]. By contrast β brass which is bcc, has poor low T ductility but increases rapidly with temperature due to greatly augmented DRV [75,77]. The hot ductility of complex brasses (60Cu-xxZn-1.5Al-yMn-zFe-wPb, y = 0.5-1.5, z = 0.1-1.3, w = 0.1-0.8) was raised as Fe content mounted above 0.8% (Figure 12) because particles of Fe-Mn formed in the melt-refined the as-cast grain size and raised the Pb tolerance from 0.1 to 0.8% [78].

In Cu-30Ni alloy, there is a ductility minimum at 625°C due to grain boundary cracking; the cracks are mainly on facets between 45 and 90° to tensile axis (10% had MnS particles) [42]. The minimum T and ε_f are lowest when the initial grain size is largest and the strain rate is lowest [42]. The ductility increases more rapidly with temperature when the grain size and strain rate are greater [42]. The workability of cupronickel decreased as Ni content was increased from 10 to 30%, and as Fe, Si, and Bi (~ 15 ppm) increased, although Mn and Al concentrations had little effect [79]; there was a ductility minimum from 527-727°C associated with GB cracking [79]. In leaded alloys of Cu-Ni-Zn-Mn tested at 750°C and 800°C, the ductility of material containing 72% β + 28% α phase was higher than that when only α phase was present [76]. In 70-30 α brass, the Pb content must be held below 0.02%, and Bi below 0.002%, for good workability [36,80]. By contrast both β brass and $\alpha + \beta$ brass, which is most ductile at 30% α, are able to tolerate up to 1% Pb [36,80]. These differences arise both because the lower dynamic recovery in the single phase α leads to greater stress concentration (hence cracking) at GB and because the presence of Pb particles impedes GB migration during DRX which usually provides ductility to α brass. In comparison to 70-30 brass, 60-20-20 nickel silver and 95-5 phosphor bronze have about one half and one fifth the tolerance for Pb and βi. The tolerance of brass could be increased tenfold by additions of Zr or U to equal 1.5 (% Pb + 5x % βi). Nickel silver had a fivefold increase from additions of misch metal but bronze was not improved by such additions [80].

STATIC RECRYSTALLIZATION

After hot working, recrystallization in Cu commonly takes place very rapidly, as compared to Al, unless effective quenching procedures are employed [10]; time for 50% SRX rose from 10 s at 810°C to 500 s at 540°C [55]. The time dependencies of restoration determined mechanically (including recovery) and microscopically exhibit S curves (Figure 13) and obey classical Avrami kinetics [47,58,81-83]. In specimens annealed at 800°C for 30 min after hot working under various conditions, the grain size was smaller when the samples were deformed at lower temperatures, at higher strain rates and to higher strains up to ε_s (about 60%) [43]. In a group of specimens deformed to the same flow stress (different combinations of T and $\dot{\varepsilon}$), the statically recrystallized grain size decreased as $\dot{\varepsilon}$ was increased and as the strain increased into steady state regime [57]. In commercial alloys of 34 to 68% Ni deformed at 1 or 10 s^{-1} to strains of 2.7, SRX was complete in 1 sec at 900 and 1000°C but only partially at 800 and 700°C [53]. When the strain was 0.5, SRX was 40-70% at 1000°C, 30-50% at 900°C and 15-30% at 800°C when cooled in 1 sec. The SRX grains were larger than the DRX ones from which they came [53].

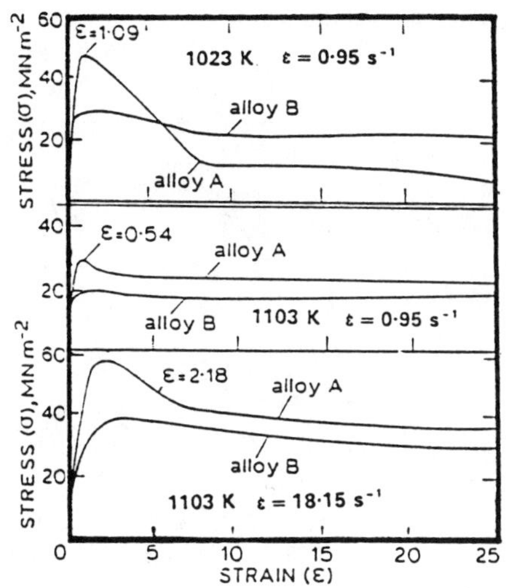

Fig. 10. Flow curves of Al-bronze exhibit a peak due to DRX when 34% α phase is present along with β (alloy A) but a simple plateau when there is only β phase which undergoes a high level of DRV (alloy B). After Granostajki and Ziemba [73].

Fig. 11. The ductility of pure Cu (a) shows a minimum between 400 and 525°C due to grain boundary sliding and cracking with the rise at higher T due to DRX. Sliding is reduced relatively at higher $\dot{\varepsilon}$ and DRX is speeded up. The presence of Al$_2$O$_3$ particles enhances cracking but inhibits DRX to eliminate the high T improvement. After Davies et al. [74]. The ductility of Cu Ni Zn Mg duplex alloy (b) is enhanced by the presence of β phase. After Ward and Helliwell [76].

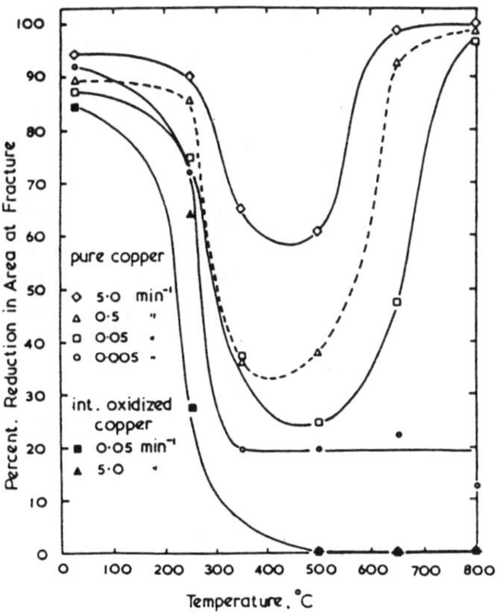

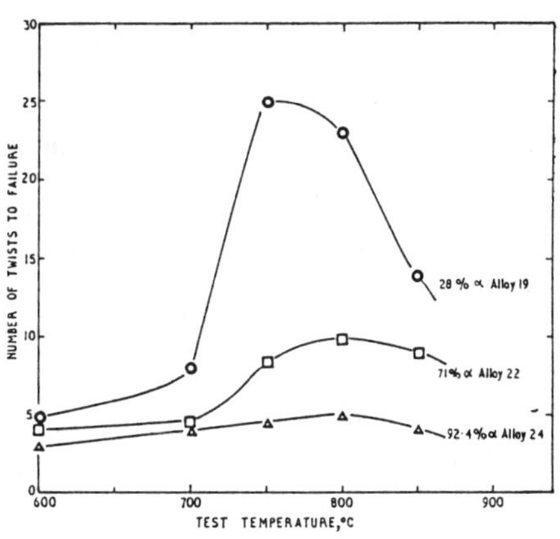

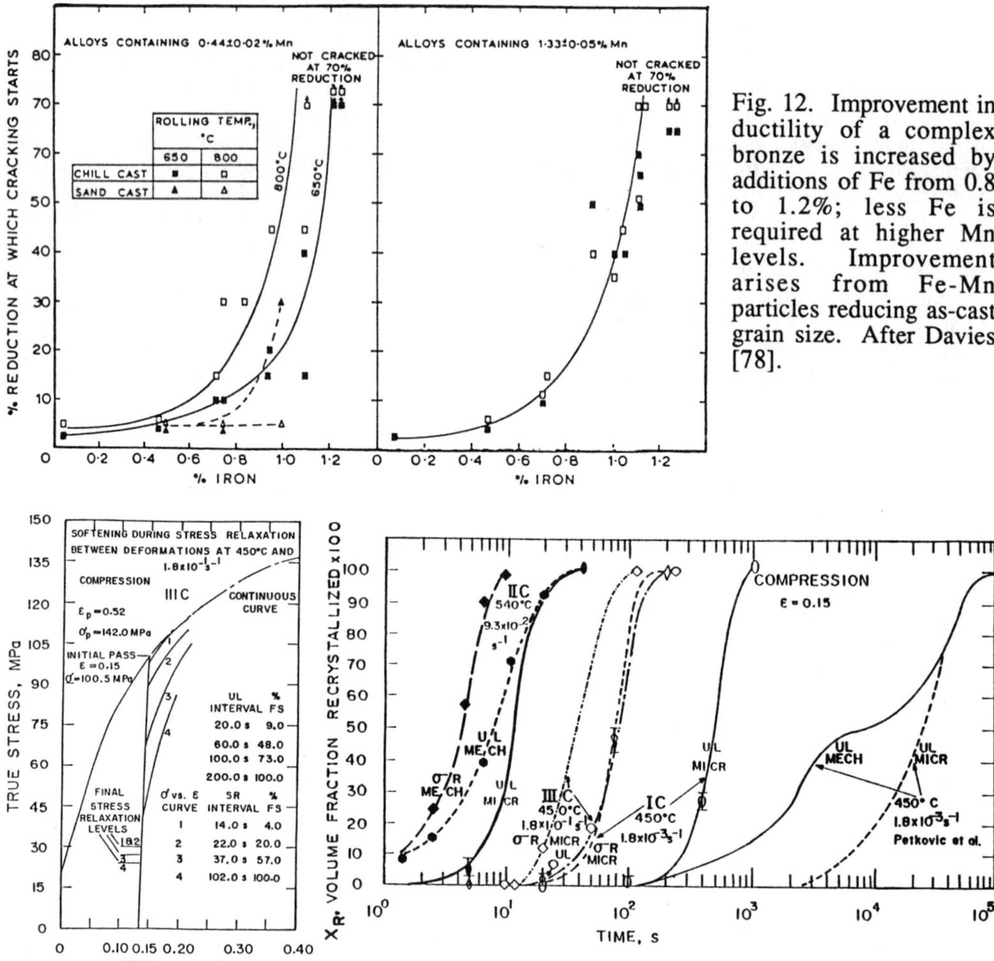

Fig. 12. Improvement in ductility of a complex bronze is increased by additions of Fe from 0.8 to 1.2%; less Fe is required at higher Mn levels. Improvement arises from Fe-Mn particles reducing as-cast grain size. After Davies [78].

Figure 13.   Duplex stress strain curves for measuring fractional softening (= σ_m-σ_R/σ_m-σ_o) after different periods of stress relaxation. Evolution of fraction recrystallized (microscopy) and of fractional softening (mechanical) for both unloading UL and stress relaxation σR for OFHC copper. After Vazquez et al. [47,58].

A precise study of static restoration kinetics in tough pitch Cu has shown that in specimens deformed less than 0.12, fractional softening of the order of 35% arises from static recovery and proceeds to saturation [84]. This occurs in two stages: first the relocation of dislocations into the sub-boundaries and then the annihilation and rearrangement within them [15,21,85]. For strains greater than 0.12 but less than the peak strain (~ 0.4), classical static recrystallization occurs whereas beyond the peak, the mechanism gradually becomes metadynamic, i.e., the continued growth of dynamic nuclei takes place without a prior incubation period [7,8,15,16,23,45,84]. At low strain rates (~ 10^{-2} s^{-1}), the new metadynamic grains appear mainly at the grain boundaries and do not completely replace the dynamic ones [84]. The metadynamic grains are usually larger than the DRX ones [8,45,46]; the density of growing nuclei is fixed since no additional dynamic nuclei are forming as they would if deformation was continuing [45]. Similar studies on OFHC copper showed SRX to be much faster in the absence of the oxide particles [47,58,81-83].

Static restoration has also been studied under conditions where the specimen is not unloaded; softening took place during stress relaxation and followed two regimes. Under stress relaxation, after high prestrain conditions, the rate of recovery appears to be reduced in the initial stages before the stress declines substantially (Figure 13) [47,58,81-83]. After low prestrain, the recovery progresses to a saturation level well before recrystallization starts; however, after high prestrain, SRV occurs simultaneously with the recrystallization which has a much shorter incubation period. When the interruption strain is high, the recrystallization is enhanced relative to that upon unloading [47,58,81,82], because the strain energy of the substructure has been less degraded by static recovery [58,82]. After low interruption strain, the rates of recrystallization are about the same when unloaded or load relaxed because the recovery has reached almost similar levels. In addition, the remaining stress is very low possibly enhancing recovery [47,58,81]. In another study, the progress of stress relaxation was used to determine the extent of softening by SRV [77]. An inflection in the curve marked the onset of SRX and then provided a measure of its progress. With increasing $\dot{\varepsilon}$, the fractional relaxation attributed to SRV rises to a maximum where SRX appears and then declines as the latter provides up to 100% softening. After the peak, softening during relaxation is attributed solely to MRX. For fine grained materially rising T raised SRV softening but induced SRX and MRX at lower strains. At constant T, rising grain size reduced SRV softening and delayed SRX and MRX to higher strains. While this technique agrees in part with the restoration during unloading [57,84,85], it does not appear to be as sensitive. In comparison to the relaxation studies above [47,58,81-83] the degree of softening is not determined by unloading, consequently being a much less precise measure.

In a simulation of deformation during continuous cooling of Cu-Al alloys, σ was lower than the isothermal value because of carryover of the more highly recovered substructure [7,35]. If the T decline were halted, considerable strain was required to establish the isothermal steady state substructure.   If the test was interrupted to produce SRX, then the flow curve on restraining exhibited normal isothermal behavior [7,35].

PRODUCT PROPERTIES

The SRX grain size was finer after higher Z deformation which gave rise to product strengthening due to the Hall-Petch effect. [28,45] If the hot work substructure was retained, the hardness was increased 60% before DRX had occurred but only by 30% after DRX where the dislocation density is lower [35]. The strengthening effect of raising Z from 0.1 to 10 s^{-1} is a change from 80 to 100 DPH for as-worked but only from 50 to 60 DPH after SRX [45]. For 800°C 10 s^{-1}, the hardness is equivalent to 40% cold work. As for other metals, the 25°C yield stress has been shown to be related to subgrain size by Equation 6 [9,28] or alternately by a similar equation with power of - 0.5 [29,49]. Wire drawn from unannealed extruded rod has improved properties due to the carryover of the substructure and grains resulting from DRX [86]; similar effects are seen in continuous processing of Al alloy wire [87].

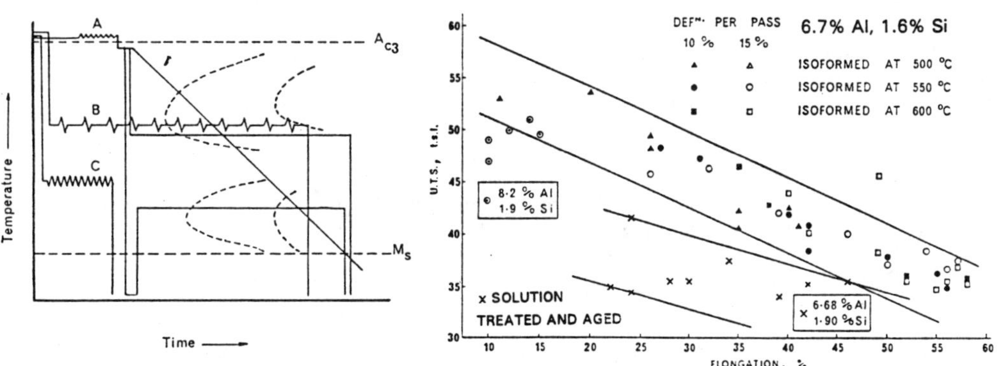

Figure 14.   TMP schedules are illustrated in (a) for (A) hot working and controlled cooling, (B) isoforming and (C) ausforming. In (b) the enhanced combination of strength and ductility resulting from isoforming are compared to those from deformation and heat treatment. After Tyler and Lockington [88].

As a result of the potential eutectoid and martensite transformations, Cu-7Al-2Si and Cu-10Al-4.5Fe-5Ni, may be subjected to steel-like TMP [88]. Isoforming (Figure 14) consists of deforming the alloy during the entire progress of the isothermal eutectoid transformation (optimally near the nose of the T-T-T diagram). This results in small spherical particles of the intermetallic phase pinning a fine substructure instead of a lamellar eutectoid. Ausforming consists of deforming the high T phase just above the martensite start (optimally in a domain of delayed transformation in the T-T-T diagram) and then quenching to martensite which retains the warm deformation substructure. From isoforming, the strength at a given ductility level was much higher than after standard heat treatment (Figure 14). The strength of Cu-Al-Si increased (with reduced ductility) as T was lowered from 600 to 500°C and pass strain raised from 10 to 15% at 65% total reduction [88]. The strength of Cu-Al-Fe-Ni was similarly affected by isoforming ranges of 650-550°C and 10 to 20% passes; however ausforming at 500°C gave an improved combination of properties [88].

CONCLUSIONS

During hot working, copper undergoes dynamic recrystallization which terminates the strain hardening domain with a peak stress at strains of 0.3 to 1.0 and then causes work softening leading to a domain where stress is independent of strain. In this steady state regime, the recrystallized grains are equiaxed, constant in size related to flow stress and contain a substructure. After hot deformation, static recovery is quickly followed by static recrystallization to a grain size which is finer as reciprocal temperature, strain rate and strain increase. However, strain has no effect with attainment of steady state after which static grains form proportionally coarser than the dynamic ones. Similar behavior is observed in solid solution alloys; however, dynamic recrystallization is slowed down leading to higher peak strains and stresses and higher steady state values also.

Ductility is increased by dynamic recrystallization with a rapid rise from a warm working domain of grain boundary cracking. Such cracking is more severe in alloys and the increase is delayed to higher temperatures and limited in extent by the retardation of dynamic recrystallization. Ductility declines rapidly as melting approaches; the temperature is lowered for alloys and additionally when there are grain boundary segregates.

The flow stress dependence on strain rate has been described by three equations: the hyperbolic sine, the exponential (high stress) and the power low (low stress); the last has been used most frequently but the first also covers that range. The Arrhenius function describes the temperature dependence; however, there is considerable disagreement in the activation energy. One group of reports places the value above 300 kJ/mol and rising with solute content. Other reports give values much lower and in some cases independent of alloy content. The latter behavior is not easily explained and disagrees with general observations in alloys of other metals.

REFERENCES

1.   J.J. Jonas, C.M. Sellars and W.J. McG. Tegart, Met. Rev., Vol. 14, 1969, 1-24.

2.   C.M. Sellars and W.J. McG. Tegart, Int. Met. Rev., Vol. 17, 1972, 1-24.

3.   C.M. Sellars, Phil. Trans. R. Soc. Lond., Vol. A288, 1978, 147-158.

4.   H. Mecking and G. Gottstein, Recrystallization of Metallic Materials, F. Haessner, ed., (Dr. Reiderer Varlag Gmbh, Stuttgart, 1978) pp. 195-222.

5.   W. Roberts, Deformation, Processing and Microstructure, G. Krausz, ed., (ASM, Metals Park, Ohio, 1983), 109-184.

6.   H.J. McQueen and J.J. Jonas, Metal Forming Interrelation Between Theory and Practice, A.L. Hoffmanner, ed., (Plenum Publishers, New York, 1971) pp. 393-428.

7.   H.J. McQueen and J.J. Jonas, Treatise on Materials Science and Technology, Plastic Deformation of Materials, Vol. 6, R.J. Arsenault, ed., (Academic Press, New York, 1975) pp. 393-493.

8.   H.J. McQueen and J.J. Jonas, J. Appl. Metal-Work, Vol. 3, 1985, 233-241.

9.   H.J. McQueen and J.J. Jonas, J.l Appl. Metal-Work, Vol 3, 1985, 410-420.

10.   H.J. McQueen, Trans. Japan Inst. Metals, Vol. 9, supp., 1968, 170-177.

11.   S.Fulop,K.Cadien,M.J.Luton and H.J.McQueen,J. Testing Evaluation,Vol.5,1977,419-426.

12.   H.J. McQueen, Can. Met. Q., Vol. 21, 1982, 445-460.

13.   H.J. McQueen, J. Sankar and S. Fulop, Mechanical Behavior of Materials ICM3, (Pergamon Press, Oxford, 1979), Vol. 2, pp. 675-684.

14.   H.J. McQueen, Defects Fracture and Fatigue, G. Sih and J. Provan, eds., (Mannus Nihoff Pub., The Hague, 1983) pp. 459-471.

15.   H.J. McQueen, Mat. Sci. Eng., Vol. A101, 1987, 149-160.

16.   H.J. McQueen, E. Evangelista and N.D. Ryan, Recrystallization ('90) in Metals and Materials, T. Chandra, ed., (TMS-AIME, Warrendale, PA 1990) pp. 89-100.

17.   T. Sakai and J.J. Jonas, Acta Metal., Vol. 32, 1984, 189-209.

18.   H.J. McQueen and D.L. Bourell, Inter-Relationship of Metallurgical Structure and Formability, A.K. Sachdev and J.D. Embury, eds., (Met. Soc. AIME, Warrendale, PA, 1986) pp. 341-348. J. Met., Vol. 39 [7], 1987, 28-35.

19.   H.J. McQueen, Hot Deformation of Aluminum Alloys, T.G. Langdon et al., eds. Met. Soc. (TMS-AIME, Warrendale, PA, 1991) (in press).

20.   H.J. McQueen and M.E. Kassner, Superplasticity in Aerospace, H.C. Heikkenen and T.R. McNelley, eds., Met. Soc. AIME, (Warrendale, PA, 1988) pp. 77-96.

21.   H.J. McQueen and E. Evangelista, Czech J. Phys., Vol. B38, 1988, 359-372.

22.   N.D. Ryan and H.J. McQueen, High Temp. Tech., Vol. 8, 1990, 27-44.

23.   N.D. Ryan and H.J. McQueen, High Temp. Tech., Vol. 8, 1989, 185-200.

24.   H.J. McQueen, J. Bowles and N.D. Ryan, Microstruc. Sci., Vol. 17, 1989, 357-373.

25.   E. Evangelista, N.D. Ryan, and H.J. McQueen, Met. Sci. Tech., Vol. 5, 1987, 50-58.

26.   N.D. Ryan and H.J. McQueen, Can. Metall. Q., Vol. 30, 1991 (in press).

27.   N.D. Ryan and H.J. McQueen, Materials Forming, Vol. 14, 1990, 283-295.

28.   R.J. McElroy and Z.C. Szkopiak, Int. Metal. Rev., Vol. 17, 1972, 175-202.

29.   D.J. Abson and J.J. Jonas, Metal Sci., Vol. 4, 1970, 24-18.

30.   M.J. Luton and C.M. Sellars, Acta Metal., Vol. 17, 1969, 1033-1043.

31.   J.P. Sah, G.J. Richardson and C.M. Sellars, Met. Sci., Vol. 8, 1974, 325-331.

32.   H. Mecking, B. Nicklas, N. Zarubova and U.F. Kocks, Acta Met., Vol. 34, 1986, 527.

33.   D. Hardwick and W.J. McG. Tegart, J. Inst. Metals, Vol. 90, 1961-62, 17-20.

34.   H. Ormerod and W.J. McG. Tegart, J. Inst. Metals, Vol. 92, 1963-64, 297-299.

35. R. Bromley and C.M. Sellars, Strength of Metals and Alloys, ICSMA 3, (Pergamon Press, Oxford, 1973) pp. 380-385.

36. R. Hannesen, U. Heubner and V. Schumacher, Z. Metallkunde, Vol. 58, 1967, 423-32.

37. H.P. Stuwe, Deformation Under Hot Working Conditions (SP108), (Iron Steel Inst., London, 1968) pp. 1-9.

38. B. Drube and M.P. Stuwe, Z. Metallkunde, Vol. 58, 1967, 499-506, 799-804.

39. L. Blaz, T. Sakai and J.J. Jonas, Met. Sci., Vol. 17, 1983, 609-616.

40. A.Korbel,L.Blaz,H.Dybiec,J.Gryziecki and J.Zasadzinski, Met.Tech., Vol.6,1979, 391-397.

41. M. Ueki, S. Horie and T. Nakamura, Mat. Sci. Tech., Vol. 3, 1987, 329-337.

42. R.W. Evans and F.L. Jones, Met. Tech., Vol. 5, 1978, 1-6.

43. L. Blaz and A. Korbel, Hot Working and Forming Processes, C.M. Sellars and G.T. Davies, eds., (Metals Soc., London, 1980) pp. 57-61.

44. A. Korbel and L. Blaz, Scripta Metal, Vol. 14, (1980), 829-834.

45. H.J. McQueen and S. Bergerson, Met. Sci., Vol. 6, 1972, 25-29.

46. H.J. McQueen, Strength of Metals and Alloy, ICSMA 6, R.C. Gifkins, ed., (Pergamon Press, Oxford) 1982, pp. 517-522.

47. H.J. McQueen and L. Vazquez, Mat Sci. Eng., Vol. 86, 1986, 355-369.

48. D.H. Warrington, European Regional Conf. Electron Microscopy Delft, 1960, p. 354-

49. P.B. Hirsch and D.H. Warrington, Phil. Mag., Vol. 6, 1961, 735-768.

50. T. Hennaut, J. Othmazouri and J. Charlier, Z. Metallkunde, Vol. 73, 1982, 744-753.

51. C.M. Sellars and W.J.McG. Tegart, Mém. Scient. Revue Métal., Vol. 63, 1966, 731-746.

52. S.K. Samanta, J. Mech. Phys. Solids, Vol. 19, 1971, 117-135.

53. S.K. Samanta, Acta Polytechnica Scandinavia ME49, Stockholm, 1970.

54. D.W. Livesey and C.M. Sellars, Met. Tech., Vol. 11, 1984, 149-155.

55. Y. Ohtakara, T. Nakamura and S. Sakui, Trans. ISIJ, Vol. 12, 1972, 207-216.

56. S.K. Samanta, Int. J. Mech. Sci., Vol. 11, 1969, 433-453.

57. R.A. Petkovic, M.J. Luton and J.J. Jonas, Met. Sci., Vol. 13, 1979, 569-572.

58. L.Vazquez, H.J.McQueen, M.Charest and G.Carpenter, Can.Met.Q., Vol.25,1987, 337-348.

59. M.B. Staker and D.L. Holt, Acta Met., Vol. 20, 1972, 569-579.

60. J.E. Pratt, Acta Met., Vol. 15, 1967, 319-

61. M. Ueki, S. Horie and T. Nakamura, Scripta Metal., Vol. 19, 1985, 547-549.

62. T. Ito, Y. Nakayama and T. Taketani, Strength of Metals and Alloys ICSMA 7, H.J. McQueen et al., eds., (Pergamon Press, Oxford 1986), Vol. 2, pp. 911-916.

63.   T. Ito, T. Taketani and Y. Nakayama, <u>Scripta Met.</u>, Vol. 20, 1986, 1329-1332.

64.   A. Korbel, W. Bochnisk, L. Blaz and J.D. Enbury, <u>Met. Sci.</u>, Vol. 18, 1984, 216-222.

65.   L. Blaz, <u>Annealing Processes - Recovery, Recystallization, Grain Growth</u>, N. Hansen et al., eds., (Riso National Lab., Roskilde, DK, 1986) pp. 221-227.

66.   A. Wantzen, P. Karduck and G. Gottstein, <u>Strength of Metals and Alloys (ICSMA 5)</u>, P. Haasen et al., eds. (Pergamon Press, Oxford, 1979) Vol. 1, pp. 517-522.

67.   G. Gottstein, <u>Met. Sci.</u>, Vol. 17, 1983, 497-502.

68.   G. Gottstein and U.F. Kocks, <u>Acta Metal.</u>, Vol. 31, 1983, 175-188.

69.   P. Karduck, G. Gottstein and H. Mecking, <u>Acta Metal.</u>, Vol. 31, 1983, 1525-1536.

70.   V.M. Sample, G.L. Fitzsimmons and A.J. Deardo, <u>Acta Metal.</u>, Vol. 35, 1987, 367-379.

71.   M. Szczerba and L. Blaz, <u>Annealing Processes - Recovery, Recrystallization, Grain Growth</u>, N. Hansen et al., eds., (Riso National Lab., Roskilde, DK., 1986) pp. 561-566.

72.   L. Blaz and M. Szczerba, <u>Arch. Hutnicka</u>, Vol. 30, 1985, 557-570.

73.   J.Z. Gronostajski and H.H. Ziemba, <u>Metal Sci.</u>, Vol. 16, 1982, 405-409.

74.   P.W. Davies, G.R. Dunstan, R.W. Evans and B. Wilshire, <u>J. Inst. Metals</u>, Vol. 99, 1971, 195-197.

75.   B.J. Sunter and N.M. Burman, <u>J. Australasian Inst. Metals</u>, Vol. 17 No. 2, 1972, 91-100.

76.   D.M. Ward and B.J. Helliwell, <u>J. Inst. Metals</u>, Vol. 98, 1970, 199-203.

77.   J. Gronostajki, E. Pulit and H. Ziemba, <u>Met. Sci.</u>, Vol. 17, 1983, 348-352.

78.   D.W. Davies, <u>J. Inst. Metals</u>, Vol. 98, 1970, 174-182.

79.   J.P. Chubb, J. Billingham, P. Hancock, C. Dimbylow and G. Newcombe, <u>J. Metals</u>, Vol. 30 No. 3, 1978, 20-23.

80.   R.J. Jackson, D.A. Edge and D.C. Moore, <u>J. Inst. Metals</u>, Vol. 98, 1970, 193-198.

81.   L. Vazquez, H.J. McQueen and J.J. Jonas, Strength of Metals and Alloys (ICSMA 8, Tanpere), P.O. Kettunen et al. eds., (Pergamon Press, Oxford, 1988), Vol. 2, 1013-1018.

82.   L. Vazquez and H.J. McQueen, <u>Strength of Metals and Alloys</u> (ICSMA 7, Montreal), H.J. McQueen et al. eds., (Pergamon Press, Oxford, 1986), Vol. 2, pp. 905-911.

83.   L. Vazquez, H.J. McQueen and J.J. Jonas, <u>Acta Met.</u>, Vol. 35, 1987, 1951-1962.

84.   R.A. Petkovic, M.J. Luton and J.J. Jonas, <u>Acta Met.</u>, Vol. 27, 1979, 1633-1648.

85.   M.J. Luton, R.A. Petkovic and J.J. Jonas, <u>Acta Met.</u>, Vol. 28, 1980, 729-743.

86.   V.L. Komarek, J. Faltus, P. Hroch and V.L. Redl, <u>Metals Tech.</u>, Vol. 6, 1979, 276-287.

87.   H.J. McQueen, E.H. Chia and E.A. Starke, <u>Microstructure Control in Aluminum Alloys</u>, E.H. Chia and H.J. McQueen, eds., (TMS-AIME, Warrendale, PA 1986), pp. 1-18.

88.   D.E. Tyler and G.W. Lockington, <u>J. Inst. Metals</u>, Vol. 99, 1971, 215-222.

As a result of the potential eutectoid and martensite transformations, Cu-7Al-2Si and Cu-10Al-4.5Fe-5Ni, may be subjected to steel-like TMP [88]. Isoforming (Figure 14) consists of deforming the alloy during the entire progress of the isothermal eutectoid transformation (optimally near the nose of the T-T-T diagram). This results in small spherical particles of the intermetallic phase pinning a fine substructure instead of a lamellar eutectoid. Ausforming consists of deforming the high T phase just above the martensite start (optimally in a domain of delayed transformation in the T-T-T diagram) and then quenching to martensite which retains the warm deformation substructure. From isoforming, the strength at a given ductility level was much higher than after standard heat treatment (Figure 14). The strength of Cu-Al-Si increased (with reduced ductility) as T was lowered from 600 to 500°C and pass strain raised from 10 to 15% at 65% total reduction [88]. The strength of Cu-Al-Fe-Ni was similarly affected by isoforming ranges of 650-550°C and 10 to 20% passes; however ausforming at 500°C gave an improved combination of properties [88].

CONCLUSIONS

During hot working, copper undergoes dynamic recrystallization which terminates the strain hardening domain with a peak stress at strains of 0.3 to 1.0 and then causes work softening leading to a domain where stress is independent of strain. In this steady state regime, the recrystallized grains are equiaxed, constant in size related to flow stress and contain a substructure. After hot deformation, static recovery is quickly followed by static recrystallization to a grain size which is finer as reciprocal temperature, strain rate and strain increase. However, strain has no effect with attainment of steady state after which static grains form proportionally coarser than the dynamic ones. Similar behavior is observed in solid solution alloys; however, dynamic recrystallization is slowed down leading to higher peak strains and stresses and higher steady state values also.

Ductility is increased by dynamic recrystallization with a rapid rise from a warm working domain of grain boundary cracking. Such cracking is more severe in alloys and the increase is delayed to higher temperatures and limited in extent by the retardation of dynamic recrystallization. Ductility declines rapidly as melting approaches; the temperature is lowered for alloys and additionally when there are grain boundary segregates.

The flow stress dependence on strain rate has been described by three equations: the hyperbolic sine, the exponential (high stress) and the power low (low stress); the last has been used most frequently but the first also covers that range. The Arrhenius function describes the temperature dependence; however, there is considerable disagreement in the activation energy. One group of reports places the value above 300 kJ/mol and rising with solute content. Other reports give values much lower and in some cases independent of alloy content. The latter behavior is not easily explained and disagrees with general observations in alloys of other metals.

REFERENCES

1.   J.J. Jonas, C.M. Sellars and W.J. McG. Tegart, <u>Met. Rev.</u>, Vol. 14, 1969, 1-24.
2.   C.M. Sellars and W.J. McG. Tegart, <u>Int. Met. Rev.</u>, Vol. 17, 1972, 1-24.
3.   C.M. Sellars, <u>Phil. Trans. R. Soc. Lond.</u>, Vol. A288, 1978, 147-158.
4.   H. Mecking and G. Gottstein, <u>Recrystallization of Metallic Materials</u>, F. Haessner, ed., (Dr. Reiederer Varlag Gmbh, Stuttgart, 1978) pp. 195-222.
5.   W. Roberts, <u>Deformation, Processing and Microstructure</u>, G. Krausz, ed., (ASM, Metals Park, Ohio, 1983), 109-184.
6.   H.J. McQueen and J.J. Jonas, <u>Metal Forming Interrelation Between Theory and Practice</u>, A.L. Hoffmanner, ed., (Plenum Publishers, New York, 1971) pp. 393-428.
7.   H.J. McQueen and J.J. Jonas, <u>Treatise on Materials Science and Technology, Plastic Deformation of Materials</u>, Vol. 6, R.J. Arsenault, ed., (Academic Press, New York, 1975) pp. 393-493.
8.   H.J. McQueen and J.J. Jonas, <u>J. Appl. Metal-Work.</u> Vol. 3, 1985, 233-241.
9.   H.J. McQueen and J.J. Jonas, <u>J.1 Appl. Metal-Work</u>, Vol 3, 1985, 410-420.
10.   H.J. McQueen, <u>Trans. Japan Inst. Metals</u>, Vol. 9, supp., 1968, 170-177.
11.   S.Fulop,K.Cadien,M.J.Luton and H.J.McQueen,<u>J. Testing Evaluation</u>,Vol.5,1977,419-426.
12.   H.J. McQueen, <u>Can. Met. Q.</u>, Vol. 21, 1982, 445-460.

13. H.J. McQueen, J. Sankar and S. Fulop, Mechanical Behavior of Materials ICM3, (Pergamon Press, Oxford, 1979), Vol. 2, pp. 675-684.
14. H.J. McQueen, Defects Fracture and Fatigue, G. Sih and J. Provan, eds., (Mannus Nihoff Pub., The Hague, 1983) pp. 459-471.
15. H.J. McQueen, Mat. Sci. Eng., Vol. A101, 1987, 149-160.
16. H.J. McQueen, E. Evangelista and N.D. Ryan, Recrystallization ('90) in Metals and Materials, T. Chandra, ed., (TMS-AIME, Warrendale, PA 1990) pp. 89-100.
17. T. Sakai and J.J. Jonas, Acta Metal., Vol. 32, 1984, 189-209.
18. H.J. McQueen and D.L. Bourell, Inter-Relationship of Metallurgical Structure and Formability, A.K. Sachdev and J.D. Embury, eds., (Met. Soc. AIME, Warrendale, PA, 1986) pp. 341-348. J. Met., Vol. 39 [7], 1987, 28-35.
19. H.J. McQueen, Hot Deformation of Aluminum Alloys, T.G. Langdon et al., eds. Met. Soc. (TMS-AIME, Warrendale, PA, 1991) (in press).
20. H.J. McQueen and M.E. Kassner, Superplasticity in Aerospace, H.C. Heikkenen and T.R. McNelley, eds., Met. Soc. AIME, (Warrendale, PA, 1988) pp. 77-96.
21. H.J. McQueen and E. Evangelista, Czech J. Phys., Vol. B38, 1988, 359-372.
22. N.D. Ryan and H.J. McQueen, High Temp. Tech., Vol. 8, 1990, 27-44.
23. N.D. Ryan and H.J. McQueen, High Temp. Tech., Vol. 8, 1989, 185-200.
24. H.J. McQueen, J. Bowles and N.D. Ryan, Microstruc. Sci., Vol. 17, 1989, 357-373.
25. E. Evangelista, N.D. Ryan, and H.J. McQueen, Met. Sci. Tech., Vol. 5, 1987, 50-58.
26. N.D. Ryan and H.J. McQueen, Can. Metall. Q., Vol. 30, 1991 (in press).
27. N.D. Ryan and H.J. McQueen, Materials Forming, Vol. 14, 1990, 283-295.
28. R.J. McElroy and Z.C. Szkopiak, Int. Metal. Rev., Vol. 17, 1972, 175-202.
29. D.J. Abson and J.J. Jonas, Metal Sci., Vol. 4, 1970, 24-18.
30. M.J. Luton and C.M. Sellars, Acta Metal., Vol. 17, 1969, 1033-1043.
31. J.P. Sah, G.J. Richardson and C.M. Sellars, Met. Sci., Vol. 8, 1974, 325-331.
32. H. Mecking, B. Nicklas, N. Zarubova and U.F. Kocks, Acta Met., Vol. 34, 1986, 527.
33. D. Hardwick and W.J. McG. Tegart, J. Inst. Metals, Vol. 90, 1961-62, 17-20.
34. H. Ormerod and W.J. McG. Tegart, J. Inst. Metals, Vol. 92, 1963-64, 297-299.
35. R. Bromley and C.M. Sellars, Strength of Metals and Alloys, ICSMA 3, (Pergamon Press, Oxford, 1973) pp. 380-385.
36. R. Hannesen, U. Heubner and V. Schumacher, Z. Metallkunde, Vol. 58, 1967, 423-32.
37. H.P. Stuwe, Deformation Under Hot Working Conditions (SP108), (Iron Steel Inst., London, 1968) pp. 1-9.
38. B. Drube and M.P. Stuwe, Z. Metallkunde, Vol. 58, 1967, 499-506, 799-804.
39. L. Blaz, T. Sakai and J.J. Jonas, Met. Sci., Vol. 17, 1983, 609-616.
40. A.Korbel,L.Blaz,H.Dybiec,J.Gryziecki and J.Zasadzinski, Met.Tech., Vol.6,1979, 391-397.
41. M. Ueki, S. Horie and T. Nakamura, Mat. Sci. Tech., Vol. 3, 1987, 329-337.
42. R.W. Evans and F.L. Jones, Met. Tech., Vol. 5, 1978, 1-6.
43. L. Blaz and A. Korbel, Hot Working and Forming Processes, C.M. Sellars and G.T. Davies, eds., (Metals Soc., London, 1980) pp. 57-61.
44. A. Korbel and L. Blaz, Scripta Metal, Vol. 14, (1980), 829-834.
45. H.J. McQueen and S. Bergerson, Met. Sci., Vol. 6, 1972, 25-29.
46. H.J. McQueen, Strength of Metals and Alloy, ICSMA 6, R.C. Gifkins, ed., (Pergamon Press, Oxford) 1982, pp. 517-522.
47. H.J. McQueen and L. Vazquez, Mat Sci. Eng., Vol. 86, 1986, 355-369.
48. D.H. Warrington, European Regional Conf. Electron Microscopy Delft, 1960, p. 354-
49. P.B. Hirsch and D.H. Warrington, Phil. Mag., Vol. 6, 1961, 735-768.
50. T. Hennaut, J. Othmazouri and J. Charlier, Z. Metallkunde, Vol. 73, 1982, 744-753.
51. C.M. Sellars and W.J.McG. Tegart, Mém. Scient. Revue Métal., Vol. 63, 1966, 731-746.
52. S.K. Samanta, J. Mech. Phys. Solids, Vol. 19, 1971, 117-135.
53. S.K. Samanta, Acta Polytechnica Scandinavia ME49, Stockholm, 1970.
54. D.W. Livesey and C.M. Sellars, Met. Tech., Vol. 11, 1984, 149-155.
55. Y. Ohtakara, T. Nakamura and S. Sakui, Trans. ISIJ, Vol. 12, 1972, 207-216.
56. S.K. Samanta, Int. J. Mech. Sci., Vol. 11, 1969, 433-453.
57. R.A. Petkovic, M.J. Luton and J.J. Jonas, Met. Sci., Vol. 13, 1979, 569-572.

58. L.Vazquez, H.J.McQueen, M.Charest and G.Carpenter, Can.Met.Q., Vol.25,1987, 337-348.
59. M.B. Staker and D.L. Holt, Acta Met., Vol. 20, 1972, 569-579.
60. J.E. Pratt, Acta Met., Vol. 15, 1967, 319-   .
61. M. Ueki, S. Horie and T. Nakamura, Scripta Metal., Vol. 19, 1985, 547-549.
62. T. Ito, Y. Nakayama and T. Taketani, Strength of Metals and Alloys ICSMA 7, H.J. McQueen et al., eds., (Pergamon Press, Oxford 1986), Vol. 2, pp. 911-916.
63. T. Ito, T. Taketani and Y. Nakayama, Scripta Met., Vol. 20, 1986, 1329-1332.
64. A. Korbel, W. Bochnisk, L. Blaz and J.D. Enbury, Met. Sci., Vol. 18, 1984, 216-222.
65. L. Blaz, Annealing Processes - Recovery, Recrystallization, Grain Growth, N. Hansen et al., eds., (Riso National Lab., Roskilde, DK, 1986) pp. 221-227.
66. A. Wantzen, P. Karduck and G. Gottstein, Strength of Metals and Alloys (ICSMA 5), P. Haasen et al., eds. (Pergamon Press, Oxford, 1979) Vol. 1, pp. 517-522.
67. G. Gottstein, Met. Sci., Vol. 17, 1983, 497-502.
68. G. Gottstein and U.F. Kocks, Acta Metal., Vol. 31, 1983, 175-188.
69. P. Karduck, G. Gottstein and H. Mecking, Acta Metal., Vol. 31, 1983, 1525-1536.
70. V.M. Sample, G.L. Fitzsimmons and A.J. Deardo, Acta Metal., Vol. 35, 1987, 367-379.
71. M. Szczerba and L. Blaz, Annealing Processes - Recovery, Recrystallization, Grain Growth, N. Hansen et al., eds., (Riso National Lab., Roskilde, DK., 1986) pp. 561-566.
72. L. Blaz and M. Szczerba, Arch. Hutnicka, Vol. 30, 1985, 557-570.
73. J.Z. Gronostajski and H.H. Ziemba, Metal Sci., Vol. 16, 1982, 405-409.
74. P.W. Davies, G.R. Dunstan, R.W. Evans and B. Wilshire, J. Inst. Metals, Vol. 99, 1971, 195-197.
75. B.J. Sunter and N.M. Burman, J. Australasian Inst. Metals, Vol. 17 No. 2, 1972, 91-100.
76. D.M. Ward and B.J. Helliwell, J. Inst. Metals, Vol. 98, 1970, 199-203.
77. J. Gronostajki, E. Pulit and H. Ziemba, Met. Sci., Vol. 17, 1983, 348-352.
78. D.W. Davies, J. Inst. Metals, Vol. 98, 1970, 174-182.
79. J.P. Chubb, J. Billingham, P. Hancock, C. Dimbylow and G. Newcombe, J. Metals, Vol. 30 No. 3, 1978, 20-23.
80. R.J. Jackson, D.A. Edge and D.C. Moore, J. Inst. Metals, Vol. 98, 1970, 193-198.
81. L. Vazquez, H.J. McQueen and J.J. Jonas, Strength of Metals and Alloys (ICSMA 8, Tanpere), P.O. Kettunen et al. eds., (Pergamon Press, Oxford, 1988), Vol. 2, 1013-1018.
82. L. Vazquez and H.J. McQueen, Strength of Metals and Alloys (ICSMA 7, Montreal), H.J. McQueen et al. eds., (Pergamon Press, Oxford, 1986), Vol. 2, pp. 905-911.
83. L. Vazquez, H.J. McQueen and J.J. Jonas, Acta Met., Vol. 35, 1987, 1951-1962.
84. R.A. Petkovic, M.J. Luton and J.J. Jonas, Acta Met., Vol. 27, 1979, 1633-1648.
85. M.J. Luton, R.A. Petkovic and J.J. Jonas, Acta Met., Vol. 28, 1980, 729-743.
86. V.L. Komarek, J. Faltus, P. Hroch and V.L. Redl, Metals Tech., Vol. 6, 1979, 276-287.
87. H.J. McQueen, E.H. Chia and E.A. Starke, Microstructure Control in Aluminum Alloys, E.H. Chia and H.J. McQueen, eds., (TMS-AIME, Warrendale, PA 1986), pp. 1-18.
88. D.E. Tyler and G.W. Lockington, J. Inst. Metals, Vol. 99, 1971, 215-222.

Promotion of copper in developing countries. The Chilean copper promotion centre: a case analysis

M. Helga Larravide V.
Chilean Copper Promotion Center (PROCOBRE), Santiago, Chile

Chile's Copper Promotion Center started its activities in February 1989. Like all analogous Centers in the world, the organization is a not-for profit one with its main function being the encouragement of the use of copper and its alloys through the promotion of its different applications.

Most of the Copper Centers, usually named as CDA's, standing for Copper Development Association, are located in industrialized countries and though Chilean Copper Center follows a similar pattern of activities in technical advice and diffusion of copper information, the promotion actions had to find its own strategies in order to get the best results. The task was not easy and specially it was hard to plan the finest distribution of the limited economical resources.

The starting of a Promotion Center in a developing country, the main obstacles overcome, how projects net had to be built up, market strategies and results are displayed and analyzed in PROCOBRE's first year of experience. Future trends and suggestions to join efforts for expansion of copper promotion in other countries of Latinamerican region and also investigation on new copper uses are within the scopes of this paper, as the Center's contribution to preserve and increase the world's copper consumption.

INTRODUCTION

In the next pages a synthesis of the activities performed by the Chilean Copper Promotion Center, PROCOBRE, are displayed following a chronological order and the obstacles that "obligated"the Center orientation.
Only the main promotion projects and actions are taken in account since details and normal copper promotion have no relevant difference with other copper promotion centers in the world whose functions are well known due to the many years of living of most of them.
Being the only spanish speaking copper promotion center in Southamerica and having a non despicable potential copper consumption in this region PROCOBRE is conscious that its working area is not limited to Chile's perimeter.
The optimum performance of promotion projects will give the desired results in Chile and the experience obtained could be projected to neighbour countries with no doubts since foreign requests indicate their needs are similar to Chile's.

BACKGROUND

Chile is essentially a mining country and within minery, copper occupies the more relevant position not only at a national level but also in the worldwide context.
In fact, Chile is the main copper exporter in the world with 1.559.500 metric tonnes of refined copper exported in 1989, that is, 33% of total world refined copper exports. It is the second refined copper producer between the occidental countries, with 1.071.000 metric tonnes of refined copper in the same year meaning this amount 10% of world copper production. Chile is also the main mine copper producer in the world, with 1.609.300 fine copper metric tonnes in 1989 and 18 % of world mine copper production.

National consumption, though, is very low: 42.900 metric tonnes during 1989 and an average for the last ten years of 36.730 metric tonnes. This amount is related to the country's existent limited copper manufacture.
In a gross approach, the main copper products consumed in the country are:

-electrical conductors	57%
-tubes	32%
-sheets and strips	6%
-others	5%

"Others" includes coining, hardware, locking and gas valves and regulators.

The main manufactured products exported are:

-copper conductors	54%
-sheets, strips and foils	14%
-tubes and fixtures	20%
-blank coins	11%
-others	1%

with a total approximate amount of 20.000 metric tonnes.(1989)

The main imported products are copper conductors, with a raise from 2000 metric tonnes in past years (1985-1988) to 5000 metric tonnes in 1989.

THE CHILEAN COPPER PROMOTION CENTER
1989

In this market reality it was created the Chilean Copper Promotion Center, PROCOBRE, in September 14, 1988, under the same premise that characterizes Copper Development Associations (CDAs): to encourage the use of copper and copper alloys and to promote their correct and efficient application.
PROCOBRE's activities started in January 1989 and the actual staff was completed same year in July. Staff is composed by:

-an Executive Secretary
-a metallurgical engineer
-a librarian
-a bilingual secretary

Above the Executive Secretary there is a Board of Directors, formed with eight Directors who belong to manufacture and copper producers enterprises.
There are five permanent Specialized Commitees whose members advise and control the development of specific projects.

When PROCOBRE initiated its labour in the promotion of the copper uses many questions worried its staff. Mainly, to discover the starting point and how to build a coherent promotion program in order to get the best results.
Efforts were made to overcome the lack of public copper information and capture all the information about copper uses that was dispersed through different organisms and some light came from the first users of the yet non-established Documentation Center whose requests indicated PROCOBRE the kind of information and needs they wanted to be satisfied and that became a priority.
It is important to mention at this point that Chile had in the past the Uses of Copper Promotion Center (CPUC) until 1975 which made a very good job. Unfortunately, most of the bibliographical material and technical documents elaborated was lost with consecuences such as complete ignorance related to copper products markets, processes, technical specifications and the unknowledge of where and how technical written information on specific subjects could be found.
All of these origined the first project presented to the International Copper Association (ICA) for financing resources:

Copper Awareness, which covered two important topics:

-National Copper Market Study, general
-National Copper in Building Market Study

Project was assigned with US$25.000.

In order to orient promotion, PROCOBRE established strong ties with the rest of CDAs in the world. It was so, that in April 1989, the Center held an international seminar under the name "International Copper Market Development and the Role of Technology and its Expansion". The seminar gathered most of the world's Copper Promotion and Development Centers, and also the International Copper Research Association (INCRA).
Copper promotion in several countries under respective CDA was reported then and PROCOBRE took experience from centers to set paths of promotion in Chile.
The ICA 1989 Market Development Program considered copper and copper alloys in Building and Construction and in this context PROCOBRE implemented the following projects:

Architectural Applications
Defense Against Plastic Tubes

with US$10.000 and US$20.000 respectively assigned by ICA.

This way, during 1989, ICA assigned a total amount of US$55.000 to chilean copper promotion for these specific projects.

THE COPPER LIBRARIES NET

Parallel to the activities of promotion and studies, PROCOBRE was retrieving copper documents, books, etc. to structure the Documentation Center. However there was not enough material to satisfy the users bibliographical demand and due to financial resources it was not possible to supply it in the short term as wished.
This situation favoured the creation of PROCOBRE's Copper Libraries Net, in October 1989.

Actually, the Copper Net joins fifteen different organizations involved in the copper area throughthe country including universities, investigation centers, mining enterprises and others.
The Copper Net is oriented to offer users the fastest and the most complete information about the subject they require. If PROCOBRE is not able to provide this by itself, PROCOBRE will get the best answer from the right organism, through the Net.
The Copper Libraries Net is one of the PROCOBRE's Documentation Center most important tool to impart copper technical information and it adds to the U.S.A. Copper Data Center connection and PROCOBRE's own Data Base.
With this initiative, members of the Net also save resources avoiding to buy books, texts, handbooks and technical magazines that already some member has been subscripted.
In less than two years PROCOBRE's Documentation Center has coated most of the copper and copper alloys bibliography existent in the country and its diffusion is achieved in PROCOBRE's Bibliographical Bulletin published quarterly in Chile and distributed to pertinent organisms and people.
PROCOBRE and the Net have been promoted through Net members bulletins, magazines and reports.
Also first Copper Net Activities Report was edited in December 1990 thanks to all members effort.

It is relevant to stand that the Net operates with no funds assigned, yet.

1990

During 1990, PROCOBRE's job continued in some of the lines established in the previous year. However,when promoting copper roofing by different mechanisms, two questions constantly were made by architects and building engineers: where they could get skilled copper roofing installers and also they wanted to know about kinds of copper roofing available for them to make the best choice.
The Center couldn't meet their requirements. By the opposite, copper roofing had a very bad image in Chile because of a very unfortunate experience in a popular shopping center in Santiago where the problem was the wrong installation.
PROCOBRE, looking for a solution, got in contact with the Ministry of Education and with its help checked the Basic Technical Schools programs and discovered there were few schools including copper installation. Electricity and sanitary installation were taught but with the employ of PVC, galvanized iron, aluminum, and copper just as another material, that is, there weren't copper specific programs and copper roofing was not in any program at all.
Main projects for 1990-91, then, were two:

-Copper Manipulation in Basic Technical Schools
-Copper Roofing Prototypes, Design and Technical Especifications

Total cost of these projects is US$164.600, and half of this amount is funded by the International Copper Association.

The first of these projects is a very wide one with several steps within a module program that can be summarized as follows:

a)Training of monitors to impart copper knowledge to Basic Technical Schools teachers

b)Teachers training and the redaction of Programmed Textbooks for students

c)Selection of some test pilot schools to apply the module program

d)Evaluation

In 1990, project completed from a) to c).
PROCOBRE invited an expert from Germany, recommended by the Deutsches Kupfer-Institut (DKI),to train monitors and workers and to advise the texts content. This expert brought with him several copper handling tools and sanitary and roofing prototypes as a donation for a better accomplishment of the project.
Three hundred students will start the module copper program in March 1991 when chilean academic year begins.

The second project will provide, besides the physical copper prototypes, all the information about the "do's and don'ts" of copper roofing handling, patinas and coatings, technical specifications of each prototype and advising in the election of a copper roofing for different climates and building designs.
Prototypes are to be done by March 1991 and several roofing manufacturers will include them in their production lines.
Roofing fabricators have been collaborating with the investigators contributing their experience and even marketing opinions on the prototypes design.
It is also important to mention that copper manufacturers sent some of their workers to the copper training classes the german teacher imparted during his stay in Chile.

These two main 1990-91 projects, though they are developed by different organizations: National Training Institute, INACAP, and Catholic University Architecture Faculty, respectively, have been in a very tight contact feeding related information to each other through their evolution.

1991

1991-92 PROCOBRE's activities will be in this net project context and mainly based in the above mentioned projects results.
The diffusion and ramification of results is to be done through seminars to professionals, training to more teachers and workers, videos and the edition of more programmed textbooks for installers.
Prototypes are to be showed in the local market and in the copper exhibitions PROCOBRE assembles throughout the year.
Within the new trends and to always continue the promotion of copper in Chile, during 1991-92 period PROCOBRE is planning to diffuse the national investigation about copper and its alloys that

universities develop. The diffusion will be done by talks exposed by the investigators themselves and the audience will join students, researchers, fabricators and any others pertinent. Doing so, universities and research institutes will directly confront the fabricators and a profitable dialogue is expected to occur in order that research really answers the fabricators needs. The goals are to establish a bridge between investigators and fabricators,to encourage engineering and architecture students to enter in the copper uses field, to encourage the new uses of copper research and in an indirect manner to stimulate students to become metallurgical engineers since in Chile they are extinguishing.

This labour can be performed through PROCOBRE's Copper Libraries Net whose members are the key tool to have the copper investigation radiography in the country.

REGIONAL SUGGESTIONS

PROCOBRE receives many copper information requests from neighbour countries such as Argentina, Brazil, Colombia, Venezuela and Bolivia.
Some of the PROCOBRE's Bibliographical Bulletins have been sent to several southamerican entities and they have been very welcome.
Southamerican fabricators visiting Chile during international exhibitions have known about the existence of PROCOBRE and become users of its Documentation Center.

Estimated copper consumption in Latin America is around 500.000 tonnes with a potential 100.000 tonnes for 1992.
To defend these amounts, promotion work must be done or substitutes will take over. In fact, aluminum estimated consumption in Latin America increases approximately 6% per year and copper only 3,5%, to mention just one of the copper competitors.

PROCOBRE makes some copper technical diffusion through its Bibliographical Bulletin, technical articles in specific magazines such as the Chilean School of Architects Magazine, brochures in the Chilean Building and Construction Chamber Bulletin and the distribution of the Copper Image Campaign 1991 Calendar, all of these with a Latin American covering, accordingly to PROCOBRE's economical resources.
It is well known this is not enough. Special copper promotion campaigns should be done with more funds assigned.
The exhibition of copper uses in countries such as Argentina, Bolivia and Colombia, for instance, is a very fruitful element to show in situ the wide possibilities for copper uses and its alloys.
To impart copper handling training to plumbers and workers related, seminars similar to the ones programmed for chilean professionals, are correct ways to wake up the copper awareness in the region and only benefits for copper consumption can be derived.
The experience PROCOBRE has gained through its Copper Libraries Net is a true example of how, with limited resources, a country can be unified through information needs. This knowledge could be employed profitably to implement a Latinamerican Copper Net and unify the region towards a better copper understanding.

CONCLUSIONS

PROCOBRE's first years main results in the promotion of copper uses have been:

-Establishment of a continuous promotion in the building area covering from the training of copper installers to copper products prototypes fabrication and advising.

-Retrieving of copper information in the country diffused through quarterly Bibliographical Bulletin.

-Copper Libraries Net creation which assures the best and more complete information to users and avoids dupplication of work.

-Strong connection achieved between researchers and fabricators through projects.

-Producers conscience about the final uses of copper

<u>REFERENCES</u>

Comisión Chilena del Cobre, <u>Memoria Anual 1989</u>, Santiago de Chile.

Comisión Chilena del Cobre, <u>Estadísticas del Cobre y Otros Minerales, Anuario 1989</u>, Santiago de Chile, Agosto 1990.

T.Grof y A. Eva,Identificación de Proyectos Específicos para la Producción, en América Latina, de Metales No Ferrosos Semiacabados,<u>Documentos ONUDI</u>, 3 de Enero de 1989.

Investigación Sobre el Uso del Cobre y sus Aleaciones en la Industria Nacional de la Construcción, PROCOBRE, 1990.

The North American initiative for copper architectural applications: an industry-wide technical market development program

P.A. Anderson
Copper Development Association Inc., Greenwich, Connecticut, U.S.A.

In 1987 the Corporate Research Center of Noranda Minerals Inc. began to study opportunities for the increased use of copper in various markets. This lead to the Noranda Minerals Inc. report on the initiative for copper in architecture published in June 1989.

The study dealt primarily with sheet copper and its architectural applications in the United States and Canada. The basic conclusion of this study was that an industrywide program solely dedicated to the development of the North American market for copper architectural applications should be created, funded and operated.

The June 1989 study stated that copper architectural use in North America was 74 million pounds per year, mostly as sheet copper for commercial and institutional building roofing applications (60%), residential flashing (7%), gutters and downspouts (6%), the balance being commercial flashing and other external and internal applications, Figure 1. This consumption level is equivalent to about 0.25 pounds per capita. The per capita consumption in some European countries is much higher, for example more than 1.5 pounds per capita in Germany and Italy---more in Austria and Switzerland--- primarily the result of significant long-established market development programs.

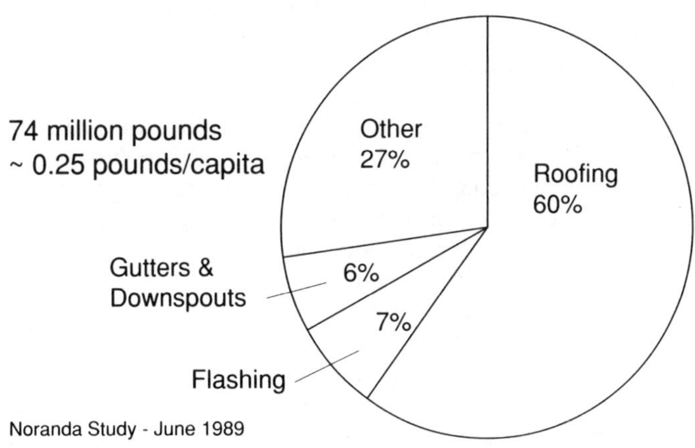

Figure 1 - Architectural use of copper in North America.

A concurrent market study by the Corporate Research Center (Union, New Jersey, USA) on metal roofing in the USA reported that copper was found to be the leading material choice in terms of customer acceptance, durability and aesthetics, accounting for 20% of all metal roofing installed. This favorable perception of copper reinforced the idea that higher copper consumption could be expected to result from an industrywide market development initiative.

The Noranda study, in fact, concluded that the North American copper market could be increased to 176 million pounds per year over a 10 year period, more than doubling from the 1988 consumption of 74 million pounds per year. Main market segment growth was identified as commercial roofing at 7%, 35% for residential roofing and 15% for commercial and residential flashing, gutters and downspouts. Consumption at these levels on the current population base corresponds to about 0.5 pound of copper per capita. While still below the 1.5 pound per capita experienced in some European countries, this is a significant increase.

In the summer of 1989 Noranda Minerals proposed to the International Copper Association the establishment and funding of a North American market development initiative for copper in architecture. The twofold strategy laid out focused on the supply and distribution side to insure availability of copper products and professional and competitive services, and on the market side concentrated on identifying opportunities and stimulating demand. Recognizing the significant opportunities that exist for increased use of copper and copper alloy products serving the building construction market in North America, the ICA consulted with the Copper Development Association Inc. and the Canadian Copper & Brass Development Association and responded favorably to CDA's proposal to develop and implement a long-term program to identify significant opportunities that would stimulate demand for copper sheet, strip and roll for roofing, flashing, gutters and downspouts and other architectural applications.

This program identified as the North American Initiative for Copper Architectural Applications (NAICAA) was organized on the basis of the Noranda Minerals proposal and a steering committee with representatives of the roofing copper fabricators and copper producers and chaired by the Noranda representative was established to provide guidance and oversee the activity of NAICAA which would be implemented and managed by CDA.

In order to carry out this multi-year program, funding was pledged by ICA, CDA, Noranda and CODELCO. CDA organized and launched the initiative and began to recruit a technical staff including a National Program Manager.

The initial thrust of NAICAA was to provide broad-based market development programs for increased copper penetration in architectural applications, targeted to double the use of copper in building construction within 10 years. Programs to be selected would define, identify and qualify market influentials (architects/designers, contractors, manufacturers, distributors, suppliers, building owners, and financiers), favorable geographic areas, availability of skilled installers/labor, legislative or jurisdictional influences, and technical/product advantages and constraints.

Early decisions by the Steering Committee encouraged CDA to proceed with a comprehensive market study for sheet copper architectural applications in order to generate solid market data which would serve as a benchmark for program guidance and progress evaluation. The committee endorsed the concept that the initiative should be organized to develop participatory value in order to attract others in the supply and installation chain, such as distributors, component and equipment manufacturers and contractors. Key programs would provide in-depth service, improved installation techniques, lower installed costs and widespread communication and promotion to the design and end use markets.

The firm of Irwin P. Sharpe & Associates conducted the basic market study to detail the use of copper sheet, strip, coil and roll for architectural and other building applications. The object was to define and quantify the North American (U.S. and Canada) market and select the prime geographic area for initiation of the NAICAA program. In order to meet these objectives Sharpe was asked to define the market as to types of existing copper products, market size in square feet and pounds, geographic influence and concentration, dynamic influence of prime and secondary copper fabricators, and to identify distribution channels and end users.

Sharpe was further asked to define and evaluate market influences including design considerations, construction segments consisting of commercial, institutional and residential, construction types including new, retrofit and restoration, suitability of current copper products and desire for new ones, availability and use of mechanical forming and seaming equipment and the influence of competitive products both domestic and imported.

Finally, the study included the role and influence of architects, specification writers, building owners, general contractors, roofing and sheet metal contractors, labor availability and qualification, and standards and codes.

The study was conducted through personal and telephone interviews with individuals who are in a position to influence the decision to use copper for roofing or architectural applications. In total, the principals or managers of 270 firms were interviewed.

The types and number of firms interviewed were: manufacturing and fabricators, 27; roofing and building products distributors, metal service centers, 42; contractors (roofing, sheet metal), 95; architects and specification writers, 92; building owners and developers, 14.

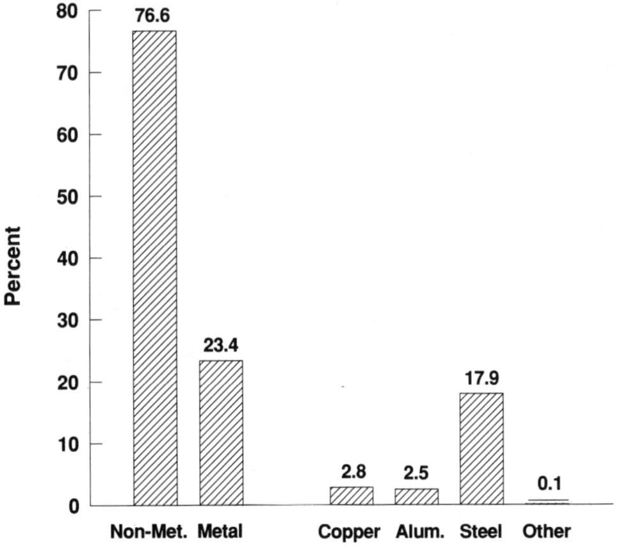

Figure 2 - Type of roofing installed by contractors surveyed. 91 million sq. ft., 1989.

The 95 contractors interviewed installed 91 million square feet of roofing in 1989 with 23.4% of that being metal roofing, Figure 2. Steel accounted for the largest segment of the metal roofing at 18% with copper at 2.8%. Fifty of the 95 contractors interviewed installed copper roofing in 1989.

Of the copper roofs installed, nearly 73% were commercial, 23% residential and the balance on institutional structures. New construction accounted for 47% of the copper roofs, re-roofing 43% and the balance on restoration projects. Copper flashing, gutters and downspouts were used on about 11% of the total roofing projects.

Building Class	Copper Roofs (Percent)
Commercial	72.8
Institutional	4.5
Residential	22.7
Total	**100.0**
Construction type	
New Construction	47.2
Re-roofing	43.4
Restroation	9.4
Total	**100.0**

Table 1 - Class of building and type of construction using copper roofing.

Color coated metal, primarily steel, is the major competitor to copper according to the majority of the contractors and 40% of the architects, Figure 3. Galvanized steel was the next most frequently identified as the primary competitor followed by aluminum and tern coated steel.

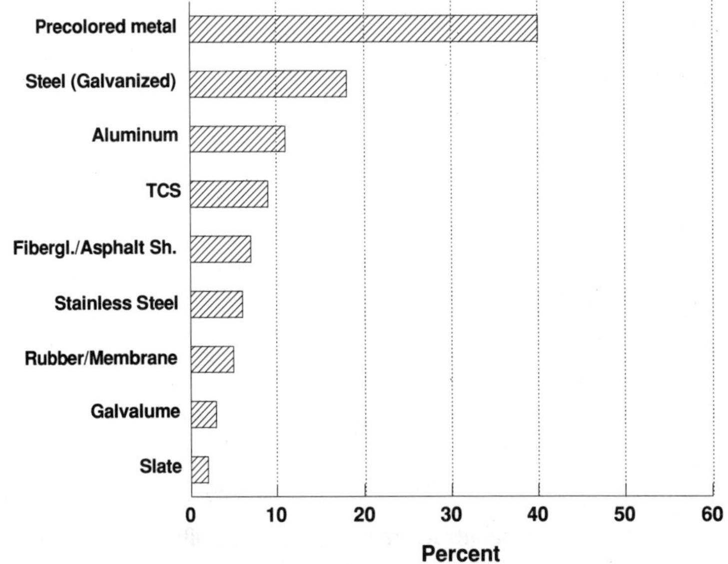

Figure 3 - Major competition to copper.

In comparison with the color coated steel roofing, copper was perceived to be better from the standpoint of life cycle cost and aesthetic values. It was considered to be worse from the standpoint of both first and installed costs, promotion and availability of information and technical services. Availability of copper was approximately the same as competing roofing materials. Factors which would convince architects and contractors to specify or use more copper were a lowered installed first cost to compete with coated steel or aluminum, the availability of skilled labor and better technical data, advertising and promotion.

It was somewhat important that new products such as shingles or corrugated textured copper and pre-form shapes be developed. New installation techniques to reduce overall installed costs were also desired.

In summary, architects are the primary influence on the use of copper in architecture but building owners/developers are also very important in making the final material selection decision based on considerations of cost and aesthetics, Figure 4. Contractors are more knowledgeable about and favorably disposed towards copper roofing than are architects. However, it is definitely a specified market.

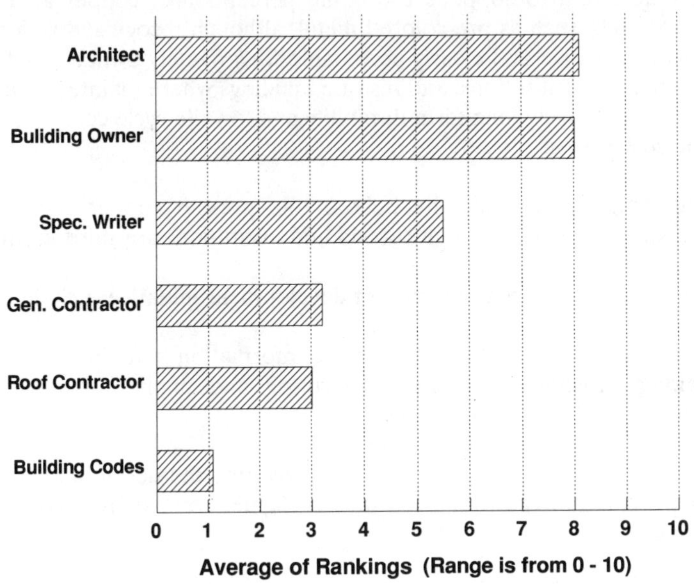

Figure 4 - Professional influence on copper use.

Aesthetics and the element of prestige appear to be the important driving forces for the use of copper roofing for privately owned commercial and residential structures. Durability is the second most important reason stated by architects. This is followed by quality image, architectural impression and matching existing roofing. Interestingly, life cycle cost is generally of least importance to architects.

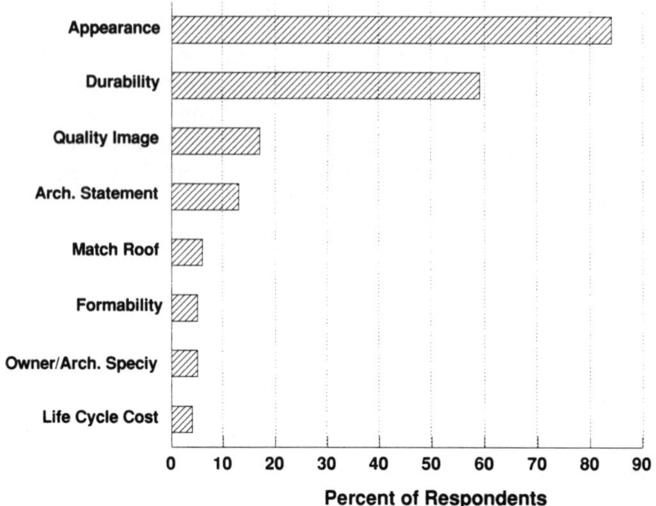

Figure 5 - Reasons for copper specification and use.

The primary deterrent to copper's use is the perception of copper as a high cost material. Newer materials, such as pre-colored metal, although recognized as less durable than copper, are less expensive and clearly acceptable for privately owned buildings. It is in publicly owned buildings, institutions, and historic buildings where maintenance is a long-term consideration, that the truly long-term durability and low life cycle costs are important motivations for the selection of copper.

Commercial structures provide the bulk of the present copper roofing market. Upscale residential structures are a distant second and institutions are third in importance.

The study recommended that our market development initiative include:

1.   A communications program to provide information to architects and owners, including effective continuing advertising and public relations aimed at architects and building owners.

2.   An educational program for architects covering the design and specification of copper, reinforced with a contractors segment on the installation of copper roofing.

3.   The development of new copper products and installation techniques to provide additional aesthetics options and lower installed costs.

The geographic areas which were recommended for inauguration of the North American Initiative included the choice of the major metropolitan areas of the Northeast, North Central and South regions if existing infra-structure was the most important criteria. If a comparatively undeveloped copper market was to be targeted and one which has the highest anticipated growth in population over the next 10 years, the far west would be the choice.

The Steering Committee of NAICAA elected to initiate the program concentrating in the West census region including the states of Alaska, Arizona, California, Colorado, Hawaii, Idaho, Montana, Nevada, New Mexico, Oregon, Utah, Washington and Wyoming and the Western provinces of Canada, British Columbia and Alberta.

While the nationwide market study was underway, CDA Inc. as program manager for the initiative conceived and commenced market development activities to establish this industrywide effort. Because of in-depth past experience within CDA in servicing the architectural market, the aptness of many of these activities proved to be confirmed by the final market study report issued by Sharpe & Associates.

Nationally, a multi-year advertising program was commenced to provide the architectural community with copper design details and specifications as used on actual buildings. The keystone of this program is production of four technical advertising inserts, one to run each quarter in Progressive Architecture magazine reaching approximately 70,000 architects per issue.

Additional advertising was placed in the Western editions of Progressive Architecture and Buildings magazines in order to influence 42,000 building owners and developers, especially in the start-up area.

A 4-page technically oriented brochure on copper roofing was produced and placed in the McGraw-Hill 1991 Sweet's Catalogue File for General Building and Renovation. This is issued each January to 25,000 architectural offices in the United States. The same brochure was carried in June 1991 in the Sweet's Canadian publication, reaching an additional 7,500 architectural firms.

A major program aimed at the emerging architect was the establishment of the first annual copper student design competition in cooperation with the American Institute of Architecture Students. This nationally recognized group of approximately 30,000 students at 130 architectural schools in the United States and Canada was very receptive to the copper sponsored design competition titled, "Where Sight Line Meet: A U.S./Canadian Gateway."

This hypothetical gateway linking the United States and Canada was described as a symbolic manifestation of peace and international cooperation. In light of the free trade agreement, it heralded the opportunities of commerce between the United States and Canada. In carrying out the competition the students explored both traditional and innovative uses of copper, including exterior wall panels, roofing, interior applications and outdoor uses such as landscape architecture and sculpture.

More than 100 student teams from 73 schools of architecture in the United States and Canada submitted entries in this unique competition. Because this program was adopted into the curricula of many of the schools including local competitions for best entries, it is estimated that over 2,500 fourth and fifth year architectural students participated in this competition.

The jury was made up of eminent architects from Canada and the United States along with an officer of a major U.S. copper sheet fabricator and a student architect. The competition was judged on April 4, 1991 at McGill University in Montreal.

First place was awarded to an entry from Texas A&M University by Brian Burke and Jeff Westhoff. These young men will present their submission to us later this afternoon. They describe their first place design as, "This kinetic gateway celebrates the common heritage and aspirations, parallel growth and cooperative spirit of Canada and the United States. The gateway consists of paralleled structures of local granite to reflect permanence of material, timber to reflect renewable material, and, of course, copper a recyclable material that is used in the cladding, roofing and connections." It is planned to continue this highly successful AIAS copper design competition in 1992 and beyond.

We are in the final stages of production of a six-part video series on copper roofing. This architect/sheet metal roofing contractor oriented video series consists of an introduction to copper roofing, construction of standing seam, batten seam, flat seam and shingle and horizontal seam roofs and gutters, downspouts and flashing details. These video tapes were started approximately a year ago in response to a strong demand for more visual representation of the use of copper in everyday construction.

The last major activity I would like to discuss aimed at education and information dissemination to all segments of the cooper sheet construction market is the development of a detailed architectural handbook for product information, design details, sources, maintenance and post occupancy evaluation. We have taken an in-depth look at existing printed material available to architects and specifiers.

Much of this material plus extensive revisions and new details will be organized in this handbook along with a companion Computer Aided Design disk for everyday use in generating drawings and specifications for copper architectural applications. The Architectural Handbook will include considerable reference to worldwide technical information and quality design examples, and ultimately, will include copper product sources, fabricator and installer lists, and machinery and competitive material information. Eventually the architectural handbook will have sections on copper curtain wall systems and other specialty areas such as restoration and renovation.

We are very pleased to date with the progress of the North American Initiative for Copper Architectural Applications. It is being closely followed as a blueprint of technical market development for end use metal application both here in North America and throughout other industrialized countries. We are extremely confident that this will be a successful program that will benefit all segments of the architectural sheet copper industry...producers, fabricators, distributors, contractor installers, architects/ designers and, of course, ultimately, building owners, the end user customer.

REFERENCES

1.   E. Gervais, and P. J. Mackey, "Copper in Architecture Proposal for a North American Market Development Initiative," Noranda Minerals, Inc., Toronto, 1989.

2.   S. Krasney, and J. Slott, "Report on Metal Roofing Market in the U.S.," Corporate Research Center, Union, NJ, 1989.

3.   I. P. Sharpe & Associates, "Roofing and Architectural Copper," Copper Development Association Inc., Greenwich, CT, 1990.

The automotive radiator market: lessons from the past and opportunities for the future

D.K. Miner
Copper Development Association Inc., Greenwich, Connecticut, U.S.A.

LESSONS FROM THE PAST

Because of its dominant position in the radiator market, the US[1] copper industry spent most of the 1960s basking on past successes rather than looking at the market changes possible in the future. It was a benign neglect of product technology developments.

Lost Ground

Early discussions during the late 1960s about the threat of aluminum radiators to the copper industry resulted in comparison of the ease of making copper and brass radiators and the excellent corrosion resistance of copper and copper alloys. Virtually no one considered any potential for aluminum to be a threat.

Therefore, the radiator market was not considered vulnerable, and no competitive effort against aluminum was felt to be necessary. The consensus in the industry seemed to be: "When we have 100% of the market, why spend money to try to improve the market." In general, having the radiator market all to itself, blinded the copper industry to any threat.

Millions for Development

The US copper industry seriously underestimated the willingness of the aluminum industry to spend millions of dollars each year to develop a suitable automotive product.

As early as 1955 the aluminum industry started promoting the virtues of the aluminum radiator. In 1963, Alcoa set up a complete radiator pilot plant to demonstrate how to make aluminum radiators and to provide prototype parts to the auto industry for evaluation. In 1962, Ford Motor Co. also set up an aluminum radiator pilot plant in the Rouge Complex in Dearborn, Michigan, at an approximate cost of $1,000,000. This Ford pilot line received substantial and constant support from the aluminum industry in the form of laboratory work on braze materials and processes, on-site engineering support and special alloy development.

---

[1] This paper presents the situation in the US radiator market and the US copper industry. The author is not as familiar with the European situation, although many of the comments in the paper may apply to Europe.

The aluminum industry demonstrated that the radiator market requires total support from the raw material suppliers that goes far beyond just selling tube and fin materials. This total marketing support continues even today.

Product Evaluation

The US copper industry did not evaluate the limitations of the copper and brass radiator in spite of known weaknesses. And it did not support any significant effort to demonstrate materials and processes to overcome these limitations. Suppliers felt providing material in response to purchase orders was all that was required to serve the radiator market.

Copper and brass materials provide outstanding performance in an automobile radiator. However, repair shop information showed that leaks do occur. These leaks were generally related to poor solder joints which the copper industry did not adequately address.

The US copper industry did not evaluate the strengths and limitations of the aluminum radiator nor the imminence of its threat.

The first high-volume aluminum radiator in service in the US was on the 1975 Volkswagen which used the low cost, mechanically assembled Sofica radiator. Although there were, and are, many limitations on the use of that type of radiator, it did introduce new technology -- using a plastic tank with a gasket to seal the tank-to-header joint. This new type of tank-to-header joint removed one of the major obstacles to the brazed aluminum radiator. Erosion of the aluminum tank was one of the failure modes that had not been solved up to that time.

The soldered tank-to-header joint on the copper and brass radiator has always had a significant failure rate. The copper industry could have utilized the same principle to improve the leak rate of the brass tank-to-header joint, but that would have meant giving away the brass tank business. So, attention was focused on saving the brass tank, and that meant risking that the market for the complete radiator assembly could be lost.

After the plastic tank was introduced on brazed aluminum radiator designs, it remained only to demonstrate the overall durability of the aluminum radiator before it was introduced into production in the US in 1982. The Volkswagen radiator had successfully demonstrated that an aluminum radiator could operate satisfactorily and give serviceable life under US road conditions and existing coolant formulations.

Weight Problems

With the then-current auto industry concerns of the 1973 and 1978 energy shortages, aluminum seemed to fill the perceived need for lightness in automobile components. This came at a time when the copper and brass radiator industry had done little, if anything, to address the weight problem.

The aluminum industry had developed and installed state-of-the-art high volume production methods for rolling thin gauge materials. Therefore, thin gauge aluminum products for the radiator were readily available. Two major brass mills supplying radiator materials were also making aluminum products. However, the brass mill rolling equipment was not as modern as that for aluminum, and the brass and copper products did not have the close tolerance and gauge control as those for aluminum strip. Therefore, the material thicknesses required for a lighter copper and brass radiator were not easily provided, and the mills generally resisted offering the thinner gauge materials.

Promotion

Although not directly related to product technology, the aluminum industry skillfully used advertising to support their technical programs. The copper industry had almost no advertising to support the radiator market.

The aluminum industry had a long running advertising theme of "Aluminum the Modern Material," which was applied to the auto radiator as well as many other auto applications that were being targeted. Younger engineers, who were not familiar with the background of copper and brass radiators, began to see aluminum as a way of getting recognition by changing to the "Modern Metal."

Auto engineers stated that the learning curve on copper and brass had come to an end after 50 or 60 years, and the learning curve on aluminum, "The Modern Metal," would enable them to improve the performance of aluminum radiators for years to come. While this statement is not true, its believability was widely enhanced by the effective aluminum advertising campaign.

High-level Contact

The US copper industry had essentially no direct communications with top management of the auto companies at that time to assess the attitude of the auto manufacturers toward aluminum. However, the aluminum industry, because it had a large number of structural and body panels, cast housings and engine parts, as well as the aluminum radiator to sell, maintained very high level contacts with each of the US auto manufacturers. Regular business and social activities were scheduled frequently with car division vice presidents as well as corporate management. The aluminum industry made sure that their story was well-known at all levels of the decision chain.

Meanwhile, the copper industry had their sales people maintaining contact mostly with the radiator materials purchasing department. Consequently, any story that the copper industry wanted to tell, seldom got beyond the purchasing manager. As a result, the copper industry did not have an effective communication line to know what plans the auto makers were considering and could not easily communicate the copper story to upper auto industry decision makers.

The strong auto interest in aluminum radiators was not apparent to the US copper industry until initial production and introduction dates were announced. Much of this important line of communication is still not in place between the upper auto management and the top management of the copper industry.

The Need to Know

Because there was only limited capability to know what the auto industry needed in support of future car programs, the US copper industry had little knowledge of the various forces working within the auto industry that influenced vehicle design, power train selections, and long-term engineering programs for each auto manufacturer.

The copper industry must have a means of getting this information early enough to keep in step with the auto industry. The auto industry is pressured by many sources about their product development programs. Customer preferences change frequently and rapidly. Congressional laws and government regulations on fuel economy and safety, environmental concerns and economic forces all combine to shape the priority list for the product developments.

Every major auto supplier industry needs to know well in advance, any major changes in the auto development priorities. However, not long ago the copper industry was still making weight savings the top priority when, in fact, the auto industry's first priority was cost.

Engine horsepower and specific design characteristics determine the cooling requirements which set the radiator design. Engine materials-of-construction have strongly influenced the coolant formulation which, again, has a strong influence on the life of the radiator. Information from the engine design programs, as well as the vehicle design, related to air flow to the radiator, must be kept up to date by the copper industry. In this way, copper and brass radiator programs can be focused with the proper emphasis and direction.

A Positive Note

The work done by Nippondenso in Japan is an outstanding example of how total market orientation can result in a superior product. Nippondenso and their material suppliers have worked closely together on all phases of materials and processes to provide their customers with an excellent, highly reliable and durable copper and brass radiator. Rather than being a one-time, short-term program, that cooperative development effort has continued.

The latest Nippondenso new production radiator appears to out-perform aluminum radiators on a BTU/lb basis. It provides improved corrosion resistance, and it is significantly lighter than previous copper and brass radiators.

When Nippondenso first started to manufacture their copper and brass radiators at their Battle Creek, Michigan, plant, they sent out inquiries for special materials to several US brass mills. Instead of responding to the materials requested, some of the US mills merely quoted on their present production materials. This gave Nippondenso the impression that the US mills were not interested in this new business opportunity. Perhaps we still have much to learn from the Japanese.

Lessons Learned

In summary, the copper industry was not prepared for the competition from aluminum in the radiator market. The US copper industry provided materials as ordered by the radiator industry without recognizing the broader requirements of the radiator market and without an effective communication path to top auto industry management. The copper industry reacted to radiator market requirements rather than being proactive to foster favorable decisions.

Under those circumstances, the US copper industry did little technology development on radiator processes or radiator designs. But now that aluminum radiator penetration has become successful, the copper industry has initiated some significant programs in an effort to stop aluminum from completely dominating the US radiator market.

OPPORTUNITIES FOR THE FUTURE

During the 1980's several major programs to improve the copper and brass radiator have been initiated by the copper industry.[2] This continuing improvement in product technology holds great promise for the copper industry.

The Splitter-Fin Design

The lightest and most effective heat transfer design for copper and brass radiators is the Splitter-Fin concept. It was designed and developed in 1975 by Granges Metallverken (Outokumpu Copper AB), Figure 1. It was first produced in 1980 by Tokyo Radiator in Japan for an Isuzu car. The radiator is still in production. It is also presently used by some Volvo cars made in Holland. The Volvo radiator is currently made by NRF Holding, B.V., Holland.[3] The system is also used in heater cores by the Chrysler Corporation in the US.

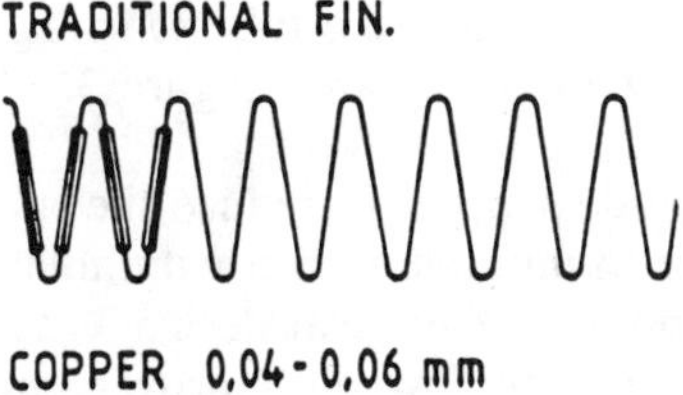

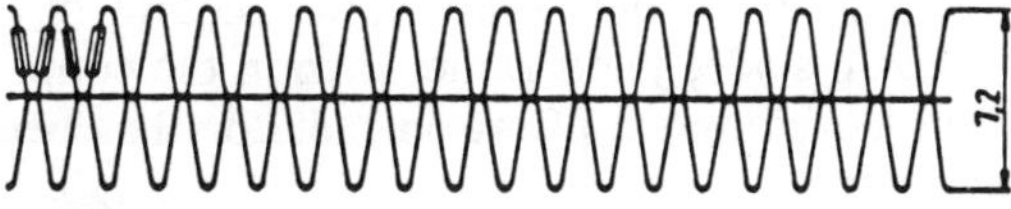

Figure 1 - Comparison of traditional-fin and splitter-fin radiator configurations.

---

[2] The copper industry radiator programs were carried out by CDA Inc., and INCRA (now known as ICA) plus a major, continuing effort by Outokumpu Copper, Ltd.

[3] SAE Paper # 890228, 1989

This outstanding design has not been readily accepted by the US auto industry for several reasons. There is only one source for the very thin strip used for the special fin construction; the special equipment to make the fin assembly is fairly complicated and not as cost-effective as an ordinary fin machine; and, there was concern for possible clogging by insects on the high density surfaces.

This last concern is not a significant problem. With so many US cars being equipped with air-conditioning, the condenser keeps the bugs and trash from clogging the radiator face. And, in actual service, there is no known experience of clogging being a problem with or without air-conditioning. Therefore, only the problems of supply sources and production equipment need to be addressed.

However, there are no known plans for this special fin design to be used by any US radiator manufacturer. This is unfortunate since the product competes so effectively against aluminum on a weight and performance basis. Perhaps this product needs a new marketing effort.

Other Design Concepts

Work done by Professor Ralph Webb of Pennsylvania State University and sponsored by the International Copper Association (ICA) has resulted in the identification of several areas for improving the durability and heat transfer capabilities of copper and brass radiators. With this information being presented at various Society for Automotive Engineering (SAE) meetings,[4,5,6,7] radiator designers have access to this information at will.

The most significant part of this effort has been the application of reformed tube ends to reduce tank-to-header stresses, Figure 2, and the tube-touching design that reduces the depth of the radiator on multiple row, CT-type cores, Figure 3.

The use of a sliding, or flexible, side support to reduce the tube-to-header joint-stresses is also an important change to improve radiator durability. Special radiators are now being assembled that incorporate these particular design features. These radiators will be run on durability tests to demonstrate the overall improvement.

New Joining Processes

As mentioned in the first section of this report, the majority of copper and brass radiator failures occur in the lead-tin solder joints rather than in the copper or brass material. For this reason, much of the development effort on radiators has been directed at improving joint strength and durability.

---

[4] SAE Paper # 850043, 1985

[5] SAE Technical Paper #870183, 1987

[6] SAE Paper #90072, 1990

[7] SAE Paper #900724, 1990

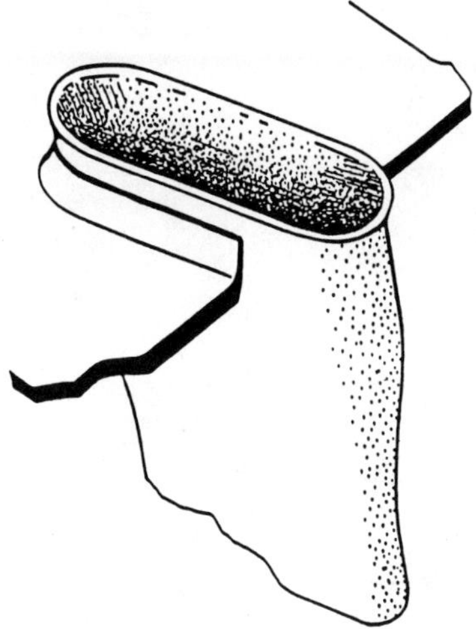

Figure 2 - Reformed tube end at tank-to header joint.

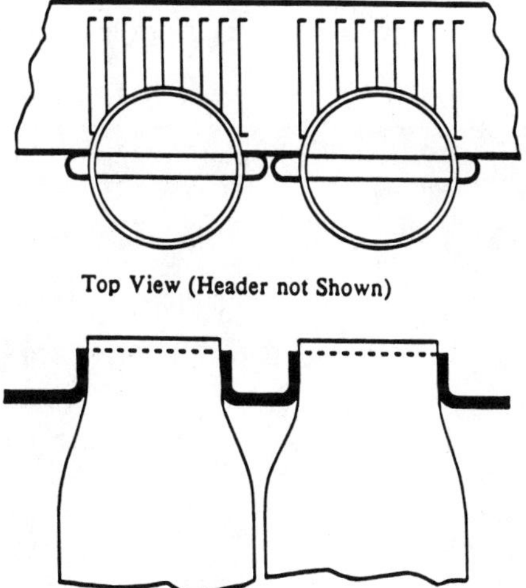

Figure 3 - Tube touching design for multiple row, CT-type cores.

To do this, three criteria had to be met. Lead-tin solder joints lose a considerable amount of strength at radiator operating temperatures, Figure 4.[8] Therefore, there was a need to find a joining material that maintained effective strength at operating temperatures.

---

[8] SAE Paper #720011, 1972

Another goal was to overcome the shortcoming of lead-tin solders which cause blooming corrosion in the tube ends under some coolant and solder-flux combinations, Figure 5. A third consideration for the new joining process was to eliminate the environmental concerns associated with the use of lead.

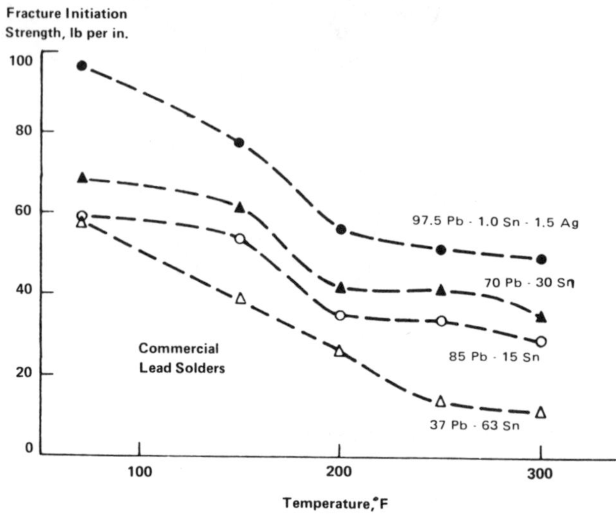

Figure 4 - Fracture strengths of lead-tin solders.

Figure 5 - Blooming corrosion.

Zinc-Based Soldering

Based on some preliminary work done in the laboratory at the US Bureau of Mines, Rolla, Missouri, the copper industry through CDA initiated a development program to determine the properties, advantages and limitations of a zinc-based soldering system. One of the reasons that zinc was chosen as a prospective solder material was that, unlike lead, it has no lasting health effect.

An effective solder composition was identified, and initial laboratory tests indicated that it maintained its joint strength at radiator operating temperatures, Figure 6. Several radiators were assembled for laboratory durability testing using the zinc solder on one end of the tube-to-header joint and production lead-tin solder on the other end. Durability testing demonstrated that failures in these radiators always occurred in the lead-tin soldered tube-to-header joints.

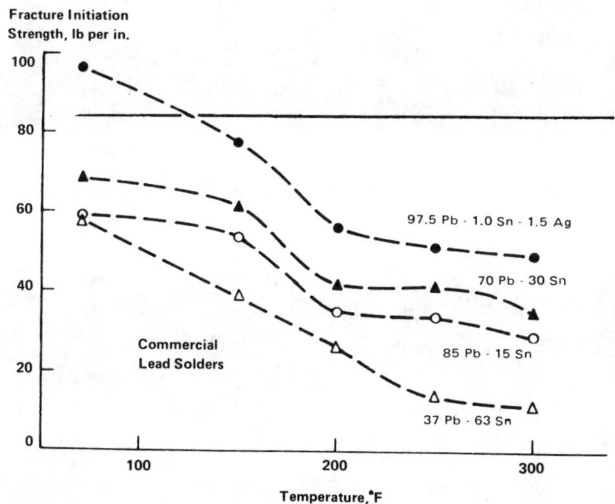

Figure 6 - Strength of new zinc-base solder ZSB1117 at typical radiator operating temperatures. Unlike conventional compositions, the new alloy's strength does not begin to fall off until around 450°F.

Following these encouraging results, 50 production radiators were made using zinc solder for the tube-to-header joints. The normal lead-tin solder remained on the tubes, and zinc solder was added to the joint in wire form. These radiators were then field tested for 2½ years with no failures. The mileage ranged from 100,000 - 160,000. As shown in Figure 7, there was no evidence of blooming corrosion after the vehicle tests.

Laboratory test results indicated that zinc-based solders met the three criteria for new joining materials. In addition, zinc-based solder has a lower density, which results in reduced core weight and a reasonably low material cost.

The next step to further demonstrate the effectiveness of zinc based solder was to make a complete zinc-soldered core. This required the use of a dedicated lock-seam tube mill with a special flux and zinc solder pot, Figure 8, installed in place of the lead-tin solder pot.

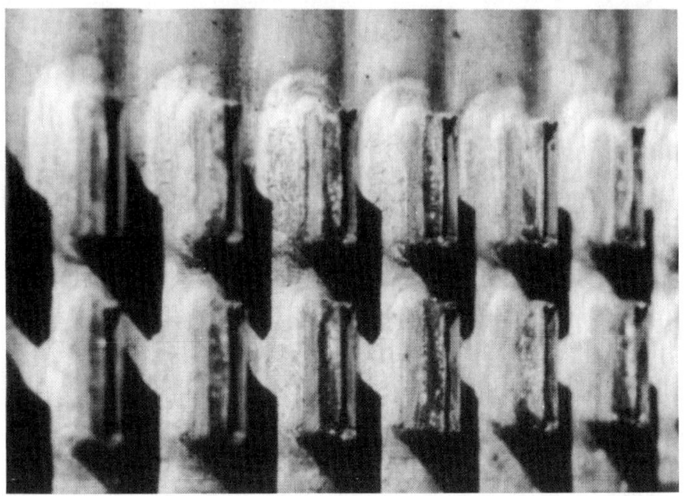

Figure 7 - Radiator joints without blooming corrosion.

Figure 8 - Lock-seam tube mill with special flux and solder pot.

A laboratory core-bake oven, Figure 9, also had to be built for the core-bake operation with the zinc-solder coated tubes.

Figure 9 - Core-bake oven for zinc-solder coated tube production.

Zinc solder picks up oxygen easily which reduces the performance of the solder. To eliminate oxygen contamination, both the tube mill solder pot and the core-bake oven used inert atmospheres. A special feature of the core-bake oven was the ability to rapidly heat the radiator core from a pre-heat temperature of 300°F to a soldering temperature of 900°F. The time at soldering temperature was a short fifteen seconds followed by a rapid cooling to 300°F in the pre-heat chamber.

The first series of zinc soldered cores has been made, and initial durability tests have been started on the Copper Development Association Inc. (CDA) radiator cycle-test stands. The tubes, fins, header plates, tanks and side supports used to assemble these cores were standard production radiator components. Only the zinc solder was new. As of the time this report was prepared, the zinc-soldered core had accumulated 3,100 hours of cycle-test time, and testing was continuing.

Low-temperature Brazing

Brazed copper assemblies have been used in special heat exchange applications for many years. Brazed copper and brass automobile radiators had never been seriously considered because the brazing temperatures normally required anneal the assembly and make the structure too soft. The joints, of course, would be very strong.

When equipment suitable for brazing aluminum radiators was developed, it became apparent that if a brazing alloy for copper could be found which melted at approximately 1,200°F, a brazed copper and brass radiator might be considered. The advantages of this brazed radiator would include very strong joints, no blooming corrosion at the tube ends, and no lead environmental concerns.

Brazing, as a process, has one advantage over soldering. The brazed radiator is a very good, reliable radiator once it passes the leak test at the manufacturing plant. If the fit and tolerances are not right and/or the process conditions are not correct, the brazing material will not completely fill the radiator joint and the radiator will fail the leak test. There are only good brazed joints or unacceptable brazed joints--no grey area in between.

With soldering, it is difficult to know whether the solder joint is very good or just good enough to pass the leak test but not able to withstand a long life in operation. This is because the solder will bridge rather wide gaps to contain the coolant. But, such a joint is not as strong as a solder joint made with the proper clearance.

The best way to ensure very good solder joint quality is to have complete statistical process control on all components of the radiator and the soldering process. The capital equipment cost for soldering is less than the equipment cost for brazing. However, the statistical process control for soldering must be more effective than most present copper and brass radiator manufacturers are doing.

ICA has funded a brazed copper and brass radiator development program since 1990. This is a five-year effort to demonstrate the effectiveness of the brazed radiator to the auto industry. Funding is authorized on a yearly basis. Program effort is directed toward three specific areas: design modifications required for a brazed radiator, copper alloy modifications required for a brazed radiator, and braze alloys that are suitable for a brazing temperature between 1,000°F -1,200°F.

At the end of 1990, a production radiator was selected as the basis for the initial brazed copper and brass construction. Permission was obtained to use available tooling and some production parts from that manufacturer. Initial design modifications were selected from the Pennsylvania State University design program and drawings were made incorporating these modifications.

An existing copper alloy, not previously used for radiator fins, was selected that provides somewhat higher anneal resistance than present fin copper. This material was rolled to a thickness of approximately 0.002" and it's properties were established. Maximum allowable brazing temperature is estimated to be 625°C (1,157°F). Conductivity of the material is approximately 90% IACS. Tensile and hardness properties are suitable after the brazing cycle if the temperature is 1,150°F and the time at temperature is less than two minutes. Evaluation of alternate tube materials will be carried out in 1991.

An available braze alloy in the CuNiSnP alloy group was found to make reasonable braze joints at approximately 1,250°F. Considerable effort has been directed at obtaining a lower brazing temperature on modified braze alloys. The results from the work in 1990 have been very encouraging, and it is expected that a suitable alloy will be available by the latter part of 1991.

During 1991, considerable effort is being directed at making a brazed radiator core with presently available materials. This will allow the brazing process to be optimized with regard to joint clearances, fixturing, process control and braze material application to the parts. By the end of 1991, the entire braze system will be ready for optimization. During 1992, tooling will be finalized, tube and fin materials produced in pilot quantities, optimized braze alloy provided in paste form, and core assemblies brazed with the optimized process conditions.

Heat-transfer test results and durability test results will be compared to the performance of the base-line radiators. This brazed copper and brass core should be ready for introduction to the auto industry early in 1993.

The Future Can Be Ours

Two new joining processes for copper and brass radiators are being developed. Zinc-based solder may be most useful for lower volume heavy-duty radiators and aftermarket radiators. Brazed copper and brass radiators are expected to be produced by OEM radiator manufacturers for higher volume cars and light truck applications. Combining these two new joining processes with the various design concepts from the Pennsylvania State University work provides the copper and brass radiator designers and manufacturers with many new tools for developing and producing new copper and brass radiators.

The new copper and brass radiators made possible by the various copper industry programs will be fully competitive with aluminum radiators as far as durability and several times better than present copper and brass assemblies. They will be superior to aluminum in overall heat transfer performance, corrosion resistance, ease of fabrication and repairability--all characteristics in which copper is inherently superior to aluminum.

It must be recognized that the copper industry can only provide new methods and concepts, not new radiators. To demonstrate these new methods and concepts requires making and testing prototype assemblies. But, the final design and construction of any new radiator must be accomplished by the auto industry and their production suppliers. The US copper industry must make every effort to ensure this happens.

REFERENCES

1. Hans Vogelaar, "Practical Experiences with Splitter Fin Products for OEM and Aftermarket Applications," SAE Paper #890228, 1989.

2. R.L. Webb, T.A. Burkett, and H.D. Foust, "Assessment of Automotive Radiator Technology and the Future of Copper/Brass Radiators," SAE Paper #850043, 1985.

3. R.L. Webb, and Wen F. Yu, "Stress Distribution and Stress Reduction in Copper/Brass Radiators," SAE Technical Paper #870183, 1987.

4. R.L. Webb, "The Flow Structure in the Louvered Fin Heat Exchanger Geometry." SAE Paper #90072, 1990.

5. R.L. Webb, and P.I. Farrell, "Improved Thermal and Mechanical Design of Copper/Brass Radiators," SAE Paper #900724, 1990.

6. R.E. Beal, "Development of New Solder Alloys for Automotive Application," SAE Paper #720011, 1972.

Tactical approaches to developing markets for copper tubular products, illustrated by programs on fire sprinkler systems and natural gas systems

A.A. Knapp, and R.J. Catterall
Canadian Copper & Brass Development Association, Don Mills, Ontario, Canada

ABSTRACT

Projects to develop new markets for materials involve several interelated activities, planned and coordinated to have an overall positive effect on the consumption of the material concerned. The activities include identification of a number of factors, such as potential markets, decision-makers involved, regulatory and code restrictions, forms of promotion needed, training required, and technical support services. Each specific area requires special skills and expertise, and a copper development association typically pulls together all of these activities into a coordinated market development initiative. Two current programs in Canada will be used for illustration. They involve the introduction of copper tube to replace steel pipe in fire sprinkler systems and in natural gas systems inside buildings

INTRODUCTION

Since mid-1989, the Canadian Copper Industry has carried out concentrated development and promotion programs on two applications. The first is automatic fire sprinkler systems for protection against the hazards of fire in all types of buildings, including single-family residences. The second is natural gas systems inside buildings, in which pipe is used to convey gas to heating equipment and appliances.

Fire Sprinkler Systems

Steel pipe with threaded joints and fittings have been used for decades for fire sprinkler systems, first to protect property, and more recently to protect against loss of life and bodily injuries. Although such steel systems have provided acceptable performance under most circumstances, efforts directed at developing more cost-effective materials identified copper tube and capillary fittings as a material system with major market potential. This potential was first promoted in Canada in 1975, and the first commercially installed copper fire sprinkler system was installed at that time in Montreal.

Natural Gas Systems

In the early 1960s, there was some use of copper tube for underground natural gas services (the service being the line from the gas main at the street to the building or house). But with the introduction of certain types of plastic pipe, interest in copper services waned.

In recent years, however, Canadian regulatory authorities agreed to permit copper tube to be used for natural gas installations above ground, which is a new application, and market, for copper. In this area, black steel pipe with threaded joints and fittings has been the predominant material used for decades. Plastic pipe is not permitted for gas installations inside buildings.

REGULATORY CONSIDERATIONS

Building Codes & Fire Sprinklers

Regulations for the design and installation of fire sprinkler systems normally fall within the jurisdiction of building codes. Additional requirements may also be contained in provincial and/or municipal legislation, regarding the need to install certain types of fire protection equipment or systems, in particular classifications of building, which are usually referred to as types of occupancies in Canada.

The National Building Code of Canada (NBC), published by the Institute for Research in Construction (formerly the Division of Building Research of the National Research Council) is the document of primary interest regarding fire detection and suppression regulations in Canada. It sets down the level of protection to be provided for various types of buildings, from single-family residences to multi-storey, multi-unit office towers, theatres, hotels and motels, and so forth. These regulations are generally used as the basis for provincial and municipal building codes, with amendments to suit local needs.

In Canada and the U.S., code-writing bodies have given virtually unanimous support to the standards issued by the National Fire Protection Association (Quincy, MA) for the design, installation and testing of fire sprinkler systems. NPPA Standards 13, 13D and 13R have become the main standards referenced in building codes throughout Canada and the United States.

NPPA Standards permit the use of copper tube and fittings for most types of buildings, with few major restrictions.

Natural Gas Installation Regulations

In Canada, CANI-B149.1, Installation Code for Natural Gas Burning Appliances and Equipment, is referenced by most provincial regulatory authorities across Canada. The Canadian Gas Association is responsible for the preparation and revision of the Code, and maintains a number of committees to provide input for the document.

In the mid-1980s, CGA revised B149.1 to permit the use of copper tube and fittings for above ground gas distribution systems. This means the lines that supply gas to a furnace, water heater, range, barbecue, or other gas-burning appliances. A number of Provinces have since accepted these sections of the Code, allowing copper to be used.

RESTRICTIVE FACTORS

With regulatory acceptance in place, it may be thought that spectacular new market growth is assured. Unfortunately, this is not usually the case.

Much laborious effort is often required to identify the factors that restrict market growth and to formulate ways and means of overcoming these restrictions.

Natural Gas Systems

On Natural Gas Systems, our market development program in 1990 focused on the following activities:

- Develop liaison with gas companies, installation contractors, and code authorities
- Publish an authoritative Installation Manual
- Participate in CEX 90 trade show
- Publish promotional folders, as needed
- Develop ASTM specification on tube for natural gas and propane

From these activities, we were able to identify a number of factors that were restricting the development of the market. They included confusion among potential users about the regulatory acceptance of copper, its installation features, fabrication techniques, as well as sourcing and purchasing of materials.

The current program includes the following activities:

- Intensify liaison with contractors, gas companies and code authorities
- Organize seminars for installation contractors
- Produce video(s) on copper systems
- Participate in CIPHEX shows in Montreal and Vancouver

Fire Sprinkler Systems

The program on Sprinkler Systems is somewhat similar, covering the following:

- Develop liaison with sprinkler contractors, fire officials and code authorities
- Identify factors restricting market growth
- Participate in CEX 90 trade show
- Publish promotional literature
- Support Operation Life Safety Canada's program
- Participate in code hearings

Here, the following have been identified as restricting factors:

- Incidents of flux corrosion
- Lack of training (fabricator skills)
- Ignorance of code acceptance
- Lack of knowledge about design and installation of copper
- Reluctance to try a "new" material

For the current year, the program has been altered to cover the following activities:

- Focus liaison on selected contractors
- Initiate actions to overcome restricting factors
- Organize training seminars for installers
- Organize promotional seminars for developers
- Include in trade show exhibits

CONCLUSION

Market development programs on Natural Gas Systems and Fire Sprinkler Systems offer significant opportunities to dramatically increase the use of copper tube and fittings in Canada. Consumption of copper in these applications could exceed 15,000 tonnes annually, and approach 25,000 tonnes if significant market penetration is achieved. This new consumption of copper will generally replace steel pipe. At the same time plastic pipe is becoming a competitor in the fire sprinkler market.

ACKNOWLEDGEMENT

The purpose of the Canadian Copper & Brass Development Association is to develop and maintain the applications for copper, its alloys and compounds. The work of the Association is funded by the Canadian Copper Industry. Additional funds for the development of the fire sprinkler and natural gas systems markets are being provided by the International Copper Association.

* * * * * * * * * * * *

Where sight lines meet—a U.S. - Canadian gateway winning competition entry

B. Burke, J. Westhoff and T.L. McKittrick
Texas A&M University, College of Architecture and Environmental Design,
College Station, Texas, U.S.A.

In conjunction with the North American Initiative's emphasis on education, a major program aimed at the emerging architect was the establishment of the first annual copper student design competition in cooperation with The American Institute of Architecture Students.

Students were invited to design a gateway at a site in the Thousand Islands area of the St. Lawrence River where a bridge between Canada and the U.S. presently exists. It was to be a symbolic manifestation of peace and cooperation, and herald the opportunities of commerce between the two nations that will result from the Free Trade Agreement.

The design could include an information center for tourists, a theatre, a gift shop with fast-food service, conference meeting rooms, places for observation of the natural environment, environmental art and landscape elements, boat docks and ancillary facilities.

On reviewing the program consideration was given to the meaning of the river, the exploration and settlement of the two nations, and the reliance on ships and boats in that process. Several metaphors were developed as sources of design inspiration including: 1) the East-to-West development of each country; 2) the largely cooperative relations between them; 3) the existence of a similar political heritage; 4) the "Melting Pot" assimilation of people from many cultures that has shaped the values of both nations; 5) the influence of water transportation on the economies of both nations, with particular emphasis on the St. Lawrence River and the Great Lakes.

The designers explored the copper alloys promoted by the sponsors of the competition for their appropriate use in the design. Of particular interest were the weathering characteristics of copper and brass which would influence later design ideas. Reference material about the use of timber in construction of large structures using brass and bronze connectors was reviewed. Consideration was given to granite as a permanent material for monumental structures in severe climatic conditions.

The following are the important features of the design:

1. The divided roadway bridge that spans the narrow water course containing the boundary line would pass through two tall parallel walls of granite.

2. Atop the walls would be counterbalanced wooden roof elements with copper roofing, designed to be pivoted from closed to open positions by the weight of time-line panels.

3.   These panels would be made by artists chosen through annual design competitions, and would depict the major historical events of the previous year in each nation. They would be of carved wood with copper or brass cladding, one material for each nation. The panels, or "tablets", would be fabricated at a workshop near the site of local timber and recycled copper and brass.

4.   There would be an annual mid-summer celebration at the site that would include several events.

5.   The first event would commence with two new tablets being hoisted onto sloping tracks near the tops of the parallel walls. The force of gravity would cause them to roll downhill on the tracks. Sections of track between wall segments would be mounted on vertical guides, and would lower on the guides when the weight of the tablets arrives on them. This would activate the roof pivoting mechanism, causing the roof panels to open, or "bloom."

6.   When the tablets reach the end of the walls and the roof panels are all open, the tablets are lowered and moved by revolving cranes to tethered floating display structures.   There are ten such structures aligned with each of the parallel walls. This enables a ten-year history of each nation to be displayed for viewing primarily from the many boats that enliven the river in this area.

7.   The revolving cranes, built of local timber with brass and bronze connectors and fittings, would sit atop granite bases. They would be powered by boats and manned by local volunteers. Their aligned forms suggest the ribs of a giant boat.

8.   A museum with an observation tower would terminate the composition at the East end. The last of the revolving cranes serves to move the tablets into a semi-enclosed space. Here the copper and brass would be stripped from the tablets for re-cycling after reaching their full patina, allowing the bare wood tablets to be seen for the next fifty years. Thus the time-line represents the approximate length of a generation.

9.   A symbolic "melting pot" would be located at the artists workshop, and could be the venue of another event, the re-cycling of the copper and brass tablet skins.

10.   The parallel walls and display structures symbolize the parallel development and mutual aspirations of the two nations, while the events associated with the revolving cranes suggest the spirit of cooperation that exists between us. The converging sight lines established by the walls, cranes, and display structures, when seen from the observation tower looking West, suggest the future of this warm, interdependent relationship.

The following are some of the jury comments for the submission:

o "This was the most sensitive in that it involved cooperation in an event through time. For me the event seemed to have its emphasis not on being monumental in scale, but in being personal and interactive."

o "A highly imaginative exploration of time, material and process that becomes a living symbol of the leadership between the two countries."

o An interesting parallel of design is that it reflects some of the process of enhancement and movement that is a part of the copper refinery industry.

The results of this competition should prove to be very useful to help Brian and Jeff gain admission to graduate schools of their choice. They are both very grateful to the Copper Development Association and the Canadian Copper and Brass Development Association for making the competition and its results possible.

Copper panels are stored in a staging area (right). Each year a panel is engraved with references to significant events in both Canada and the United States. The panel is then removed from the staging area and transferred by the first of many crane-like mechanisms (center) to a pier (two are shown on the left) from which the panel is suspended.

Each succeeding year that panel is moved by one of the cranes to the next pier in line, and a new panel is removed from the staging area. Once all the piers have a copper panel suspended, they will present a chronology of the current decade's events.

After the 11th year, the last crane in line will retire the last panel in line to this museum where it will remain on display for a year. The copper panel is then recycled to make a new panel that will be stored in the staging area until needed again.

New manufacturing technology as a means to new product development — the automotive radiator story

R. Sundberg
Outokumpu Copper, Research and Development, Västeras, Sweden

ABSTRACT

The production of thin gauge copper and brass strip used for automotive radiators goes back to the beginning of the 1950's in Outokumpu Copper (former Metallverken). As a matter of fact the first precision cold rolling mill was designed for rolling of nickel-silver strips for springs to electro mechanical telephon exchanges, typical thickness 0,4 mm. Close thickness tolerances are very important for the performance of the nickel-silver springs. The mills were of a special design, so called pre-stressed and constructed and built within the company. They were soon found possible to use even for thinner gauges with excellent thickness tolerances. At that time the automotive industry in Sweden had started to grow. Swedish production of automotive parts as radiators was already in business and Metallverken supplied some strip material. Through contacts with the radiator manufacturers it became evident that thin gauge strip with close thickness tolerances and controlled properties was needed. The development of the new product and its production equipment was closely linked to the needs of the customer, the radiator manufacturer. The knowledge of the final product, the radiator, also grew. The high conductive fins were found not to be efficiently utilized. They should be much thinner. The splitter fin design using 25 um strips was developed and proved that thinner gauge material could be used without any performance loss. Modern radiator design is now using much thinner fin material.

PRODUCTION PROCESS DEVELOPMENT

Copper and brass are generally very ductile materials which permit high reductions at cold rolling. Combining 4 or 5 of the pre-stressed rolling mills after each other made it possible to roll effectively with high reductions.

Fig. 1 - 4-stands cold rolling mill.

All inter annealings were very early in the development left out for copper. The strips were cold rolled from hot rolled slabs to annealing dimension before final rolling, reduction 98,7%. That high reduction made it necessary to control the annealing texture which will be dealt with in the next part of the presentation. The work hardening for brass is much higher than for copper.

One interannealing after the breakdown, rolling to 3 mm had to be done but then the strips were rolled to annealing point with a reduction of 95%.

In the middle of the 1960's Metallverken decided to start production of brass strip within the European common market in Holland. As the mill should be fairly small a large hot rolling mill was out of the question and strip casting process was not developed. Because of that a new process starting with centrifugal casting of a ring and continously cold rolling of that ring was developed.

Fig. 2 - Ring rolling of centrifugal cast brass rings.

     In that way it was possible to have fairly long lengths
without welding and a relatively cheap investment. The casting
and ring rolling process proved to be very demanding and the
company had a long learning period. Today 120 mm thick rings
are rolled to 3 mm, reduction 97,8%, annealed and rolled to
final annealing dimension, reduction 95%.

     With time it became evident that strip casting was the
process of the future for production of copper and copper
alloys. After evaluation of the conventional horisontal casting
process it was decided to develop our own vertical process with
high efficiency (1). A production unit for copper was started
in 1982 in our Finspong plant for production of thin gauge
copper strips. The as cast strips are cold rolled with 99,6%
reduction before annealing and final rolling.

     It was not only the casting and rolling processes that
were developed. A continouos brigth annealing process for
copper was introduced at an early stage instead of the very
problematic coil annealings that was standard in the industry.
Strand annealing of thin gauge brass was also used from the
beginning. Finally, slitting machinery of own design was taken
into use. More or less all of the equipment used in the
production of radiator strips is of our own design and
manufacture.

MATERIAL DEVELOPMENT

     The thin copper strips are used for fins which conduct the
heat from the tubes to the air, fig. 3

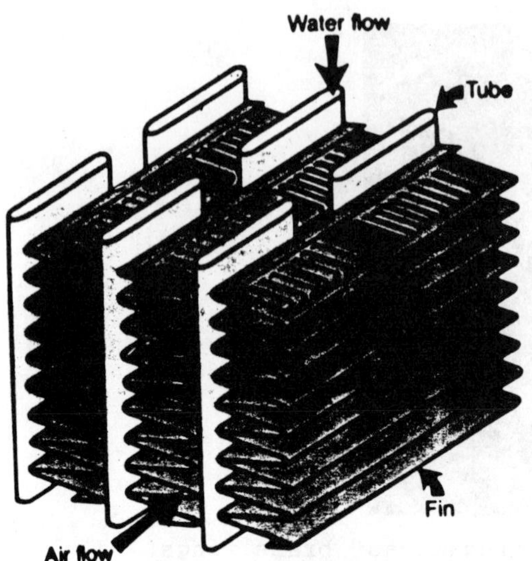

Fig. 3 - Radiator core with serpentine fins and flat tubes.

High conductivity and high softening temperature are needed. The later because the fins have to retain their strength after the soldering operation which is used to join the radiators. A third property texture had, due to the high cold reduction, to be taken into account. Conductivity, softening and texture can be influenced by small alloy additions. Fig. 4 shows how conductivity and softening temperature depend on amount of Ag, P, Cd, Sn, Zn in solid solution in copper according to (2).

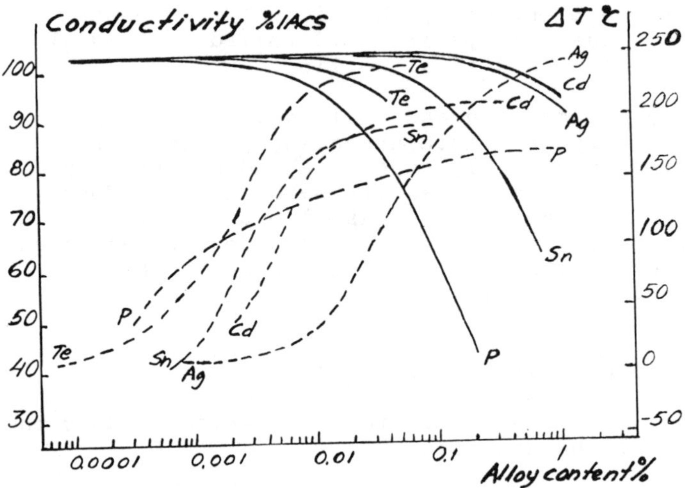

Fig 4 - Conductivity, % IACS and increase in 50%-softening temperature, °C, as a function of alloy additions to copper.

Silver was the alloy element which was used to reach the desired
properties. As it is an expensive element as low content as
possible, 0,05-0,08% Ag, was used. The influence on conductivity
is neglible but the softening temperature is lower than for
cadmium or tin. It is obvious from fig. 4 that cadmium is the
best choice. Phosphorous is an element that is present in most
copper radiator strips. It is used to deoxidize, and remaining
content after casting has to be as low as possible, not to
reduce the conductivity too much.

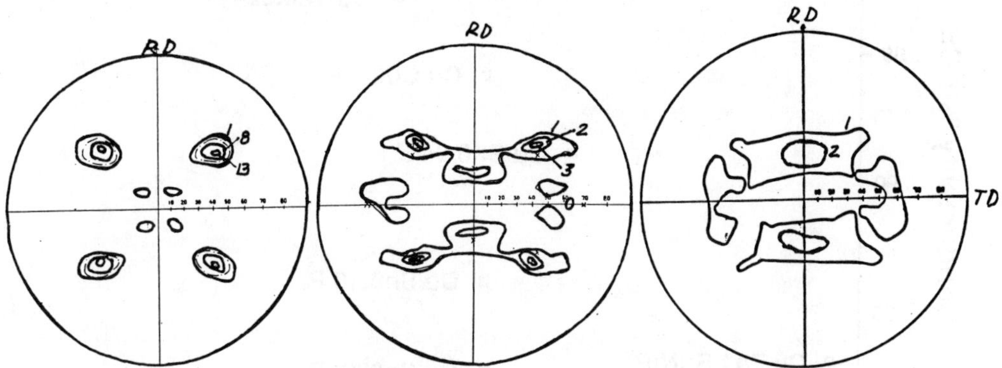

a) Cu hot rolled and   b) D.O. CuO,2%Cd. c) Strip cast and cold
   cold rolled 95%.   rolled 99,7%.

Fig. 5 - (III) polefig. for annealed copper strips.

     A strong rolling texture is developed in copper after
reductions above 90%. The rolling texture is for pure or low
alloyed copper changed into cube texture, fig. 5 a, during
recrystallisation. It gives the strip mechanical properties with
large anisotropy. That has to be avoided. Additions of cadmium
and phosphorous suppress the cube texture, fig. 5b.

     Such a small amount as 0,003% P can reduce the cube texture
dramatically, fig. 6.

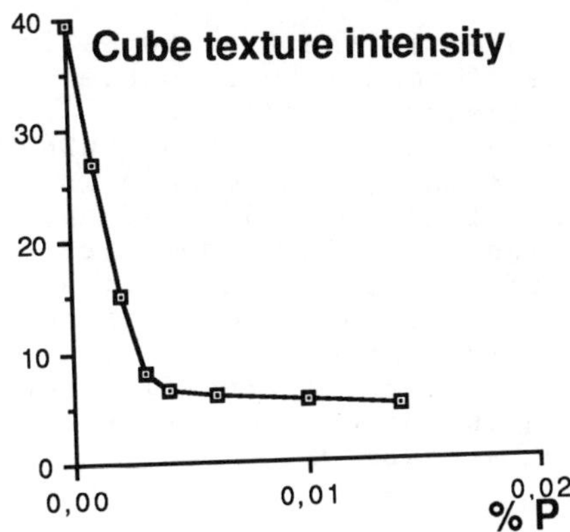

Fig. 6   - Influence of phosphorous on cube texture in
            annealed copper.

Cadmium became very early the chosen alloy element in the copper radiator strips. The only but unfortunately growing problem with cadmium is its poisonousness. That has lead to search for other elements. Tellurium has in combination with the strip casting process been found to give still better combination of properties, fig, 7.

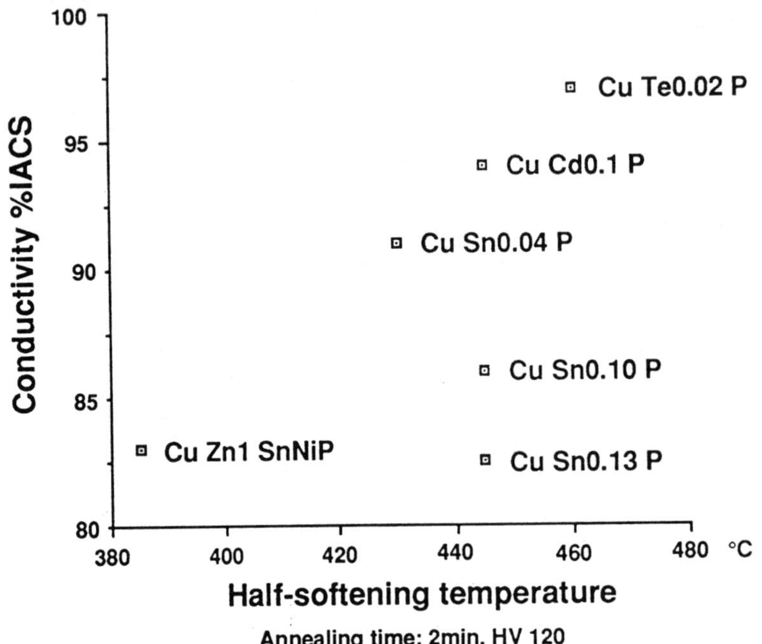

Fig. 7 -   Half softening temperature versus conductivity for
           fin copper alloys.

Cold rolling of casted strips results in a modified rolling texture and consequently the cube texture is completely suppressed, fig. 5c. It is with this type of strip with close dimension and strength uniformity possible to form the fins with all details, fig. 3.

The thin gauge brass strips are used in the flat tubes in the radiators, fig. 3. They are either formed with a lockseam, which is filled with solder, or high frequency induction welded. A number of different mainly -brasses are used. CuZn30 is standard in USA, CuZn30 or CuZn36 in Europe and CuZn35 in Japan. The strips are either used in a fine grained, 5-10 um, or annealed and cold rolled condition. The corrosion properties of the tubes have in recent years been of interest. Dezincification and intercrystalline corrosion have occured when radiators have been exposed to salt and airpollution (3). Arsenic and phosphorous additions are known to improve the dezincification resistance. Inhibited brass have been introduced. Addition of 0,02% P and fine grain size has given a brass with good properties in all aspects, table 1.

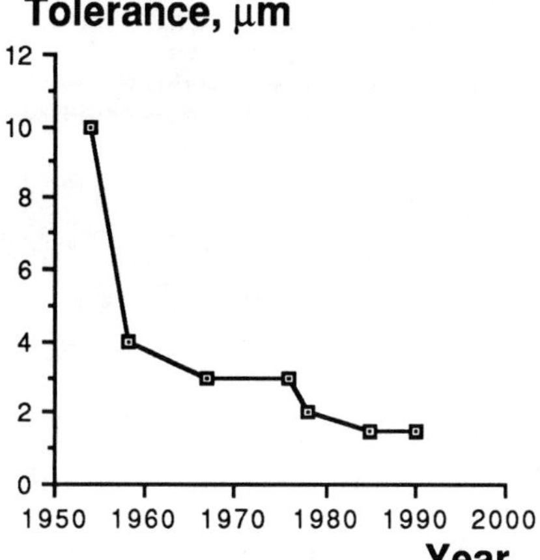

Fig. 8 - Development of thickness tolerances.

The average thickness can, thus be decreased by using strips with closer tolerances, i.e. for a certain weight a longer strip is available. This can in most cases be done without changing the performance of the radiator.

The knowledge of which variables that determines the performance was growing gradually through the contacts with the customers. It became very interesting to know which is the thinnest fin that can be used. Rough calculations indicated that a reduction to less than half the normal thickness could be possible.

Thin fins of normal design are difficult to handle. A specialtype, so called splitter fin was therefore developed, fig. 9 using 25 um strip.

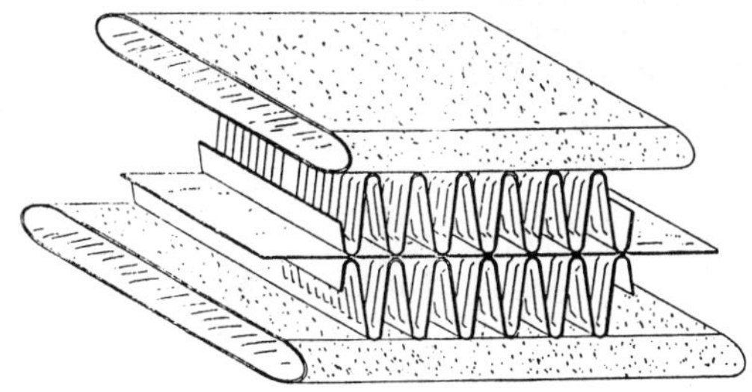

Fig 9 - Splitter fin design.

It was introduced for the first time at a CIPEC conference (4) 1975. The production line and the design has since then been developed further. It is used in some radiators and a lot of heaters. The efficiency of the new fins in a radiator core had to be known. Wind tunnel testing was needed. A long time cooperation with the technical center of the swedish air plane manufacturer started.

The knowledge of the interaction between material and radiator
design grew. In depth discussions with the customers was
possible. It became such a valuable sales tool that a wind
tunnel of our own was built some years ago.

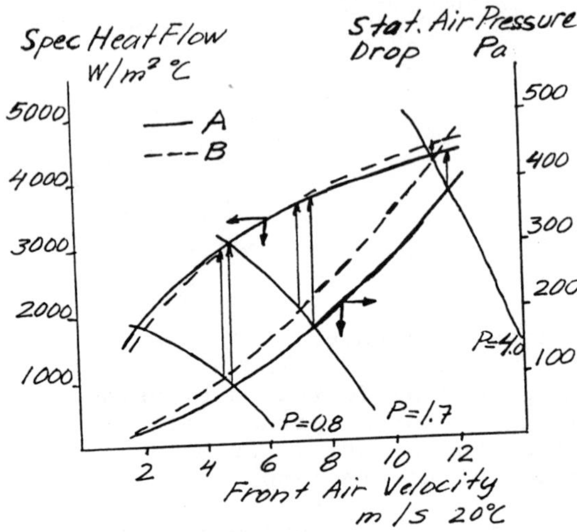

Fig. 10 - Heat performance and air pressure drop determined in
      wind tunnel.

      Fig. 10 shows the curves of heat performance and static
pressure drop versus air velocity that are normally determined
in a wind tunnel test. We often use a method suggested by Modine
Manuf. Co. (5) to evaluate the results. When comparing different
radiators a rating is given at three different air velocities.

      The competition from the copper/brass substitute aluminium
became from the middle of the 1970's very fierce due to the
first oil chock and a very high copper price, fig 11.

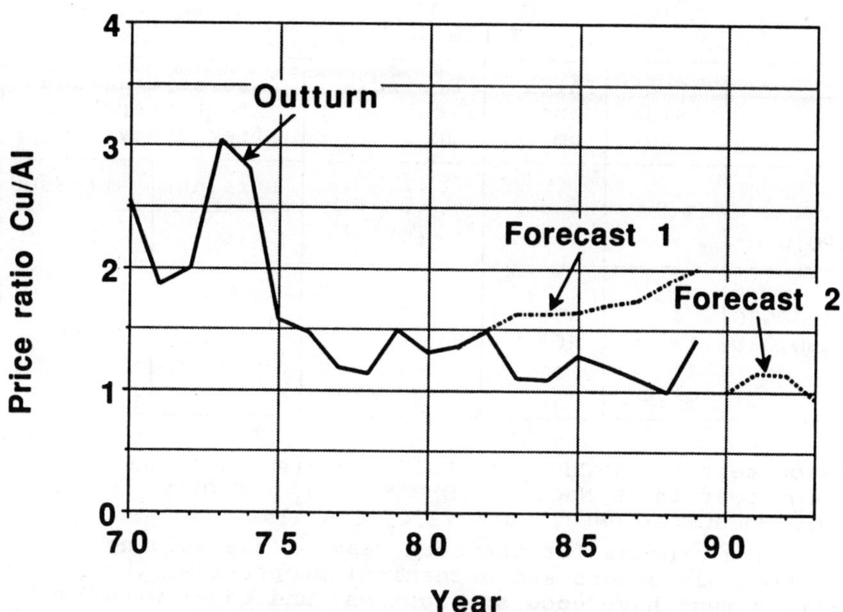

Fig 11 - Price ratio Copper/Aluminium.
         Forecast 1, Chase Econometrics 1980.
         Forecast 2, Brook Hunt, Drexel, RSI, CRU, 1989.

    Aluminium radiators were lighter than the normal
copper/brass ones. The image of copper was not very good either
when compared to aluminium which was looked upon as new and high
tech. The splitter fin design was used to prove that copper/brass
did not need to be heavier than aluminium. A great number of
radiators were tested. Fig. 12 shows a summary of the results.

Table 1

Sample	Grain size um	Thickness um	Dezinci- fication um I	Intercrystalline attack, depth um after 3 days II	
				straight	stressed
CuZn30PO,020	5	110	(5)	6	20
CuZn30AsO,030 CuZn30AsO,030	4 19 + rolled	178 105	(6) (9)	6 20	50 50

I   Immersion test in 1% $CuCl_2$ at 25°C see also (11) and (3).
II Immersion test in 1M NaCl + 0,05M Na_2 SO_3 + 0,025M CuCl + 0,02-0,05M CuCl at pH 3,5 and 22°C, see also (12) and (3).

The forming process for the lock-seam or the welding depends on even dimensions and mechanical properties. The slitted strips must have good straightness and edges with low burrs, which implies a high quality slitting operation. Strips to be slitted must be flat and free from uneven internal stresses, i.e. a good rolling result is a necessity.

RADIATOR DESIGN AND MANUFACTURE

It became very early in the development process evident that the forming and joining processes used in the manufacture of radiators had to be known in order to understand how to develop the strips. The forming of tubes and fins depends on even properties regarding hardness and dimension. Surface roughness and cleanliness can also be of importance for the forming as the strips are passing a lot of rolls when shaped into tubes or corrugated to fins. The surface properties are still more important for the joining of the radiators through soldering. As the years have gone by we have gathered a lot of knowledge regarding the interaction between the radiator production and the quality of the strips. This is of great importance as a lot of the radiator manufacturers are small companies with limited resources.

The rolling process development made it possible to decrease the thickness tolerances.

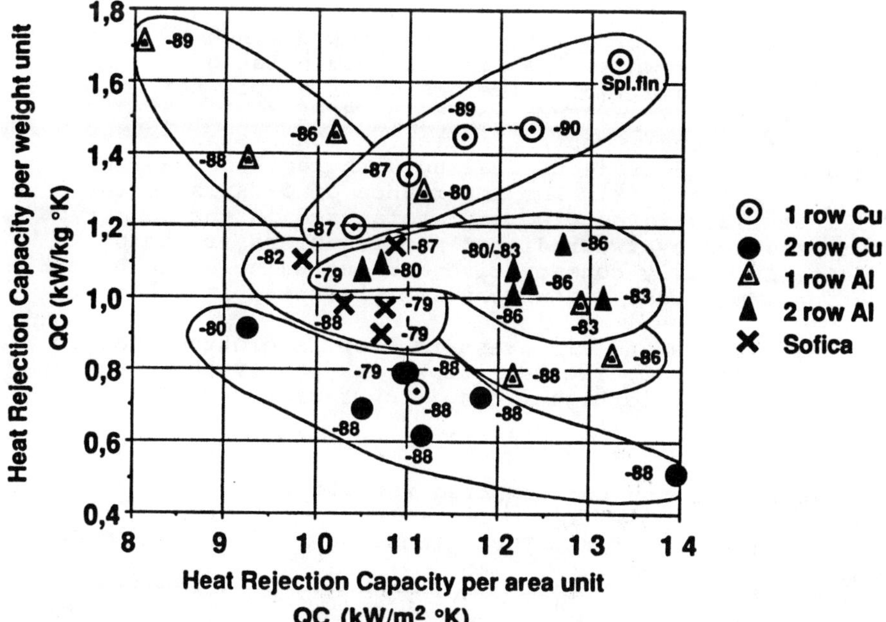

Fig. 12 - Heat rejection capacity per weight unit versus heat rejection capacity per area unit for a number of radiator cores. Results from Outokumpu Copper.

It is clear that for one row radiators (one row of tubes) a copper/brass core can be as light as aluminium. It is interesting to see that the heat rejection capacity per area unit of core is higher for copper/brass. The splitter fin core is among the best ones but other modern designs are also very good. Splitter fin was as said introduced to show that it is possible to use much thinner fin material than normal. The development of the average strip thickness delivered from our rolling mills has, as shown in fig. 13, decreased with >15% since 1980.

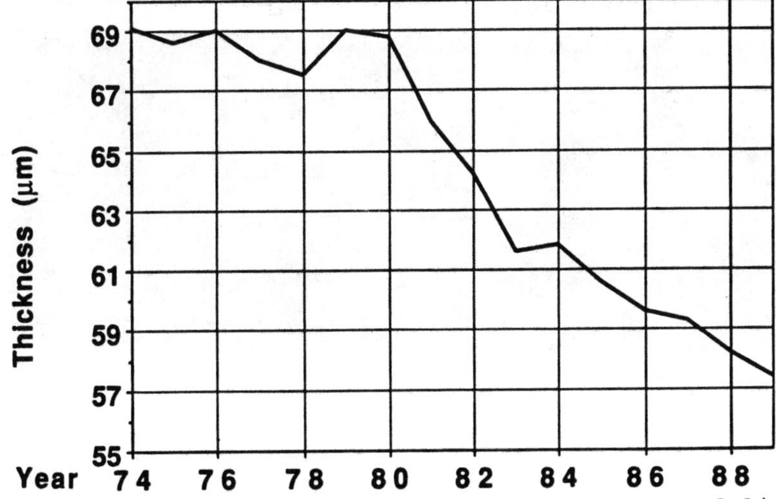

Fig. 13 - Average copper fin material thickness delivered from Outokumpu Copper Radiator Strip i Finspong

The interest for calculation of the performance of radiators has also grown as the results from a lot of measurements became available. The design of a splitter fin core that matches a conventional can now be predicted with good accuracy. International Copper Association ICA has in recent years had a number of projects dealing with radiator design problems, (6), (7) and (8). The importance of details in the design will be illustrated. Two radiators used in the same type of car but produced by two radiator manufacturers are being compared. One is fairly conservative and one using a light weight design. Important dimensions are presented in table 2.

Heat performance and air pressure drop is plotted against air velocity in fig. 10. The heat flows are nearly the same but the air pressure drop is higher for 1-row radiator B. This is astonishing.

Table 2 - Dimensions for two compared radiators.

Radi-ator	QC kW/m^2·C	Core weight kg/m^2	Tube dimension mm	Fin dimension mm
A	11,1	17,9	2 rows 16,0x2,3x0,19	40x10 x0,050
B	11,0	8,5	1 row 18,6x2,2x0,093	22x7,2x0,030

Fin count Fins/inch (FPI)	Fin surface m^2/m^2 core.
15	47
18	31

Radiator A has 50% larger fin surface, table 2. Still the heat performance is not any better. The explanation is found when looking at the louver shape, fig. 14a and b

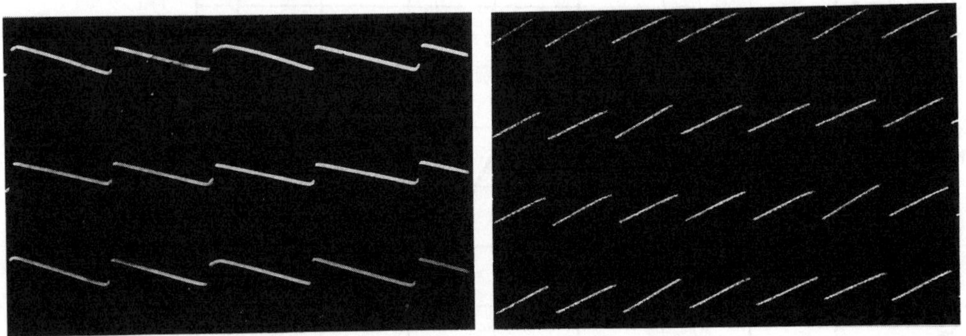

Fig. 14- a) Louvers with burrs angle 15°.
       b) Clean cut louvers, angle 25°.

Radiator A has low louver angles and burrs at the edges. This means that most of the air does not pass through them with a reduced efficiency as result.

Corrosion and corrosion protection of copper/brass radiators has in recent years become an issue. All aspects of corrosion shall not be covered here (3). Only the parts that are related to the radiator manufacture will be dealt with. The Sometimes so called non residual flux is used which eliminates the rinsing. Good geometrical tolerances are needed to get a good soldering result. This is sometimes not the case.

Another problem is flux residues either due to ineffective rinsing or wrong conditions during soldering with a non residual flux. Flux residues can accelerate corrosion either in storage or after the radiators are taken into use, (3).

A good customer service and consistent and uniform strip quality has proved to be very important to help, especially smaller radiator manufacturers to avoid problems.

DISCUSSION

The development of production processes for thin gauge strips used in copper/brass radiators and heaters has been very successful. The reason for this is not easy to explain, but some important areas will be discussed. An early understanding of the radiator design, production and function made it possible to direct the development towards important goals. The close thickness tolerances are an important sales argument but also consistent and uniform properties. They mean less problem with running in new forming tools and much less scrap in current production. The geometry of the different parts in a radiator is of importance for a good soldering result and a good final product. This is of course much easier to achieve with a consistent strip material.

Our own design and construction of production equipment was probably a necessity for the successful development of the strips. Inhouse capability to change and modify made the development much easier. A close collaboration between process and material development was also important. The possibilities that the process can give to improve the material properties could be used at an early state. In some cases on the other hand, the material had to be adjusted to the process.

The knowledge of radiator design has also been used to introduce a new improved construction, the splitter fin. It has not become a great success but has served its original purpose to show that much thinner fins can be used without loosing capacity. This has been a way to try to prevent the change from copper/brass to aluminium. It is hard to estimate the influence but our company has seen no decrease in production so far, fig. 15.

Production of Radiator Strip, kton

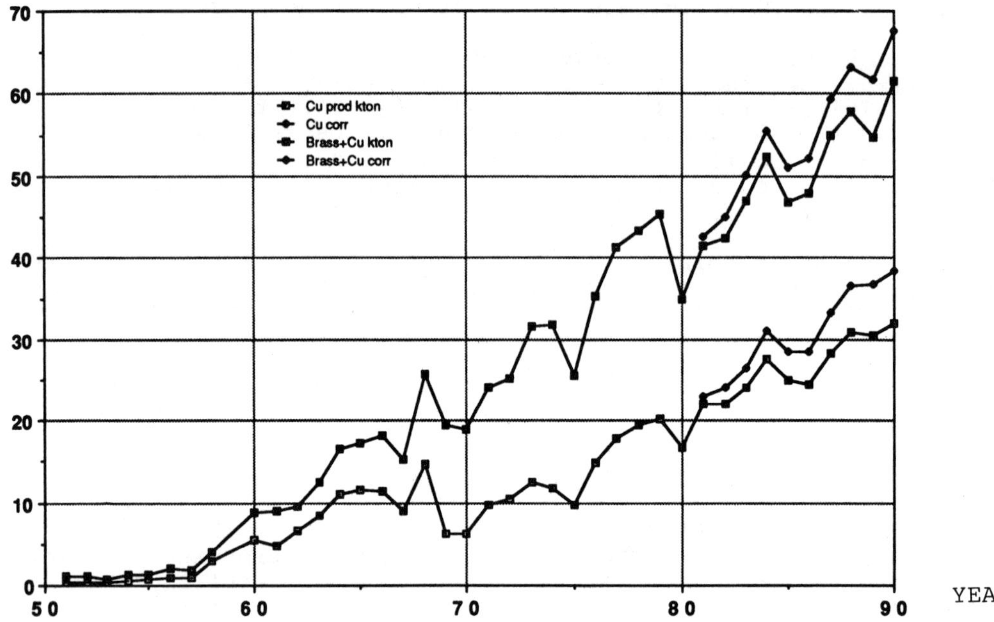

Fig. 15 - Production of radiator strips of Outokumpu
         Copper.

Knowledge of the competing material aluminium and the
properties of the aluminium radiators must also be mentioned. It
is valuable to be able to discuss with the customers even about
that (9). No mayor technical differences excist between the two
materials. A weak point for copper/brass is although joint
strength. Aluminium is brazed and copper/brass soldered.
Development work to make a brazed copper/brass radiator has
started by ICA. Another drawback is the image of copper. That
has to be tackled by the copper industry. Copper versus
aluminium price is finally maybe the most important question.
For copper, very optimistic predictions were recently made (10).

REFERENCES

(1).   Carlsson R, Kamf A, Sundberg R, Östlund S.Vertical Strip
        Casting - Materials Properties.
        Copper 90, The Inst of Metals, Västerås 1990.

(2).   Hutchinson B, The effect of alloying edditions on
        recrystallisation behaviour in Copper - literature
        review. Inst for Metals Research Report IM - 2003
        Stockholm 1985.

(3).   Sundberg R, Holm R, Hassel L.
        Corrosion and corrosion protection of automotive heat
        exchangers - Comparison between Copper/Brass and
        Aluminium.
        SAE Techn Paper 870108, Detroit 1987.

(4).   Kupfer in Internationalen Zusammenspiel.
        Pro Metal (1975):2, 30.

(5).   Modine Manufacturing Co Racine Wi, Quality control rating
        test, No 6545, 1966.

(6).   Webb R L, Yu W F Stress distribution and stress
        reduction in Copper/brass radiators. SAE Techn Paper
        870183 Detroit 1987.

(7).   Webb R L The flow structure in the louvered fin heat
        exchanger geometry. SAE Techn Paper No 900722 Detroit
        1990.

(8).   Webb R L, Farrell PA Improved thermal and mechanical
        design of copper/brass radiators. SAE Techn Paper No
        900724 Detroit 1990.

(9).   Anger S, Sundberg R Copper and brass - material for
        automotive heat exchangers - comparison between competing
        material. Copper 90 The Inst of Metals Västerås 1990.

(10).   de J Osborne R A re-energized copper industry
        perspective and outlock.
        Copper 90 Inst of Metals Västerås 1990.

(11).   Corrosion of metals and alloys - Determination of
        dezincification resistance of brass ISO-Standard 6509,
        1981.

(12).   Mazza F Torchio S Factors influencing the susceptibility
        of intergranular attack, stress corrosion cracking and
        de-alloying attack of aluminium brass Corrosion. SC
        23 (1983) 10:1053-1072.

Copper rod quality and its effect on wire drawing performance

T.A. Jones
Canadian Wire and Cable Limited, Toronto, Ontario, Canada

Abstract

At the time of its introduction, continuous cast copper rod yielded immediate benefits to wire manufacturers especially in terms of drawability and surface quality. However, pressures on rod users for improved productivity, lower costs and reduced scrap have been reflected in demands for copper rod capable of even higher levels of performance.

This paper describes the typical benefits experienced by wire producers as a result of the introduction of continuous cast rod. In addition, recent product testing and process upgrades made at Canada Wire and Cable's Montreal East copper rod mill are discussed. Particular emphasis is placed on the effect of these improvements on copper rod performance in the wire mill.

INTRODUCTION

As with all manufacturers, electrical wire manufacturers have experienced pressures for improved productivity, reduced scrap, lower product costs, and higher product quality. This has led to the widespread introduction in recent years of higher speed drawing machines, and the tandemization of operations (e.g. drawing/enamelling in the magnet wire industry). Naturally following from this has been a demand for copper rod capable of higher levels of performance in the wire mill.

The copper industry's first response to this challenge was the development in the late 1960's and early 1970's of continuous casting and tandem hot rolling for the production of copper rod. This technology allowed the manufacture of a product with more consistent properties, and much improved performance.

However, as the rod quality improved and user requirements continued to tighten, it became apparent that the traditional measures of copper rod quality such as dimensions, chemical composition, and mechanical and physical properties (i.e. tensile strength, elongation, and electrical conductivity) no longer provided sufficient information to the user on the properties of his raw material. Of increasing importance are details on the surface quality and drawability of the rod.

This paper examines the impact that continuous cast copper rod has had on the manufacturers of copper wire, especially in terms of its performance in the wire mill. In addition the various methods of defining or predicting the wire drawing performance of copper rod which have been presented in the literature are discussed. Finally, recent product testing and processing upgrades made at our Montreal East rod mill are described in terms of their ability to provide performance-relevant information and improved quality rod to the user.

INTRODUCTION OF CONTINUOUS CAST ROD

Brief Description of Process

In 1982 we commissioned a 33 tonnes/hr. Krupp-Hazelett continuous casting and rolling mill for the production of 8mm and 10mm diameter electrolytic tough pitch copper rod. This process features a twin-belt casting machine in-line with a 13 stand rolling mill. Sulphuric acid pickling is used to remove the surface oxide formed on the rod during the hot rolling operation. The major features of the line are shown in Figure 1.

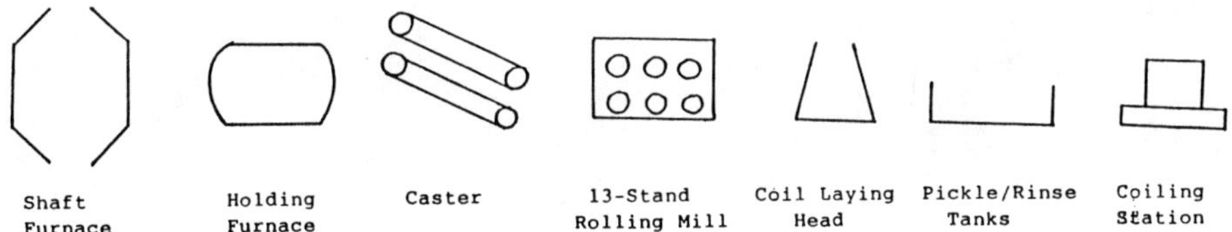

Figure 1 - Block diagram of Montreal East
rod mill.

Immediate Improvements in Drawability

Drawability can be defined as the ability of rod to be reduced in diameter in a drawing machine without breakage. Wire breaks during drawing can be caused by a variety of factors, but their causes can generally be divided into two broad categories:

       1) Drawing related
       2) Rod related

Drawing related causes of wire breaks include worn dies, poor lubrication, tangled input wire, incorrect area reduction between dies, mechanical abrasion, and poor weld between rod coils. The control of these types of breaks is, of course, the responsibility of the wire producer.

The actions of the rod producer, however, have a direct influence on the frequency of rod related breaks. The causes of this type of break include metallic inclusions, slag/oxide inclusions, porosity, and seams, slivers and other surface defects.

The introduction of continuous cast rod eliminated (or at least greatly reduced) two of the most frequent causes of wire breaks among wire drawers. The frequency of weld breaks was reduced by a factor of at least 30 simply as a result of the change in coil size. The old 300 lb. wire bar rod coil was replaced by a 10,000 lb. coil. Similarly, the lack of the "surface set", characteristic of wire bar rod, eliminated wire breaks caused by high oxygen content at the rod/wire surface.

The improvement in drawability, as measured by the number of pounds of wire drawn per wire break, resulting from the use of continuous cast rod is illustrated in Figure 2. This data represents drawing experience at one of our wire mills six months prior to and six months after their use of continuous cast rod began.

The roughly six-fold reduction in the frequency of wire breaks resulting from the introduction of continuous rod is especially significant when one considers that each wire break costs approximately 30 minutes of machine downtime.

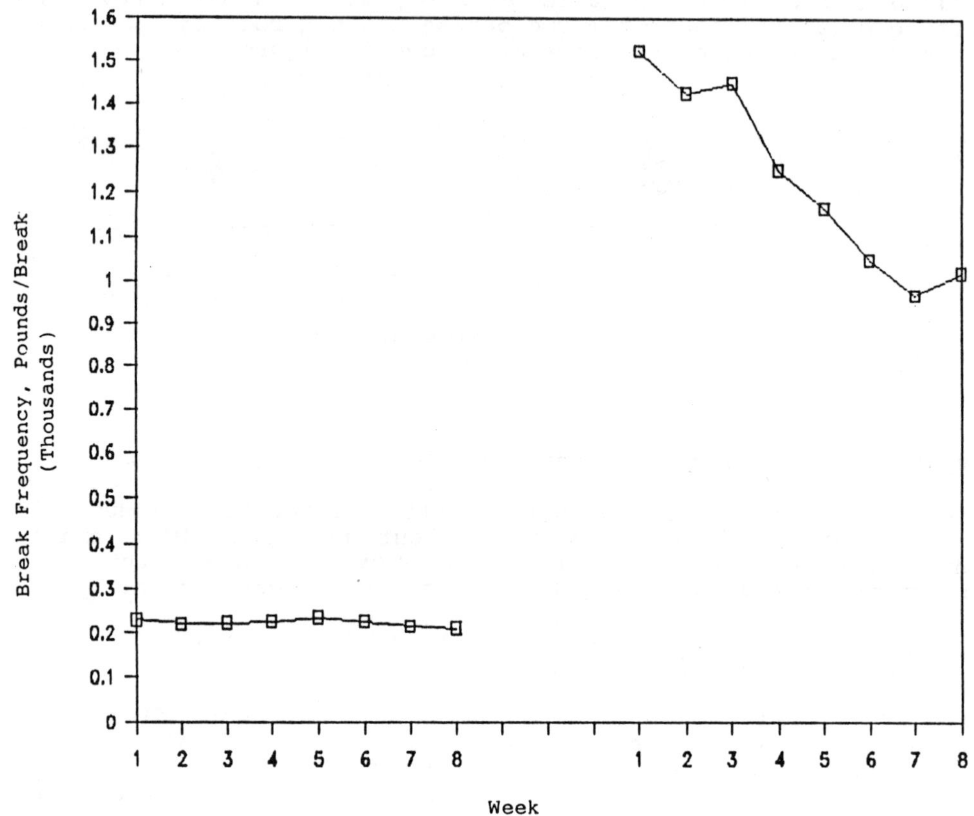

Week

Figure 2 - Immediate effect of introduction of continuous cast
rod on drawability of 30 AWG (0.25 mm) wire. Wirebar
rod performance at left, continuous cast rod performance
at right. Breaks of all types considered.

TIGHTENING OF USER REQUIREMENTS

Shortly after the universal acceptance of continuous cast rod by
wire drawers, a combination of factors resulted in a re-
examination of the measures of copper rod quality. First of all,
the higher level of product quality resulting from the new rod
making technology meant that it was almost impossible to not meet
the existing industry standard for electrolytic tough pitch
copper rod (ASTM B49). Prior to 1990 this specification required
that only the most fundamental properties be met (1).

Secondly, as wire manufacturers searched for ways to reduce
production costs the cost of inferior quality copper rod became
obvious. With trends toward increased drawing speeds and the
tandemization of drawing and insulating processes, the costs of
machine downtime resulting from wire breaks increased
dramatically. Also, the cost of scrap generated as a result of
rod related surface defects on enamel coated wire (for use in

motor windings) was of particular concern for the magnet wire industry.

As a result of this re-examination, copper rod users were no longer satisfied with rod conforming to existing ASTM requirements. Instead, demands were being made for rod meeting a more stringent, and meaningful, set of requirements.

As shown in Table I ASTM B49 was revised in 1990 to include much tighter requirements for chemical composition and a new surface oxide requirement (2). However, despite the best efforts of the ASTM committee the critical performance criteria of surface quality and drawability remain inadequately defined/quantified in the revised standard. This is largely a result of the lack of a standardized test that can be performed on rod to fully determine or predict its surface quality or drawability performance in the wire mill.

Table I - E.T.P. Copper Rod
Requirements as Defined
By ASTM

Property	ASTM B49-78	ASTM B49-90
Chemical Composition*		
Copper, min.	99.90%	99.90%
Tellurium	–	2
Selenium	–	2
Bismuth	–	1.0
Group Total	–	3
Antimony	–	4
Arsenic	–	5
Tin	–	5
Lead	–	8
Iron	–	10
Nickel	–	10
Sulfur	–	15
Silver	–	25
Oxygen	–	100-650
Maximum Allowable Total	–	65
Tensile Elongation (min. % in 250mm)	30	30
Electrical Resistivity (max. $\Omega.g/m^2$)	0.15328	0.15328
Diameter Variation (mm)	±0.38	±0.38
Surface Oxide	–	1000 A
Surface Quality	As per good commercial practice	As per good commercial practice
Drawability	–	–

* In ppm, max. unless specified otherwise.

ASSESSMENT OF ROD QUALITY

One of the biggest challenges facing copper rod producers then, is to identify a test, or a series of tests, which will accurately predict the rod's level of performance in the wire mill. To this end, many tests have been proposed and evaluated by both rod users and producers. Table 2 briefly describes the more common of these tests which have been proposed in the literature as measures of the drawability and/or surface quality of copper rod (3).

Table II
Twist Tests to Evaluate Drawability or Surface
Quality of 8 mm Copper Rod

Test	Procedure	Evaluation
Twists to Failure	-twist an 8 in. sample of rod in one direction -count the number of turns before rod breaks	-the higher the number of twists to failure, the better the drawability.
Reverse Twists to Failure	-twist an 8 in. sample of rod 25 turns in one direction and then reverse the twisting direction until failure -count the number of reverse twists before the rod breaks	-the higher the number of reverse twists to failure, the better the drawability
10 x 10 Twist Test	-twist a 10 in. sample of rod 10 turns in one direction and 10 turns in the reverse direction	-examine rod surface for cracks, slivers, flakes and fines -indicator of surface quality
3-Die Twist Test	-draw rod through 3 dies, each die reducing the cross-sectional area by 21% -twist a 10 in. sample of the drawn wire 14 turns in one direction and 14 turns in the reverse direction.	-examine wire sample at 7X magnification for cracks, slivers, flakes and fines. -indicator of surface quality

Unfortunately, all of these tests have significant flaws. Those tests designed to evaluate the surface quality of the rod, and its likelihood of causing surface flaws on the drawn wire, suffer from the subjective nature of the test. In all cases the tester is required to judge the severity of the surface defects produced on the rod's surface as a result of the twisting. This, of course, makes it extremely difficult to standardize.

In addition, all of these tests are performed on short (20-25 cm) samples of rod. This represents an extremely small portion of the approximately 10 km in a single rod coil, and inevitably results in discussions between rod producers and users about the representativeness of the sample.

Finally, while examples of isolated correlations between these test results and rod drawability and drawn wire surface quality exist, a more general, long-term correlation is difficult to find (4).

NON-DESTRUCTIVE ASSESSMENT OF ROD QUALITY

In light of the shortcomings associated with the torsion type tests used to assess the surface quality and drawability of rod, we opted for continuous, non-destructive methods to evaluate the quality of its rod. On-line eddy current testing is used to determine surface quality, and a ferrous inclusion detector is being evaluated as a tool to assess the rod's drawability.

Eddy Current Testing

The Montreal East rod mill uses a Dr. Foerster Defectomat C eddy current tester for the detection of surface flakes, seams, slivers and cracks.

The theory behind this type of testing has been thoroughly described in the literature (5), so only a brief outline of its operation will be given here. On exit from the last roll stand the rod passes through the centre of an encircling coil. This coil produces eddy currents in the rod which are sensitive to inhomogeneities on the rod's surface. Disturbance of these eddy currents is measured electronically. Those disturbances, or fault signals, which are larger than a determined threshold value are automatically counted and recorded for each rod coil.

For a given set of eddy current tester settings/test parameters, the size of the fault signal is only roughly proportional to the severity of the defect. Factors such as the shape of the defect, and the presence and nature of any included foreign material will affect the size of the signal. Despite these variables it has been possible, through close communication with a rod user, to develop a set of eddy current tester settings which give a good, nonsubjective indication of the surface quality of the copper rod.

The rod mill's tester settings were developed in conjunction with our Simcoe, Ontario plant. This plant manufactures magnet wire – the most demanding application of copper rod especially in terms of surface quality. The Simcoe plant also has an eddy current tester (similar to the unit used in the Montreal East rod mill) installed on their rod breakdown/shaving machine. As the result of much operating experience, they have been able to determine a frequency and amplitude of fault signals at this stage of drawing which is likely to impact on the processability and/or quality of the material in downstream operations.

By comparing the results of the rod mill's surface analysis with Simcoe's eddy current results and processing experience, a set of

tester settings were determined for the rod mill. These settings
are shown in Table III.

Table III - Eddy Current Tester
Settings for 10 mm Copper Rod

Parameter	Setting
Frequency	30 kHz
Sensitivity	57.0 dB
Filter	14
Threshold Amplitude	30%

The correlation between the Montreal East and Simcoe testers is
shown in Table IV. This table shows the percentage of rod judged
to be of superior surface quality by the two facilities and also
illustrates one of the drawbacks of this type of testing.

Although a good correlation is evident for three of the seven
months tabulated here (months 1, 2 and 7), there is a wide
discrepancy for the remaining four months. In these cases the rod
mill has judged the surface quality more harshly than the wire
mill. The discrepancy is believed to be caused by the shaving
operation performed at the Simcoe plant prior to eddy current
testing, and the use of single threshold testing at the rod mill.

Shaving is a standard operation performed by magnet wire
producers in which an outer layer of 5 - 7 mils thick is removed
(shaved) from the rod. This operation is designed to remove the
majority of shallow surface defects. Consequently, the surface
tested at the wire mill is a different surface than the one
"seen" by the eddy current tester at the rod mill.

In addition, by using single threshold testing at Montreal East
the total number of fault signals indicates only the number of
indications more severe than the threshold setting. It does not
indicate what proportion of these defect indications just barely
exceeded the threshold and what proportion greatly exceeded the
threshold. The approximate proportionality between the amplitude
of the defect indication and the defect size is not taken
advantage of. Single threshold testing tends to err on the side
of caution, then, with small defects counted the same as large
ones.

Referring again to Table IV, it appears that the rod produced in
months 3,4,5 and 6 contained many small defects that were easily
removed by shaving.

The solution to this problem is to use multi-threshold testing.
This allows defect indications to be counted at several severity
levels simultaneously. Experimental work is currently underway at
the rod mill at Simcoe plant to develop a correlation for this
type of testing.

Table IV - Percentage of 10mm Rod
Judged To Be of Superior Surface Quality

Month	Montreal East	Simcoe Plant
1	30	34
2	20	19
3	16	82
4	8	66
5	3	43
6	15	57
7	59	58

Ferrous Inclusion Detector

To compliment the continuous, on-line, non-destructive detection of surface defects on copper rod, we are experimenting with the continuous, on-line, non-destructive detection of ferrous inclusions. As previously mentioned one of the major causes of rod related wire breaks is metallic inclusions. Typically these inclusions are small pieces (usually less than 3 mg) of rolls or guides which become embedded in the surface of the rod during the hot rolling process. On drawing to wire the copper matrix deforms differently than the inclusion material. This inhomogeneity results in a wire break and costly machine downtime.

Commercial inclusion detectors are available which operate on the basis of the behaviour of a magnetic field in the presence of a magnetic particle. A magnetic field is created by a permanent magnet mounted onto a standard eddy current test coil. In the presence of a non-magnetic material, such as copper rod, this field is undisturbed. However, when a magnetic material is introduced, the field is disturbed and a voltage is induced. This voltage is amplified by the detector to produce a defect signal.

Laboratory experiments have been performed with a Dr. Foerster Ferromat ferrous inclusion detector to determine its effectiveness in detecting a variety of sizes and types of inclusion materials. Simulated inclusions were created by affixing pieces of magnetic materials onto a short length of copper rod. The copper rod (and inclusion) was then passed through the test coil by hand and the signal intensity recorded. The test results are shown in Figure 3.

Although the defect signal intensity is quite low in the range of particle weights of interest (i.e. approximately 3 mg), it is believed that this problem will be remedied in on-line operation. This is because in addition to the inclusion size, its magnetic permeability, and its location within the rod, the signal intensity is also a function of the speed of the inclusion through the magnetic field. The faster the line speed, the larger the disturbance of the magnetic field and, consequently, the larger the signal intensity (6).

On-line trials of this detection technique are currently underway and while initial results are encouraging, they are as yet inconclusive.

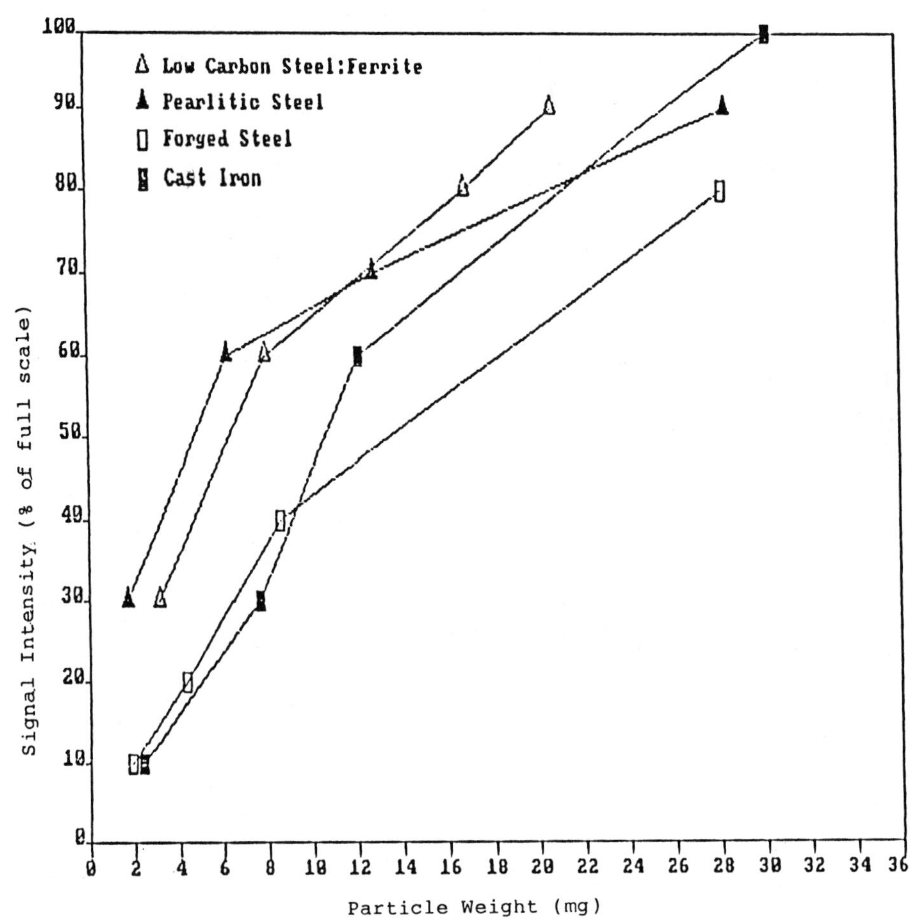

Figure 3 - Particle weight vs signal intensity for Dr. Foerster Ferromat. Laboratory trials.

ROLLING EMULSION CONTROL AND ITS EFFECT ON DRAWABILITY

Our rod mill uses a water/alcohol/lubricant emulsion to provide lubrication and cooling to the copper rod and tool steel rolls during the rolling process. It has been found that proper maintenance of this emulsion is critical for achieving a consistently high level of copper rod drawability.

In March of 1990 a program was instituted to improve the control and maintenance of the emulsion. This program included the following changes to normal operating practice:

1) Small, regular additions of lubricant, alcohol and emulsifier rather than large, infrequent additions.
2) Regular measurements of bacteria content, pH, lubricant content, conductivity, alcohol content, and surface tension.

3) The addition of a finer emulsion filter for use when the line is down (eg. on weekends). This extra filtration operation compliments the normal filtration during normal mill operation.

Figure 4 shows the drawability figures obtained at one of our plants which manufactures communication cables. The data is presented in terms of the frequency of inclusion type wire breaks and is expressed in MCF (millions of copper feet) per inclusion break. This particular plant produces wire mainly in the 22 AWG to 26 AWG (0.40-0.64 mm) size range.

The effect of this emulsion maintenance program is obvious. The regular addition of emulsion components helps to create a stable emulsion. This in turn reduces degradation of the mill rolls (a major source of inclusion particles). In addition, the fine filtration on weekends removes many ferrous particles from the emulsion which otherwise would be recirculated and possibly become embedded in the rod.

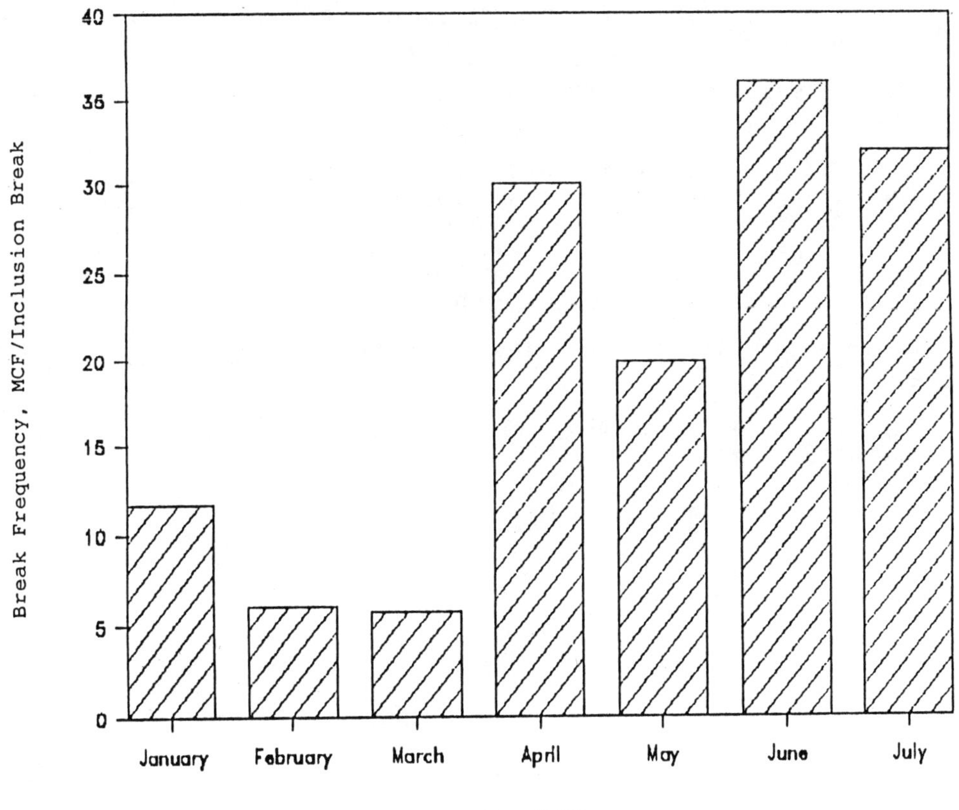

Figure 4 - Effect of improved rolling emulsion control (started in late March) on drawability of rod.

CONCLUSIONS

1) The introduction of continuous cast copper rod had a significant positive impact on the performance of the rod in the wire mill, especially in terms of drawability.

2) Continuous, on-line, eddy current testing can be used to evaluate the surface quality of copper rod, and to give an indication of its likely performance in the wire mill.

3) Although laboratory test results are encouraging, additional rod mill/wire mill trials are requred to determine the effectiveness of the ferrous inclusion detector as a predictor of ferrous inclusion wire break frequency.

4) Proper maintenance of the rolling mill emulsion is an effective means of significantly reducing the frequency of inclusion type wire breaks in the 0.40 - 0.64 mm size range.

REFERENCES

1. ASTM B49-78 "Standard Specification for Hot-Rolled Copper Redraw Rod for Electrical Purposes".

2. ASTM B49-90 "Standard Specification for Hot-Rolled Copper Redraw Rod for Electrical purposes.

3. E.H. Chia, R.D. Adams and J. Kajuch, "Torsional Stress Tests for Copper Rod: Indicator of Quality and Performance", _Wire Journal International_, June 1986, 57-67.

4. H. Pops and D. Hobbs, "A Fine Wire Drawability Case Study", _Wire Journal International_, June, 1990, 22-35.

5. B. Faith, "Eddy Current Inspection", _Wire Industry_, April, 1987, 245.

6. T. Einspanier, "Testing for Ferrous Inclusions", _Wire Industry_, February 1988.

High-performance copper alloys for electrical and electronic applications—an integrated approach to meet customers needs

D.E. Tyler and J. Crane
Olin Metals Research Laboratories, New Haven, Connecticut, U.S.A.

ABSTRACT

Olin has focused the development of new alloys over the past 25 years on the electrical and electronic segments of the copper alloy strip market. Major usage areas have been leadframes for semiconductor packaging and connectors or terminals.

The commercial success of this thrust was realized by translating the customers' needs into alloy performance targets, offering competitively priced alloys, and integrating marketing, manufacturing, and R&D operations to provide timely development and delivery of new materials to the marketplace.

This paper reviews the factors involved in each of the foregoing aspects. Issues related to coordination of marketing, manufacturing, and R&D functions consistent with the needs of the marketplace are addressed. The fine lines between advancing understanding of product performance, and not being responsive to customer needs, between extending manufacturing capability, and not being able to produce the desired product consistently within cost targets, are examined.

Finally, the case history of a recently commercialized alloy, C7025, used both as leadframe and as connector material, is followed to illustrate the problems and rewards of each stage of the process, starting with defining customer-driven targets to dealing with international markets and standards.

INTRODUCTION

Olin Brass has invested a major portion of its resources for process improvements aimed at producing quality strip at competitive prices. A significant effort has also been directed toward coupling an enhanced production facility with high performance products including new alloys aimed predominantly at electronic and electrical market segments. These products encompass fabricated strip, welded tubing, composite strip, etc. made largely of standard alloys but also including some Olin-developed alloys.

This paper offers some insights on the ingredients of an alloy development program. The emphasis was, is, and will always be, on tailoring the product to the marketplace. It will be evident through this discourse, however, that meeting the long-term needs of the marketplace is a complex and demanding goal.

The relationships of the alloy developer with the end user, and with the marketing and manufacturing groups which make and market those alloys, are also described.

THE MARKET

In 1965, the total U.S. copper alloy strip consumption was about 1 billion pounds. This is approximately the same volume consumed in 1990. Copper alloy strip used for connectors in 1965 was largely for automotive and appliance terminals and household electrical receptacles. Copper alloy leadframe for IC packaging was unknown and copper alloy used for discrete devices was small.

Major growth of electronic connectors did not occur before the late 70s; the mid to late 80s marked the advance in automotive electronics. Leadframe for transistor outline packages reached large volumes by the early 80s and comparable numbers were achieved for copper-base IC leadframe by mid 80s. Application Specific IC (ASIC) leadframe reached measurable but small volume only within the last few years.

A list of the alloys which serviced the electrical and electronic connector markets in 1965-70 is shown in Table I. The primary alloy was cartridge brass, used in most automotive connectors, household receptacles, and a large portion of appliance terminals. Phosphor bronze, in addition to brass, was a major material for telecommunications connectors. Beryllium copper was the material of choice for connectors in high reliability telecommunications and military applications and selected premium-grade receptacles. These alloys still dominate the connector market. The leadframe market was serviced exclusively by nickel and iron-nickel alloys. While iron-nickel alloys still enjoy wide usage today, copper alloys are a major component of the leadframe market.

Table I - Preferred Copper Alloys Used for Electrical and Electronic Connectors

Alloy		Application
C230 C260	Brass	Automotive connectors, household receptacles, appliance terminals
C510 C521	Phosphor Bronze	Telecommunications, computer
C172 C170 C175 C173	Beryllium Copper	Military, aerospace, computer, telecommunications

Olin's strategy was to develop new alloys which offer enhanced performance and are competitive with the higher volume alloys in electrical and electronic markets. The goal of this strategy was to provide the best combination of properties required for any specific application coupled with good fabricability for the customer. This approach required a critical analysis of the actual demands of the individual application.

In 1965, properties of strip used for connectors were largely specified by ultimate tensile strength and elongation, sometimes yield strength. Designer modelling equations used Young's Modulus and longitudinal 0.2% offset yield strength to determine contact force. On this basis, higher yield strength and Young's Modulus were the desired performance targets. Before developing new alloys, it was necessary to establish the validity of these criteria. This required understanding of the material's performance in end user applications and transposing this information to materials characteristics measurable in laboratory tests. Similarly, it was essential to establish criteria for identifying fabricability. This objective demanded more in-depth knowledge of materials behavior during various customer manufacturing operations. In some cases test parameters already existed for "manufacturability" criteria, in others, existing tests had to be modified, or new tests developed. New product targets including cost, fabricability and properties would then be established by Olin Marketing. Marketing also provided the opportunity for dialogue between the customer and Olin R&D personnel regarding the materials interaction with service demands.

APPLICATIONS ANALYSIS AND PROCESSIBILITY

The user must be concerned about the ease of converting strip to a final product because cost and part function can be adversely impacted by this conversion. Most customers are aware of the primacy of system costs, and are alert to metal per pound savings that are lost by excessive wear of stamping tools, plating problems, etc. Similarly, benefits attributable to improved alloy performance can go unrealized if the new material fabricates poorly. As a consequence, an understanding of the alloy and process factors affecting bend formability, drawability, stamping, blanking, plating, tool wear, etc., was required to assess fabricability (1,2). In the case of semiconductor packaging, additional alloy characteristics like adhesion to polymers and low cycle fatigue had to be determined to predict likely behavior during assembly.

Over the years, performance characteristics, once focussed on strength versus conductivity, have been extended to define specific application needs. For bulk behavior of connectors, load deflection or secant modulus, and stress relaxation in bending, have largely replaced simple strength criteria to assess spring performance, i.e., contact normal force (2,3). Conductivity though still very important is now coupled with stress relaxation resistance to accommodate the thermal interdependence of Joule heating with the alloy's ability to withstand relaxation at temperature (4). For a complete profile of connector performance, surface attributes reflecting contact resistance as a function of environment chemistry and micro and macro wear have to be evaluated to separately assess their effects on connector contact resistance (5,6).

For semiconductor packaging applications, leadframe performance potential is assessed by evaluating strength; expansion mismatch with silicon, encapsulating resins and circuit boards; compliance; thermal and electrical conductivity to assess thermal dissipation; and stability in a range of thermal and mechanical environments (1). Surface characteristics are usually related to compatibility with coatings like nickel, gold, silver, solder, and tin. Most recently, adhesion of copper alloys to encapsulating resins has become another surface characteristic of concern as manufacturers of low profile packages seek ways to maintain structural integrity and reduce permeability of the package to moisture (8,9).

The analysis of performance in terms of materials characteristics usually requires simulation tools, but ultimately cooperative experiments with the end user are critical. This typically is an iterative process where an array of materials characteristics are evaluated until the optimum product performance is obtained. This process should lead to a database and design equations which provide predictability of performance. The cooperative nature of this effort is required because of the complex interaction of various components and the expense of specialized equipment to evaluate performance.

For examples, let's examine two significant electronic applications. It is well known that plastic encapsulated semiconductor devices dissipate heat better when supported by copper alloy leadframes in contrast to lower thermal conducting iron-nickel alloys. Conductivity can be measured in the laboratory to rank alloys, but the actual performance benefit is not linearly dependent on conductivity of the leadframe. The payoff to the end user is the ability to dissipate heat which is measured as Θ_{JA}, and reflects the junction temperature of the powered chip. Figure 1, a plot of Θ_{JA} versus leadframe conductivity, indicates, for the system tested, the real utilization of leadframe conductivity to the end user (10). The data in Figure 1 are for Plastic Dual-in-Line Packages (P-DIPs) and will differ for surface mounted packages.

Another illustration of the benefit of supplier/customer collaboration to establish real performance gains, is to assess the interaction of components in a connector. Figure 2 shows the key components in an assembled connector. The heart of the connector is the metal spring component which maintains the required normal force to minimize contact resistance of the circuits connected. The load/deflection characteristics of potential contact materials and their stability at a given load, temperature, time, etc. can be measured and compared (2). However, the contact resistance is also impacted by the surface coating on the contact and potentially by the plastic housing. The ability of the connector to meet its functional goal--to be transparent to the circuitry it connects--is therefore also dependent on the behavior of these other components and their compatibility with the copper alloy contact material. The extent to which this material system meets the end user's goal will be measured by elaborate instrumentation that determines the changes in system resistance over time under whatever environmental conditions are appropriate.

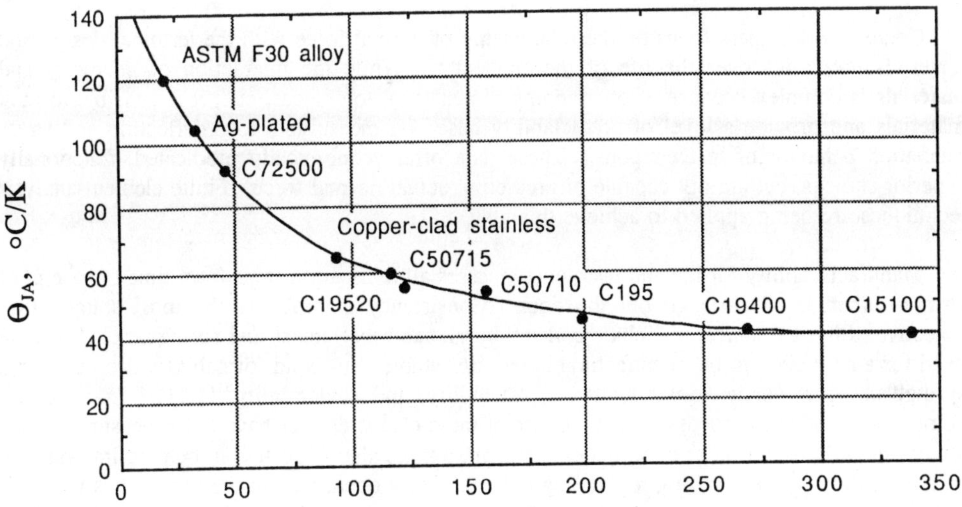

Thermal Conductivity, W/m-K

Figure 1 - Plot of thermal conductivity of leadframe alloy
versus Θ_{JA} of 14-lead PDIP.(10)

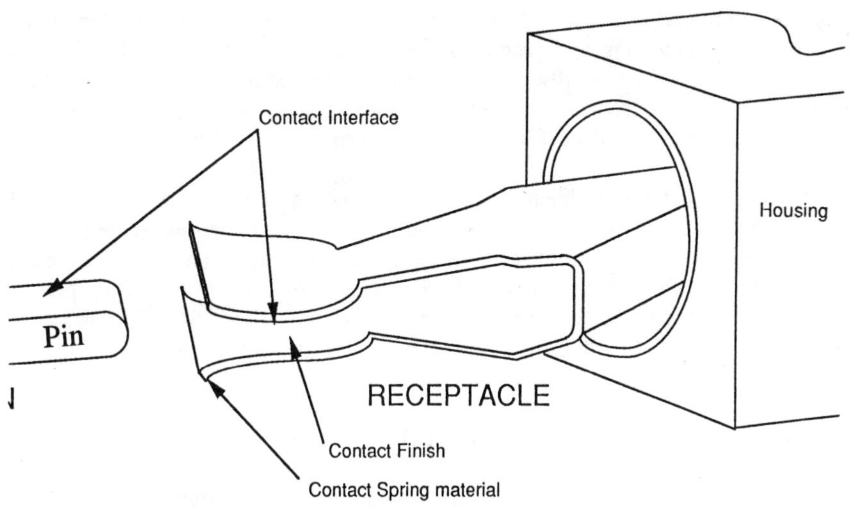

Figure 2 - Connector Components

Connector designers focus on the relationship of normal force with the terminal design and materials properties over the life of the connector. While the interaction of geometry and materials is complex because of overlapping dimensional tolerances, etc., it is possible to rank materials and provide a level of predictability based on measurable load/deflection and stress relaxation behavior of test coupons. These data offer economically predicated, functionally superior choices, but are not capable of predicting actual normal force. Finite element analysis techniques are being applied to achieve this goal.

Manufacturability for the end user, encompasses all fabrication steps from time of receipt of strip to when good parts go out the door. Consistency is probably the most sought after objective. Physical concerns include coil set, burrs, camber, flatness and springback. Thus, the need for a new alloy to be manufacturable requires studies in: bend formability, deep drawing, springback, burr versus tool tolerances, plateability, and solderability, (1,11). In all cases, composition and strip processing dependence of these characteristics have to be measured. Such studies provide useful input to alloy development, but are also important in helping introduction of a new alloy when adjusting to existing manufacturing operations in the customer's plant.

STRIP PROCESSING

Alloy development proceeds through various stages. After performance/fabricability targets are demonstrated to be technically feasible, the emphasis is focussed on the ability to commercialize the alloy within the constraints of existing strip manufacturing capability. This approach offers the best opportunity to minimize processing costs.

Processing technology, drawing from process metallurgy and engineering disciplines and experience, is directed first at making the alloy with in-house equipment. If this is not possible the choices are to adjust alloy composition or to develop alternative "unconventional" processing or modifications to existing processes. Alloy development by its very nature usually extends existing process technology but hopefully does not require new equipment.

In this phase of development, it is not uncommon to compromise properties achievable in the laboratory to permit good commercial strip processing. Here the interrelationship with plant engineering, process engineering and production personnel is critical for developing an alloy which can be made routinely eventually. This is a delicate balance because the property benefits can not be so badly compromised as to make the alloy non-distinguishable. Nor can a "unique" alloy be successful in the marketplace if it cannot be made to customer requirements consistently.

While there is no handbook to identify the best tradeoffs, a flexible and controlled process technology, coupled with committed production and process engineering personnel, has served to optimize this aspect of commercialization. The commitment to process research for the overall alloy mix has benefitted this optimization; and, conversely, solutions to new alloy processing dilemmas have fed back to enhance processing of the overall alloy mix. This has been an ongoing interaction through all stages of processing from melting and casting, composition control, hot and cold rolling, annealing and cleaning.

TECHNOLOGY

The technologies associated with alloy development start with physical and mechanical metallurgy related to alloy design and proceed through those associated with strip processing, customer fabrication, assembly, and finally to performance in service. Table II traces this path

Table II - Technologies Required for Alloy Development

ALLOY DESIGN & STRIP PROCESSING

Alloy Design	Physical and mechanical metallurgy/phase transformation, metal deformation
Melting & Casting	Physical metallurgy/solidification
Hot & Cold Rolling	Mechanical & physical metallurgy/metal deformation, recrystallization
Annealing, Cleaning	Physical and chemical metallurgy/recrystallization, oxidation, chemical and electrochemical dissolution and surface chemistry

APPLICATION

Connector Fabrication	Leadframe Fabrication
Stamping/forming - mechanical metallurgy/metal deformation, lubrication, residual stress, tool wear	Stamping/etching - mechanical and chemical metallurgy, metal deformation, lubrication, residual stress, tool wear, metal dissolution, surface chemistry
Cleaning - chemical metallurgy/surface chemistry and electrochemistry	Cleaning - chemical metallurgy/surface chemistry and electrochemistry
Plating - chemical metallurgy/chemistry and electrochemistry	Plating - chemical metallurgy/chemistry and electrochemistry
Storage - chemical and physical metallurgy/corrosion, diffusion	Storage - chemical and physical metallurgy/corrosion, diffusion
Assembly - mechanical and chemical metallurgy/deformation, residual stress	

Leadframe Assembly (Packaging)

Die attach - physical and mechanical metallurgy/diffusion, adhesion

Wirebond - physical and mechanical metallurgy/deformation, diffusion

Encapsulation - chemistry/polymer chemistry, adhesion

De-flash - physical and chemical metallurgy/corrosion and dissolution

Curing - physical metallurgy/diffusion, polymer chemistry

Lead finish - chemical and physical metallurgy/chemistry and electrochemistry, diffusion

Trim and form - mechanical metallurgy/deformation, residual stress

Test and burn-in - mechanical, physical and chemical metallurgy/diffusion

Board assembly - residual stress, oxidation, corrosion, fatigue

from alloy design to end use and indicates the associated technologies for typical connector and semiconductor applications.

It is evident from Table II that getting the alloy produced involves only about half the technology disciplines required by the alloy development to produce the final working part. The importance of understanding the customer's demands to make, assemble and utilize the end product is a vital part of the alloy developer's technology portfolio.

CASE HISTORY OF C7025 DEVELOPMENT

This copper-nickel-silicon-magnesium alloy has application in both connector and leadframe markets. A review of its development and commercialization illustrates the importance of all phases of the development of a new alloy. The alloy is aimed at the high performance end of both market segments. For connector applications, this includes good formability and high stability at modest temperatures; conductivity was targeted mid-range to accommodate the electrical and electronic markets and insure that elevated temperature stress relaxation resistance would not be compromised by Joule heating. ASIC leadframe requiring high strength coupled with good formability and moderate conductivity was the other application area targeted.

Table III lists the properties targeted for this state of the art alloy for both markets. This comprehensive list which reflects both performance and fabricability requirements is in strong contrast to the strength, elongation, conductivity targets of the 1960's.

Table III - Properties Targeted for Connector and Leadframe

Connector
      UTS, YS, elongation
      Load/deflection or secant modulus vs stress
      Stress relaxation resistance as a function of temperature
      Conductivity - electrical, thermal
      Bend Formability
      Stress corrosion resistance
      Corrosion resistance
      Plateability
      Solderability, solderability shelf life
      Tool wear
      Springback

Leadframe
      UTS, YS, elongation, modulus
      Conductivity - electrical, thermal
      Bend formability as a function of width and thickness
      Leadbend fatigue
      Softening resistance
      Plateability
      Solderability, solderability shelf life, solder adhesion
      Stress corrosion resistance
      Corrosion resistance
      Springback

The research phase of this alloy development identified that a precipitation-hardening system was required to meet the combination of strength, formability, and conductivity targets. Technical feasibility of using a Cu-Ni-Si alloy was established after extensive research defined the trade-offs among the aforementioned properties and relaxation resistance.

The development phase defined the processing and composition controls required to provide ranges of properties that could be met using commercial equipment which offered user consistency regarding both fabricability and performance. The effects of annealing on properties and surface quality, and the relationships of process routes to potential tool wear, solderability, plateability, and cosmetics issues were rigorously defined.

Processing was identified for a range of product types, both leadframe and connector. Plant processed metal, reflecting what would be commercially produced was then subjected to a complete series of tests to verify that property targets were achieved. Other characteristics aside from those targeted in Table III, were measured for information purposes, e.g., interdiffusion with certain coatings, fatigue strength, thermal expansion. Effects of gage, width and directionality of properties were also determined to provide more in-depth understanding and to assess new applications.

After the initial field evaluation, some process revisions were required to accommodate specific customer needs and extend the impact of the alloy in the marketplace. There was particular interest for improved bend formability, particularly at the higher tempers. Process revisions achieved these improvements in formability and also enhanced elevated temperature stability. Figure 3, a plot of bend formability versus stress remaining after 3000 hours at temperatures from 105-175°C reflects these improvements.

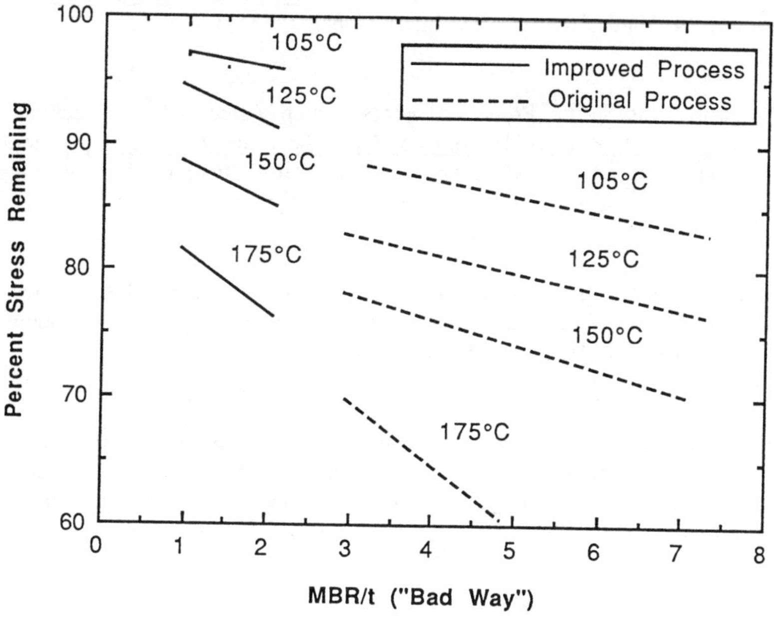

Figure 3 - Stress remaining vs bend formabililty of C7025.

The current global nature of the electrical/electronic industries demands that metal suppliers be aware of the various standards used in different countries. There are a number of material evaluation tests that differ in methodology and duration. In addition, there are cultural differences--preferences--among the various host countries that process, test or assemble the metal enroute to its final destination. These logistics and cultural considerations cannot be ignored.

In commercializing C7025, many of these "local factors" have required resolution of measurement methods and some customizing to accommodate end-user processing.

SUMMARY

Commercialization of new copper alloys has changed significantly over the past two decades. The miniaturization of components and increasingly stringent thermal, electrical, and environmental conditions have extended the alloy performance criteria that have to be satisfied. At the same time, demands of productivity and system quality by end users of copper alloys require improved and more consistent manufacturability of these materials(12,13).

The global industrial community now expects these improvements in performance and fabricability to be achieved more rapidly and competitively. These factors necessitate intimate supplier/user relationships. For Olin, this has confirmed the philosophy of close linkage among the alloy developer, marketing, manufacturing, and the end user.

REFERENCES

(1)   F. Mandigo and J. Crane, "Forming of Copper Alloys," <u>Metals Handbook, Forming and Forging</u>, Volume 14, Ninth Edition, ASM International, 1988, pp 809-823

(2)   J. Breedis, "Formability/Spring Property Factors Which Affect Alloy Selection for Connectors", <u>Materials Issues in Electronic & Opto-Electronic Connectors,</u> Minerals, Metals, and Materials Society, BookCrafters, 613 E. Industrial Drive, Chelsea, Michigan, 1991, pp 43-52.

(3)   J. Crane, "Performance and Fabricability Requirements for Copper-Base Alloys in Electrical and Electronic Connectors", Technical Paper, Society of Manufacturing Engineers, 20501 Ford Rd., Dearborn, Michigan, MS76-363.

(4)   C. Andersson, A. Kamf, and R. Sundberg, "Stress Relaxation of Some Copper Alloys for Connectors - Comparison Between Ni and Fe in Combination With Sn - and Influence of Conductivity", <u>Materials Issues in Electronic and Opto-Electronic Connectors</u>, Minerals, Metals and Materials Society, BookCrafters, 613 E. Industrial Drive, Chelsea, Michigan, 1991, pp 29-42.

(5)   R.S. Mroczkowski, "Materials Considerations in Connector Design", Paper presented at ASM World Materials Congress, Chicago, Ill., Sept. 1988.

(6)   M. Antler, "Survey of Contact Fretting in Electrical Connectors," <u>Proceedings of the 12th International Conference on Electric Contact Phenomena"</u>, Sept. 17-21, 1984, pp 3-22.

(7)   J. Crane, J.F. Breedis and R.M. Fritzsche, "Leadframe Materials", <u>Electronic Materials Handbook</u>, Volume 1 Packaging, ASM International, Materials Park, Ohio, Nov. 1989, pp 483-492.

(8)   T.O. Steiner and D. Suhl, "Investigations of Large PLCC Package Cracking During Surface Mount Exposure", IEEE/CHMT, 1987, Volume, CHMT-10, pp 209-216.

(9)   Proposal - "Impact of Moisture on Plastic I/C Package Cracking", IPC-SM-786, available from Institute for Interconnecting and Packaging Electronic Circuits, 7380 N. Lincoln Ave., Lincolnwood, Ill 60646.

(10)   T. Sakomoto, "Leadframe Materials and Market Trends, <u>"Leadframe Handbook</u>, Dec. 1988, Chapter 5.

(11)   A.D. Romig, Jr., Y.A. Cheng, J.J. Stephens, D.R. Frear, V. Marcotte, and C. Lea, "Physical Metallurgy of Solder-Substrate Reactions", <u>Solder Mechanics - A State of the Art Assessment</u>, D.R. Frear, W.B. Jones, K. Kinsman, Editors, Minerals, Metals and Materials Society, 1991, pp 29-104.

(12)   R.S. Pokrzywa, "Materials Challenges and Mechanical Requirements for Future Connectors", <u>Materials Issues in Electronic and Opto-Electronic Connectors</u>, Minerals, Metals and Materials Society, BookCrafters, 613 E. Industrial Drive, Chelsea, Michigan, 1991, pp 1-16.

(13)   Y. Matsuda, "Recent Environmental Demands on Automotive Connectors", <u>Materials Issues in Electronic and Opto-Electronic Connectors</u>, Minerals, Metals and Materials Society, BookCrafters, 613 E. Industrial Drive, Chelsea, Michigan, 1991, pp 17-28.

Development of mould flux for continuous casting of 65/35 brass

M. Yamamoto, T. Shimada, N. Kimura
Kurami Works, Nippon Mining Co., Ltd., Kanagawa, Japan

M. Iwase
Department of Metallurgy, Kyoto University, Kyoto, Japan

ABSTRACT

In order to evaluate uniformity of heat removal through the mold wall, slabs of 65/35 brass were continuously cast by using a continuous casting machine of production-scale. Fluctuations of temperature during con-casting process were evaluated through a number of thermocouples embedded in a water-cooled copper mold. It was observed that the fluctuation of temperature increases with increasing viscosity of the mold flux.

INTRODUCTION

Wrought products of 65/35 brass, in the forms of sheet and strip, are widely applied for the production of automotive and electronic parts such as terminals and connectors. In producing slabs of 65/35 brass by continuous casting, mold fluxes are routinely utilized, filling on to the molten metal in the mold. From the viewpoint of the product quality of the slab and its yield, the selection and usage of the mold flux plays an important role in the continuous casting. In general, the mold flux provides lubrication between the solidifying shell and the mold wall, uniform heat removal, prevention of reoxidation of brass by insulating it from the atmosphere and the forth. It has been pointed out that these functions can be controlled by adjusting physical properties of the mold fluxes such as viscosity, crystallization temperature and melting rate. In the field of steelmaking, numerous investigations in terms of mold powders and fluxes for continuous casting have been conducted in conjunction with the measurements of these physical properties (1,2). However, there are only a few studies (3) on the continuous casting of copper and its alloys. Within the authors' knowledge, no literature has been published on the mold fluxes and powders for con-casting of 65/35 brass.

In this study, slabs of 65/35 brass were prepared by using a continuous casting machine of industrial scale, using mold fluxes with different viscosities, and fluctuations of temperatures were measured via a number of thermocouples embedded in the mold wall. Effects of viscosity of the mold fluxes on the uniformity of heat removal through the mold wall were, then, evaluated.

EXPERIMENTAL ASPECTS

Mold Flux

Three types of mold fluxes, whose viscosities ranged from 1 to 100 centi-poise at 1,000 $^{\circ}$C as given in Table I, were prepared for this experiment. Melting rates of the mold fluxes, quantitatively evaluated at 1,000 $^{\circ}$C[4], were almost similar to each other. Before application in the in-plant trials, the mold powders were dried at 200 $^{\circ}$C for more than 2 hours in order to remove moisture.

Table I   Viscosities of mold fluxes used at 1,000 $^{\circ}$C.

Powder A	1 centi-poise
Powder B	10 centi-poise
Powder C	100 centi-poise

Production of Slab and Temperature Measurement

The production unit for 65/35 brass slab, schematically shown in Figure 1, consists of the induction melting furnace, the tundish and the casting machine. The installation used in this experiment is capable of producing 5 m long slabs of 3.2 ton in weight on seven strands. The mold assembly of the casting machine installed below the tundish comprises the water-cooled copper mold oscillated, 65 cm x 15.6 cm x 30 cm deep, with the secondary water cooling spray. Submerged-nozzle with two side orifices is used in order to prevent from entrainment of air during casting.

Molten alloy prepared in the induction furnace was poured into the mold through a tundish, which had seven graphite nozzles in its bottom with stopper needles. In order to maintain a constant meniscus level of the molten metal during casting, the stopper needles were controlled by hand-operated screws. The level of molten metal in the mold was measured during the casting period. At the beginning of casting, relatively small amounts of the melted flux was placed on the molten metal surface, in order to minimize oxidation of the molten metal as soon as possible. After filling with the molten brass in the mold, the powder was supplied to cover completely over the liquid metal surface. The alloy solidified in the mold was, then, vertically withdrawn from the bottom of the oscillating mold. Sprays of secondary cooling water, installed just below the mold, impinged on the descending slab. The casting was terminated when the length of slab became more than 4 meters. After the casting, the flux remained in the mold was carefully removed, and its weight was measured in order to estimate the consumption of the flux used during casting. The detailed operating conditions in this experiment are given in Table II.

Table II - Casting conditions of 65/35 brass ingot by continuous casting

Casting temperature	1050 °C
Casting speed	130 mm/min
Frequency of oscillation	100 cpm
Oscillation stroke	3 mm
Flow rate of cooling water	2,100 l/min
Mold size	156 mm x 650 mm
Ingot size	150 mm x 630 mm x 4,000 mm

Twelve thermocouples of alumel-chromel, 1.6 mm in diameter, were embedded in the water-cooled copper mold parallel to the withdrawal direction. The cross section of casting mold and positions of the thermocouples embedded are schematically illustrated in Figure 2. Thermocouples of No.1 through 5 were carefully adjusted at the position of 3 mm from the mold wall surface and 40 mm from the top of mold such that temperatures below the meniscus can precisely be measured. The meniscus level during casting was normally controlled to 20 mm from the top of the mold. Thermocouples of No.6 and 7 were placed at the positions of 24 mm and 39 mm, respectively, from the mold wall surface with the same height (40 mm) from the mold top. Thermocouples of No.8 through 12 were located at the positions of 20, 30, 40, 50 and 60 mm, respectively, from the mold top, while the distances from the mold wall were remained at 3 mm.

RESULTS AND DISCUSSION

A record of the temperature fluctuations, monitored with the thermocouples embedded in the mold, is illustrated in Figure 3. Mold temperature rose up to more than 100 $^\circ$C for a few minutes from the beginning of casting and then continued to rise moderately until the end of casting, generating fluctuations of temperature repeatedly. This moderate temperature ascent, which would correspond to the steady-state, is presumably due to consumption of mold flux flowing down between the solidified shell and the mold wall during casting, although some amounts of mold flux are supplied in order to cover molten metal in the mold perfectly.

The temperature fluctuation, however, can be affected by two factors; one is the changes in meniscus level and the other is non-uniformity of heat removal. If the meniscus level vigorously fluctuated as shown in Figure 4, then the temperature measurements would not evaluate the extent of uniformity of the heat removal. Fluctuations of the meniscus level of molten metal during casting, therefore, are to be carefully controlled. In the present study, the meniscus level was kept constant within a range of $\pm$ 10 mm.

Even if the meniscus level was kept constant, heat removal can be affected by the viscosity of mold fluxes. The changes of mold temperature with the mold flux of various viscosities are illustrated in Figure 5. It seems that the fluctuation of mold temperature increases with increasing viscosity of the mold flux, although the temperature rise with casting time at steady-state is of the same tendency for all the mold fluxes.

To evaluate the extent of uniformity of the heat removal through the mold wall, fluctuations of mold temperature must be analyzed in more detail, using data obtained during casting. Mold temperature was measured at an interval of 8 seconds with an aid of a strip-chart recorder connected with the personal computer. As illustrated in Figure 6, the fluctuation of mold temperature is defined as T_{fi} (= $T_{i+1} - T_i$). Values for T_{fi} at the steady-state measured for various mold fluxes are shown in Figure 7, where the term, "steady-state", corresponds to the period between 10 to 25 minutes after the beginning of withdrawing. It is shown that the fluctuation of mold temperature increases with increasing viscosity of mold flux.

The difference, ΔT_i, between the maximum and minimum values of temperatures over time interval, t_i, is obtained in Figure 8. Nakano et al.(1) defined the mold heat removal non-uniformity index ΔT, which varies with the viscosity of mold flux and casting speed, as the following equation:

$$\Delta T = \sum_i^n T_{fi} / n \dots\dots\dots\dots\dots\dots\dots\dots\dots\dots\dots\dots\dots\dots\dots\dots(1)$$

The mold heat removal non-uniformity indexes ΔT as a function of time intervals from 8 to 64 seconds for different mold fluxes are shown in Figure 9. ΔT increases with increasing viscosity of the mold flux and the time interval t_i. ΔT as a function of analyzing time span at a fixed time interval, t_i = 64 sec, is also shown in Figure 10. The mold flux A, which has the lowest value of viscosity (1 centi-poise), has the lowest of ΔT.

The difference ΔT_i between maximum and minimum temperature for each 8 seconds for 15 minutes at steady-state was counted in order to evaluate uniformity of heat removal from the mold as another method. Relation between frequency and T_{fi} for each 8 seconds for various mold powders used is shown in Figure 11. It is obtained that frequency near ΔT = 0 for powder A (the lowest viscosity mold flux) is standing out. The standard deviations of the temperature fluctuation for various time spans were calculated in order to obtain a quantitative general tendency of the uniformity of heat removal through the mold. The standard deviations are plotted against the analyzing time span for the mold fluxes used in Figure 12. The standard deviations   are also plotted against viscosities of the mold fluxes in Figure 13. It is made clear that heat removal through the mold is homogenized by decreasing the viscosity of the mold flux, ranged from 1 to 100 centi-poise, being independent of the time span and the time interval in the steady-state.

CONCLUSION AND SUMMARY

In the continuous casting for 65/35 brass slab, temperature fluctuation of the mold during casting was measured, using thermocouples embedded in the mold. Evaluation for uniformity of heat removal was made in various ways, using mold fluxes with different viscosities. The results obtained are summarized as follows;

(1) Mold temperature rises up to more than 100 $^{\circ}$C in a few minutes from the beginning of casting, and, then, continues to rise until the end of casting. Short-period temperature fluctuations were detected repeatedly during the course of the steady-stage of the temperature rise.

(2) The fluctuation of temperature increases with increasing viscosity of the mold flux.

(3) The standard deviation of ΔT_i at the steady-state can be a quantitative measure of the uniformity of heat removal through the mold, and is independent of time span and time interval of temperature measurement.

ACKNOWLEDGMENTS

The authors wish to thank Dr. T. Ogura and Messrs. K. Yanagimachi and K. Kitanaka of Kurami Works, Nippon Mining Co., Ltd. for their helpful discussions in carrying out the present experiments.

REFERENCES

1.   T. Nakano, T. Kishi, K. Koyama, T. Komai and S. Naitoh, Trans. ISIJ, Vol. 24, 1984, 950
2.   R. V. Branion, Iron and Steelmaker, Vol.13, No.9, 1986, 41
3.   H. S. Bhamra, J. F. Hill, S. Garber and V. Kondic, Metals Technology, 1976, 54.
4.   M. Yamamoto, T. Shimada, N. Kimura and M. Iwase, to be published.

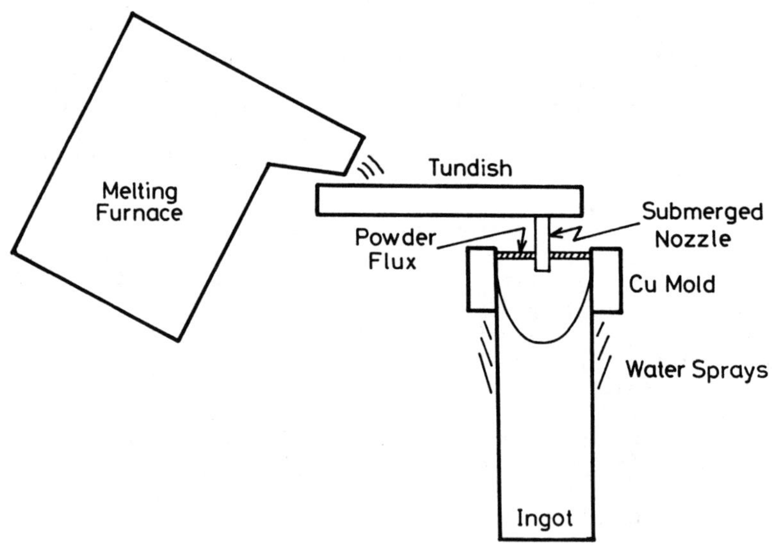

Figure 1 - Schematic illustration of continuous casting for slab of brass

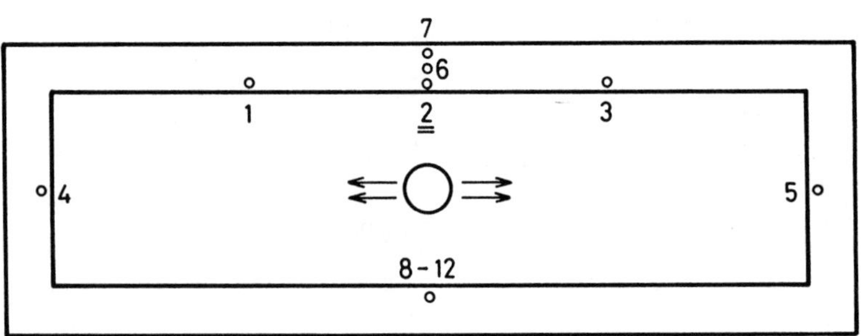

Figure 2 - Positions of thermocouples embedded in the mold

Distance from the mold   40 mm : 1 - 7
                          20 - 60 mm : 8 - 12

Distance from the mold surface   3 mm : 1 - 5, 8 - 12
                                   24 mm : 6
                                   39 mm : 7

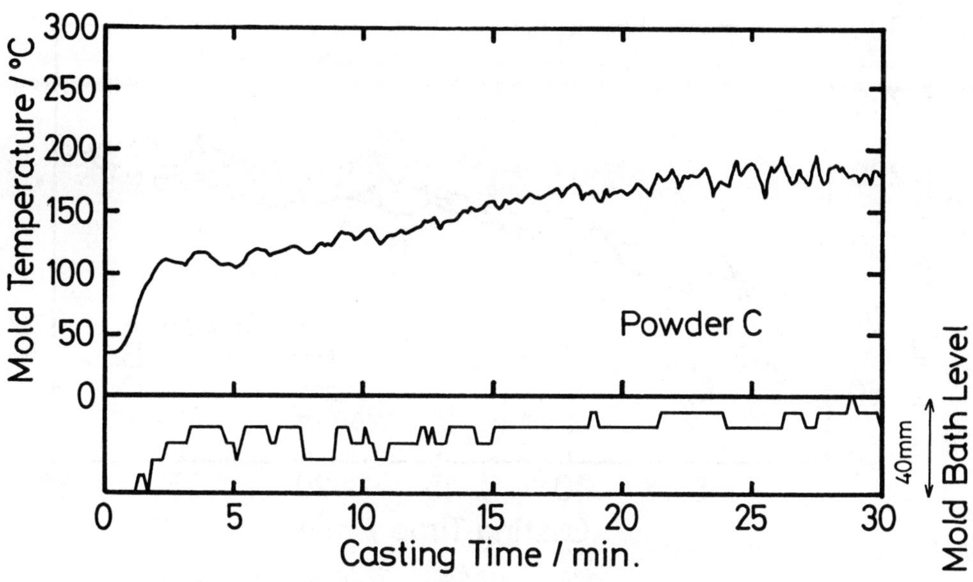

Figure 3 - Changes for temperature of the mold and meniscus level by using powder C
when small fluctuation of meniscus level took place

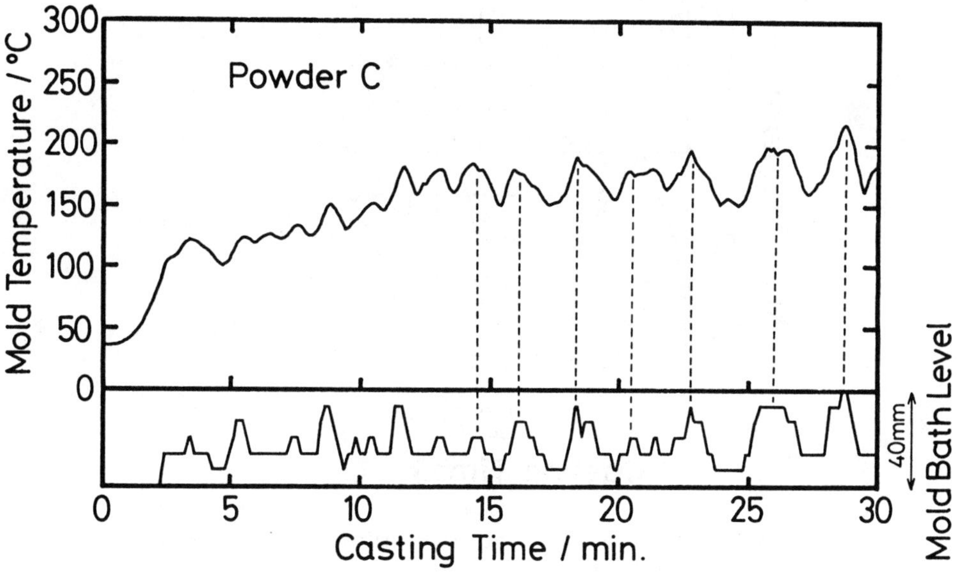

Figure 4 - Changes for temperature of the mold and meniscus level by using powder C
when large fluctuation of meniscus level took place

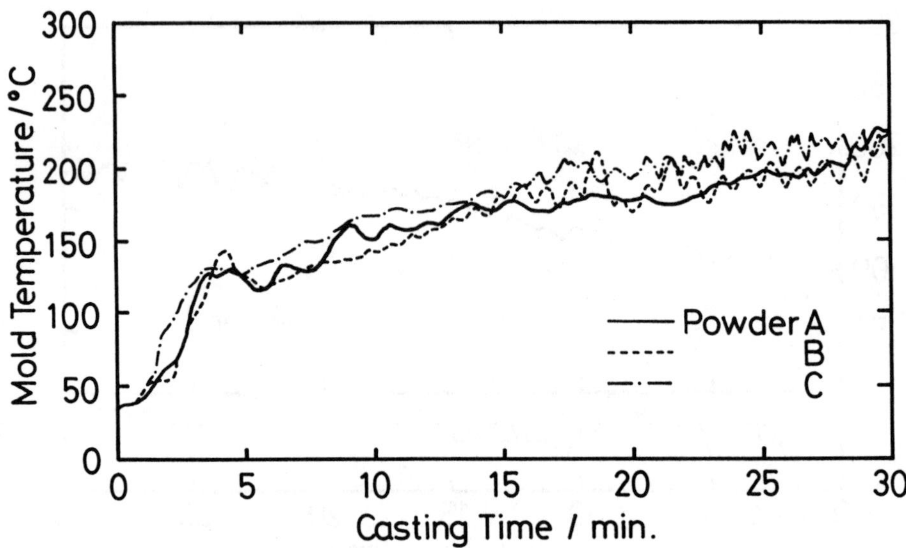

Figure 5 - Comparison of temperature fluctuations of the mold for various powder
fluxes

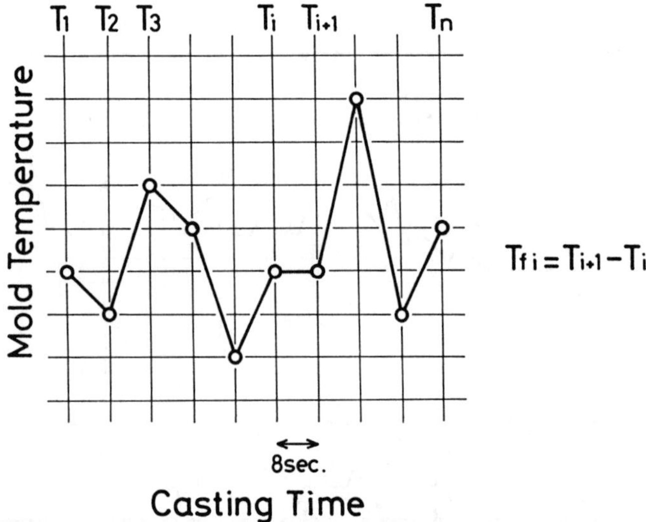

$$Tf_i = T_{i+1} - T_i$$

Figure 6 - Schematic illustration of temperature fluctuation measured and
definition of Tf_i

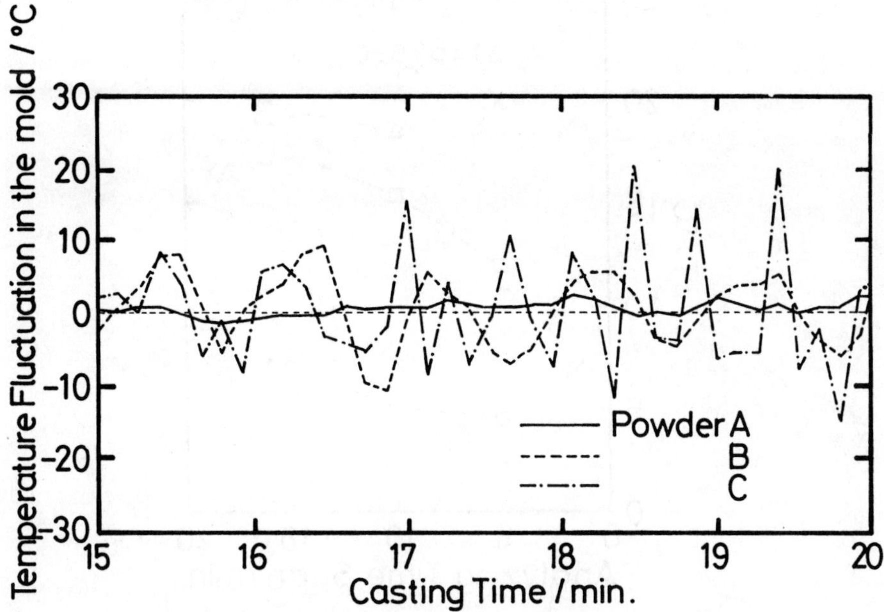

Figure 7 - Temperature fluctuations for various powder fluxes at the steady-state during continuous casting

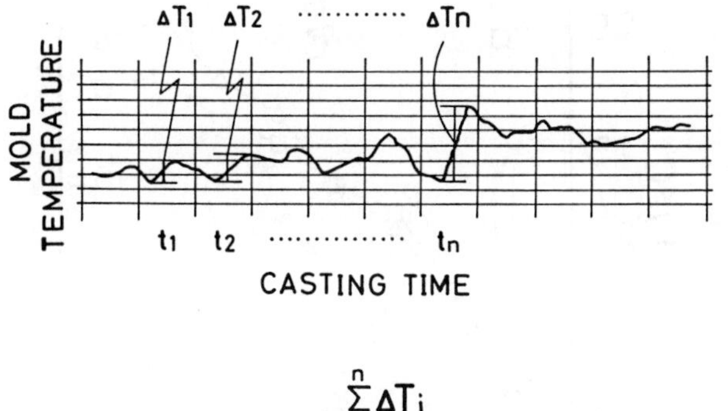

$$\Delta T = \frac{\sum_{i}^{n} \Delta T_i}{n}$$

NAKANO et al. 1984

Figure 8 - Schematic illustration of temperature fluctuation and definition of ΔT

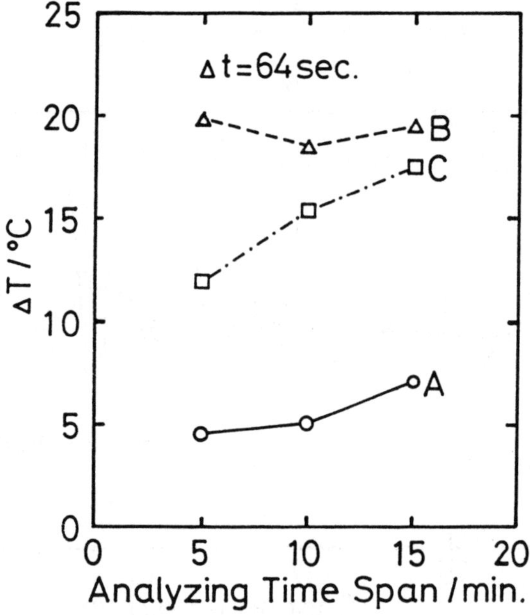

Figure 10 - Relationship between ΔT and analyzing time span for various powder fluxes

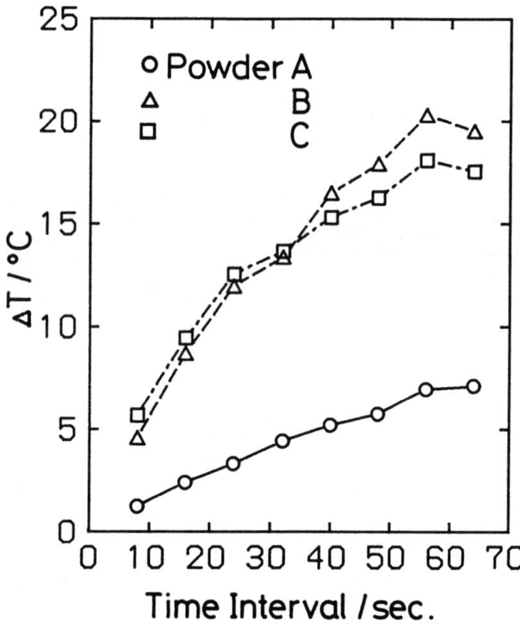

Figure 9 - Relationship between ΔT and time interval for various powder fluxes

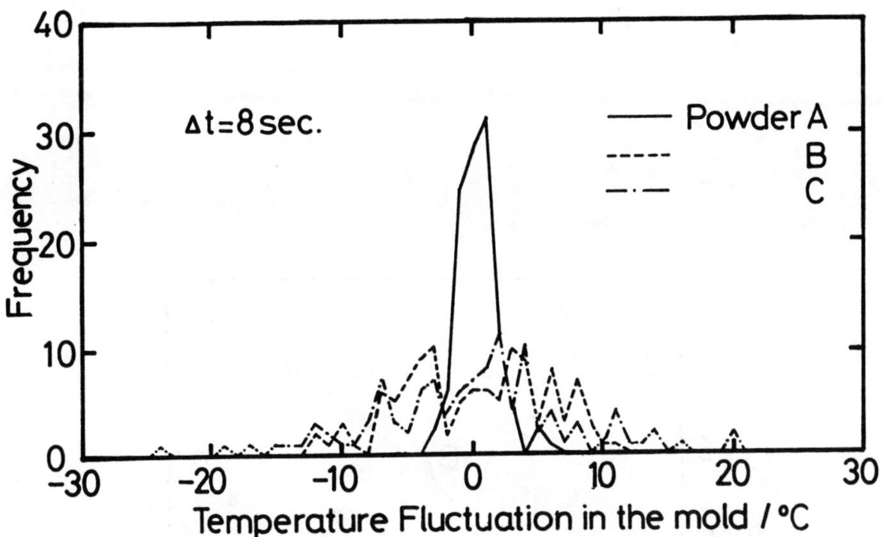

Figure 11 - Frequency vs. temperature fluctuation in the mold for various powder
fluxes

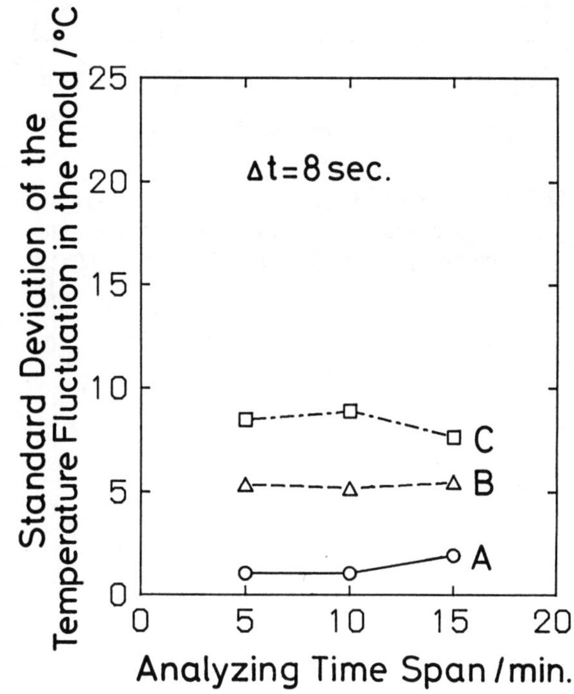

Figure 12 - Standard deviation of the temperature fluctuation
in the mold vs. analyzing time span

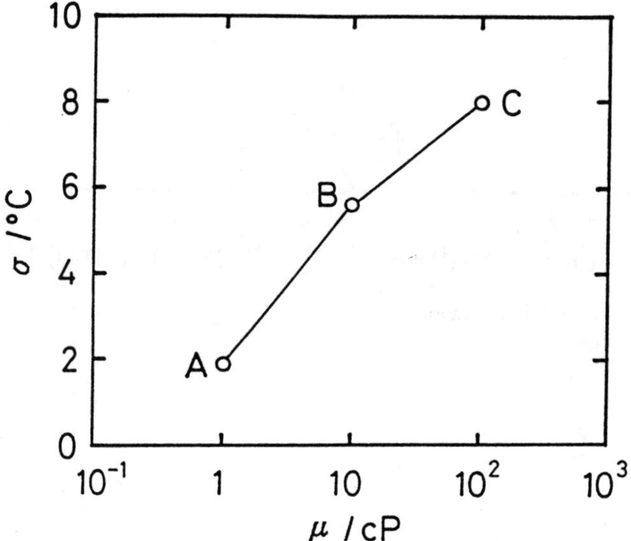

Figure 13 - Relationship between standard deviation of
the temperature fluctuation (σ) and absolute
viscosity of molten flux (μ) at 1,000°C

Entrainment of molten flux during continuous casting of copper alloys

M. Yamamoto, T. Saito, T. Shimada
Kurami Works, Nippon Mining Co., Ltd., Kanagawa, Japan

M. Iwase
Department of Metallurgy, Kyoto University, Kyoto, Japan

ABSTRACT

Entrainment of molten fluxes and gas bubbles to the casting ingot during continuous casting will affect the surface quality of the final product. Water model studies were conducted for better understanding on the role of the stream of molten metal teemed from submerged nozzle into the mold. If the depth of submerged nozzle can be kept at about 40 mm from the meniscus, neither internal defects nor surface defects were detected, in accordance with the results of the water model studies.

INTRODUCTION

In the continuous casting of brass, the flow pattern of the molten metal teemed into the mold can affect significantly the surface quality of the final products. For example, scab cracks would arise from the entrainment of molten fluxes and gas bubbles. The molten brass through the inverse-T shape submerged nozzle with two exit openings, installed at the bottom of tundish, is teemed into the mold in which the surface of liquid metal is covered with molten flux. These flow conditions of molten metal would significantly depend upon casting conditions such as casting speed, types of submerged nozzles and meniscus level. Water model studies can be employed to provide qualitative information for the flow pattern in the mold and/or the tundish. Numerous investigations in regard to water model studies for steel con-casting have been, therefore, undertaken to simulate the behavior of stream from ladle to tundish and tundish to mold(1,2). Water model studies for continuous casting of brass are, however, very limited.

Casting conditions of brass refer to Table I. The casting conditions of liquid brass, e.g., casting speed, pouring temperature and mold size, are quite different from those of steel con-casting. Figure 1 illustrates schematically the flow pattern of molten brass within the mold, in comparison with that for liquid steel. In general, the submerged nozzle for steel casting is in a shape of inverse-Y, while inverse-T shape nozzle would commonly be applied for brass casting. The exit end of the nozzle is located about 200 mm and 50 mm below the meniscus for steel and brass con-casting, respectively. These differences in nozzle design would cause a difference in flow pattern between molten steel and liquid brass. The flow of liquid metal below meniscus for brass continuous casting would presumably be directed from the center to the wall of the mold.

Table I - Casting conditions of brass ingot by continuous casting

Casting temperature	1,050 $^{\circ}$C
Casting speed	130 mm/min
Frequency of oscillation	100 cpm
Oscillation stroke	3 mm
Flow rate of cooling water	2,100 l/min
Mold size	156 mm x 650 mm
Ingot size	150 mm x 630 mm x 4,000 mm

Water model studies with a particular emphasis upon the application to the continuous casting of brass were, therefore, conducted for better understanding on the role of flow patterns in the entrainment of molten fluxes as well as gas bubbles during relatively stable sequence of casting. Influence of casting conditions on the entrainment of molten flux will also be discussed with results of both water model studies and in-plant trials at the continuous casting machine.

WATER MODEL STUDIES

The objective of water model studies prior to in-plant trials is to demonstrate how flow patterns in various teeming conditions can affect entrainment of molten fluxes and gas bubbles. Consideration was given to the following dimensionless variables, by comparing physical properties of copper with those of water:

$$\text{Froude number} = (v^2) / (g\,L) = (\text{inertia force})/(\text{gravity force}),$$

$$\text{Reynolds number} = (v\,L) / V = (\text{inertia force})/(\text{viscous force}),$$

$$\text{Weber number} = (p\,v^2\,L)/ q = (\text{inertia force})/(\text{surface tension force}),$$

where terms, v, g ,L ,V, p and q, are velocity of the stream, gravitational constant, characteristic length, kinematic viscosity, density and surface tension, respectively. The physical properties of copper and water were listed in Table II. The ratios of these parameters for water(w) to those of copper(c), Fr(w)/Fe(c), Re(w)/Re(c) and We(w)/We(c), were obtained in Table III. As shown in the Table, it is considered that a full scale model was constructed on the basis of Froude criteria or Froude-Reynolds criteria(3).

Table II - Physical properties for water and molten copper

		Liquid copper (1,200 $^{\circ}$C)	Water (20 $^{\circ}$C)
Absolute viscosity, v	(centi-poise)	4.50	1.01
Density, p	(g / cm^3)	8.90	1.00
Kinematic viscosity, v/p	(centi-stokes)	0.51	1.01
Surface tension, q	(dyne/cm)	135.0	72.8

Table III - Ratios of dimensionless variable, Fr, Re and We of the water model

Fr(w) / Fr(c)	1.0
Re(w) / Re(c)	0.5 to 1.0
We(w) / We(c)	0.2

The equipment for water model studies is schematically illustrated in Figure 2. A full scale mold, modeled of a 150 mm x 650 mm slab mold, was made in order to examine the factors affecting the behavior of stream teemed into the mold through submerged nozzle. The location of solid-liquid interface within the metal phase at the vicinity of the mold wall was previously determined(4). Based on such determinations, the mold of the water model was designed to have tapered walls on its short sides.

By placing paraffin on the surface of water for simulating molten fluxes, it has been possible to observe and record photographically the entrainment of paraffin into water associated with various flow patterns. To facilitate the observations of flow patterns within the moldeled mold, a grid made of vinyl film was attached on the long side wall of the modeled mold as shown in Figure 3.

The amounts of water pouring into the mold, carefully controlled with the valve installed in the water pipe, were measured more than 5 times by using a flow meter and a measuring cylinder. A valve installed on the bottom of the modeled mold was used to maintain the level of water in the mold at a constant height. Observations, with an aid of a cinematographical equipment, were conducted for a period of 10 minutes after the water stream was brought to the steady-state.

The amount of water teemed into the mold, i.e., volumetric flow rate, the depth of submerged nozzle from the level of water surface in the mold, and the diameter of the submerged nozzle are altered as the key factors which would influence the flow pattern. Ranges of these factors are described in Table IV. If the submerged nozzle is placed above the water surface, the value of the depth of the submerged nozzle is symbolized with minus (-).

Table IV -   Experimental conditions for water mold studies.

Volumetric flow rate	50 to 400 cm^3 / sec
Depth of submerged nozzle	- 45 to 165 mm
Outlet diameter of nozzle	8 to 20 mm
Experimental temperature	18 ^0C

RESULTS OF WATER MODEL STUDIES

Water model experiments were, firstly, conducted with a fixed volumetric flow rate of 200 cm^3/sec and a nozzle diameter of 12 mm. These values were chosen to simulate the practical brass casting machine. The effects of the depth of the submerged nozzle on the entrainment of liquid paraffin were investigated, and the results are given in Figure 3.

If the nozzle position is above water level such as the case of - 25 mm, liquid paraffin was vigorously entrained into water, due to the collision of water pole from the nozzle with water surface, and finally forced to the bottom of the mold, Figure 3 a). Air was also entrained into water.

When the depth of the submerged nozzle was relatively shallow below the water surface level in the case of 15 mm, the fluctuation of the interface between water and liquid paraffin in the vicinity of the short side mold wall took place due to strong vortices, Figure 3 b). Entrainment of air, however, did not occur due to existence of liquid paraffin on water. This fluctuation tremendously decreased with an increase in the depth of the submerged nozzle from the meniscus, Figure 3 c).

Second, effects of the diameter of submerged nozzle on the entrainment of paraffin were investigated, at a fixed nozzle depth of 35 mm and a fixed volumetric flow rate of 200 cm^3/sec, Figure 4. If the diameter of submerged nozzle is less than 8 mm, liquid paraffin is vigorously entrained into water because of the higher linear flow rate, Figure 4 a).

Since changes of casting speed in the actual continuous casting corresponded to those of volumetric flow rate in this water model, effects of volumetric flow rate are also investigated at flow rates ranged from 100 cm^3/sec to 400 cm^3/sec, Figure 5. Entrainment of liquid paraffin as well as fluctuation of liquid paraffin near the mold wall increases with increasing volumetric flow rate.

A parameter, dc, was defined as the critical value of the depth of the submerged nozzle above which the entrainment of liquid paraffin did not take place. Values of dc are plotted against volumetric flow rate in Figure 6: dc increases with increasing volumetric flow rate.

Relation between dc and linear flow rate is shown in Figure 7, where linear flow rate equals to volumetric flow rate divided by the area of exit opening. Values of dc increase linearly with increasing linear flow rate, being independent of nozzle diameter.

An important comment pertaining to casting operation can be made with the aid of Figure 7. In practical continuous casting, even at a constant casting speed, changes in volumetric flow rate would arise from ripples of molten metal level, resulting in enhancement of fluctuation in the depth of submerged nozzle. During casting operations, the surface level of molten metal may accidentally enter to the entrainment region, associated with a decrease in nozzle depth, see Figure 7. An operational practice to be taken would be to raise the meniscus level. However, care must be taken in raising the meniscus level, since sudden changes of the molten metal level can, in turn, cause entrainment of molten fluxes.

RESULTS OF IN-PLANT TRIALS

In-plant trials were conducted in order to understand the state of entrainment of molten flux during casting, comparing with the results of the water model studies. Brass ingots of 3.2 tons at the same cross section of the water model studies were produced, using a continuous casting machine. The meniscus level was continuously measured during casting. The brass ingots were subsequently hot- and cold-rolled to 5 mm in thickness. The number of internal defects in the brass sheet was then counted by the ultra-sonic inspection. The number of surface defects due to entrainment of molten flux was also counted by the visual inspection as well.

Relationships between fluctuation of molten metal level, and internal and surface defects, are shown in Figures 8 and 9, respectively. If the depth of submerged nozzle was kept at about 40 mm from meniscus level with a fixed linear flow rate of 100 cm/sec, neither internal defects nor surface defects, which would be due to entrainment of molten flux, were detected, in accordance with the results of the water model studies.

SUMMARY

Water model studies were conducted in order to understand how the casting conditions such as the depth of submerged nozzle, nozzle diameter and volumetric flow rate of molten metal can affect the entrainment of molten flux into the ingot cast. Results of the water model studies were compared with those of in-plant trials. The results are as follows;

(1) If the nozzle position is above water level, liquid paraffin along with air is vigorously entrained into water due to the collision of water pole from the nozzle with water surface.

(2) Entrainment of liquid paraffin increases with decreasing the nozzle diameter and increasing volumetric flow rate.

(3) Values of dc increase linearly with increasing linear flow rate, being independent of nozzle diameter.

(4) If the depth of submerged nozzle can be kept at about 40 mm from meniscus in the conventional continuous casting condition, neither internal defects nor surface defects were detected, in accordance with the results of the water model studies.

ACKNOWLEDGMENTS

The   authors wish to thank Dr. T. Ogura, and Messrs. K. Yanagimachi and K. Kitanaka of Kurami Works, Nippon Mining Co., Ltd. for their comments and constructive discussions in regards to writing this paper.

REFERENCES

1.   W. J. Maddever, A. McLean, J. S. Luckett and G. E. Forward, <u>Can. Met. Quart.</u>, Vol. 12, 1973, 79.
2.   J. Little, M. Van Oosten and A. Mclean, <u>Can. Met. Quart</u>., Vol. 7, 1968, 235.
3.   J. Szekely, J. W. Evans, and J. K. Brimacombe, <u>The Mathematical and Physical Modeling of Primary Metals Processing Operations</u>, John Wiley & Sons, Inc. 1988, 199.
4.   M. Yamamoto, to be published.

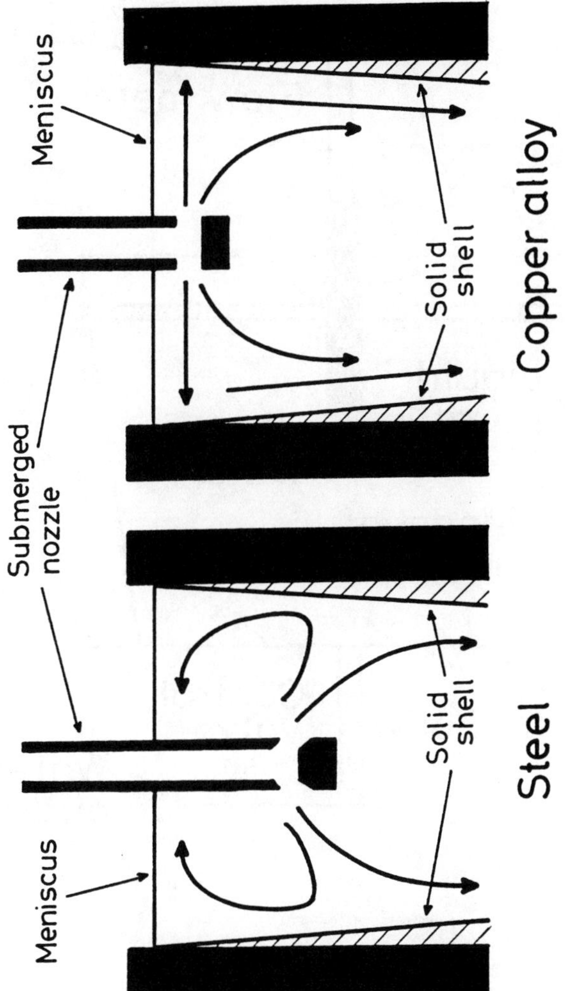

Figure 1 - Comparison of schematic flow patterns for copper alloy and steel

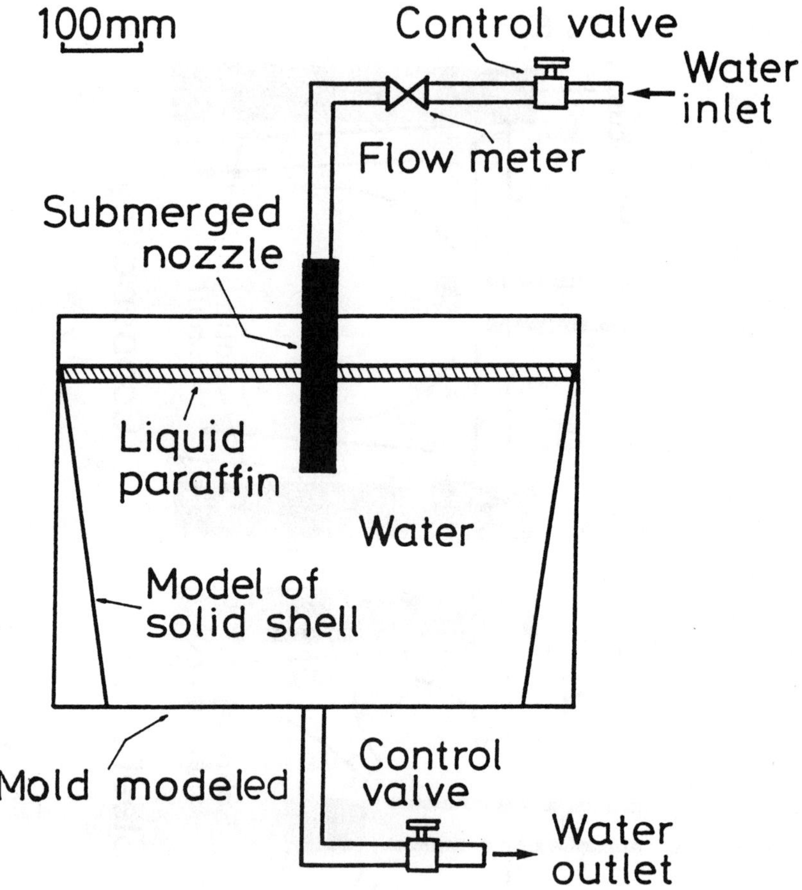

Figure 2 - Experimental apparatus for water model studies

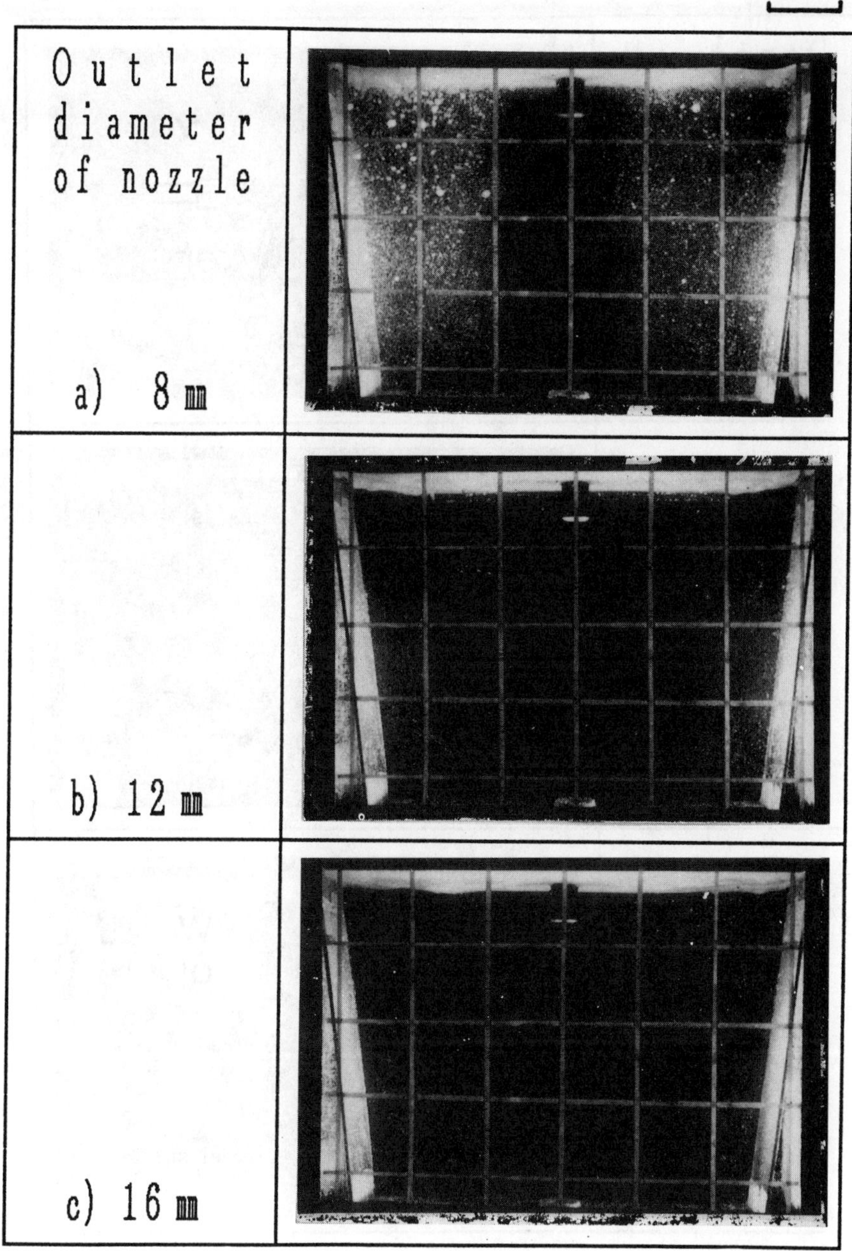

Figure 4 - Effects of nozzle diameter for entrainment of liquid paraffin ; depth of submerged nozzle : 35mm , volumetric flow rate : 200cm^3/sec

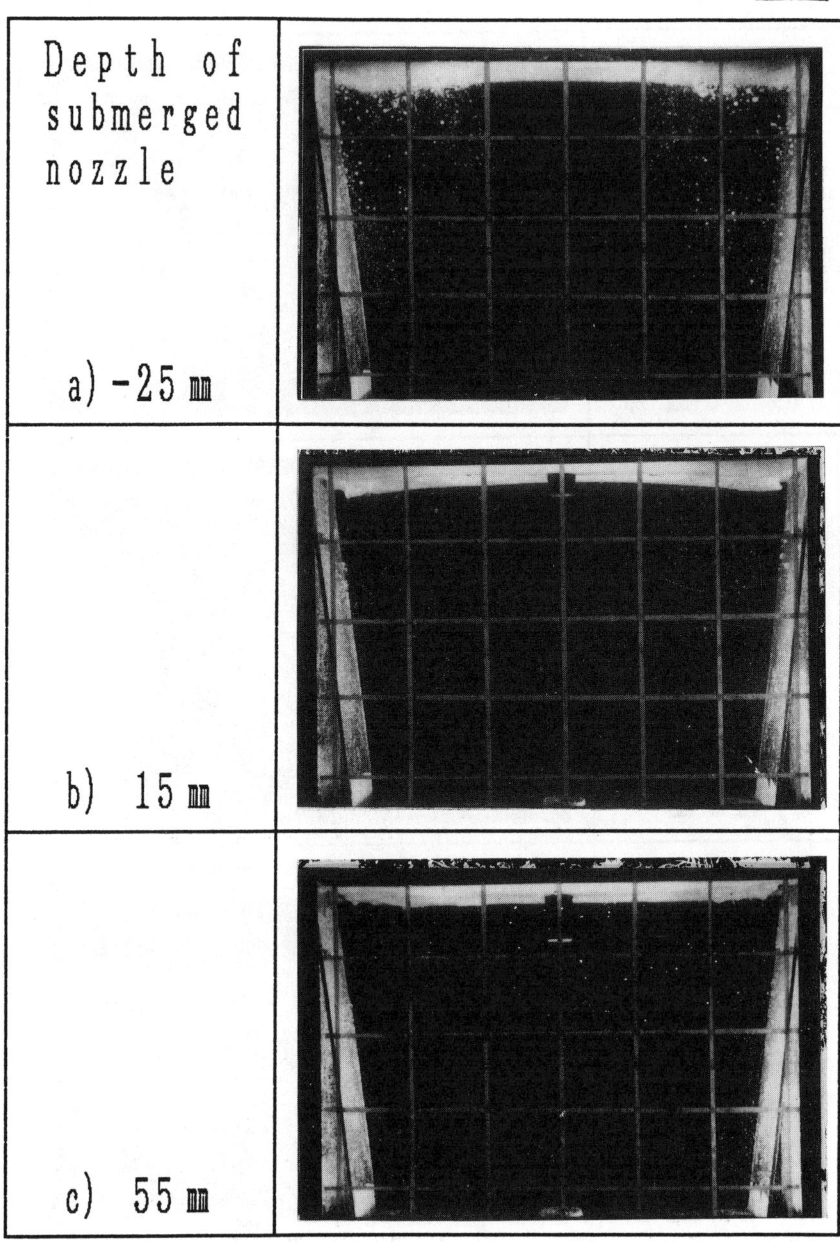

Figure 3 - Effects of the depth of submerged nozzle for entrainment of liquid paraffin ; nozzle diameter : 12mm , volumetric flow rate : 200cm³/sec

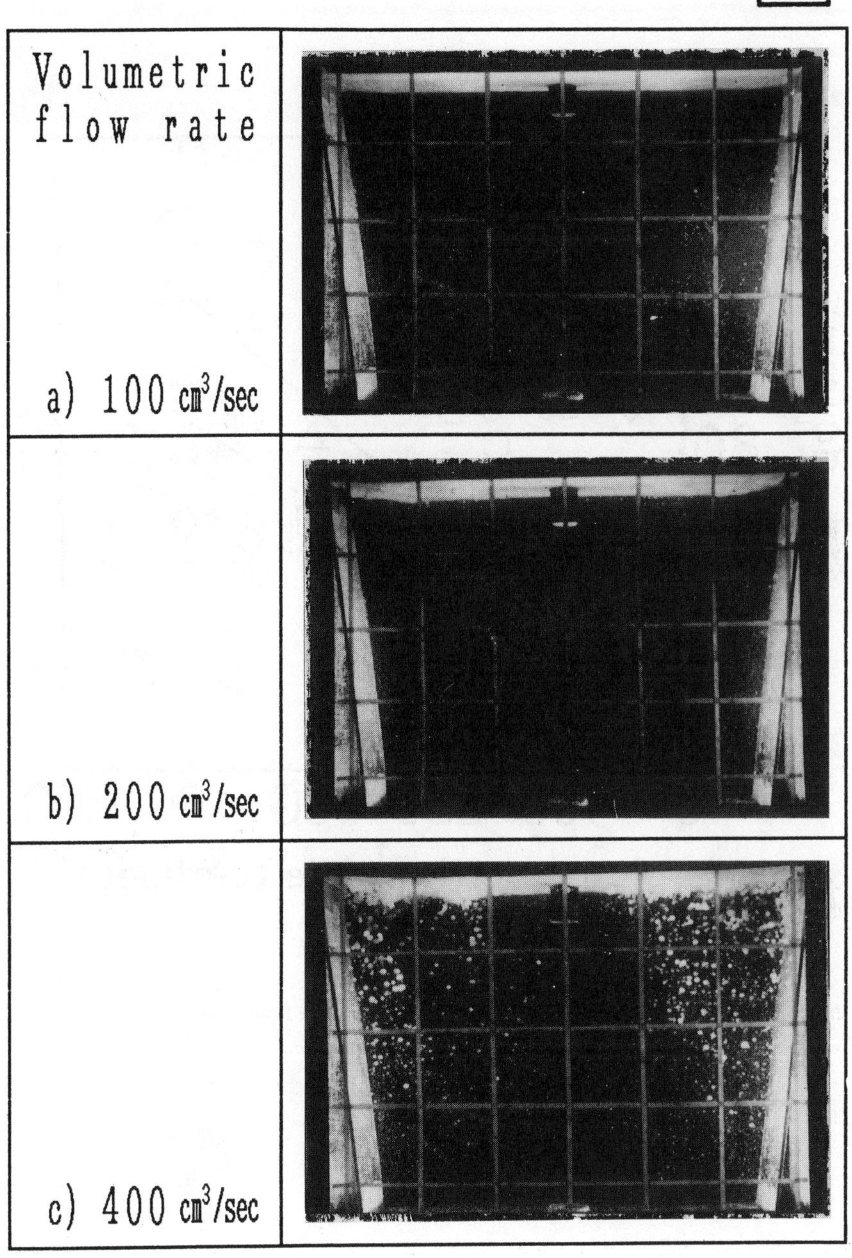

Figure 5 - Effects of volumetric flow rate for entrainment of liquid paraffin ;
depth of submerged nozzle : 35mm , nozzle diameter : 12mm

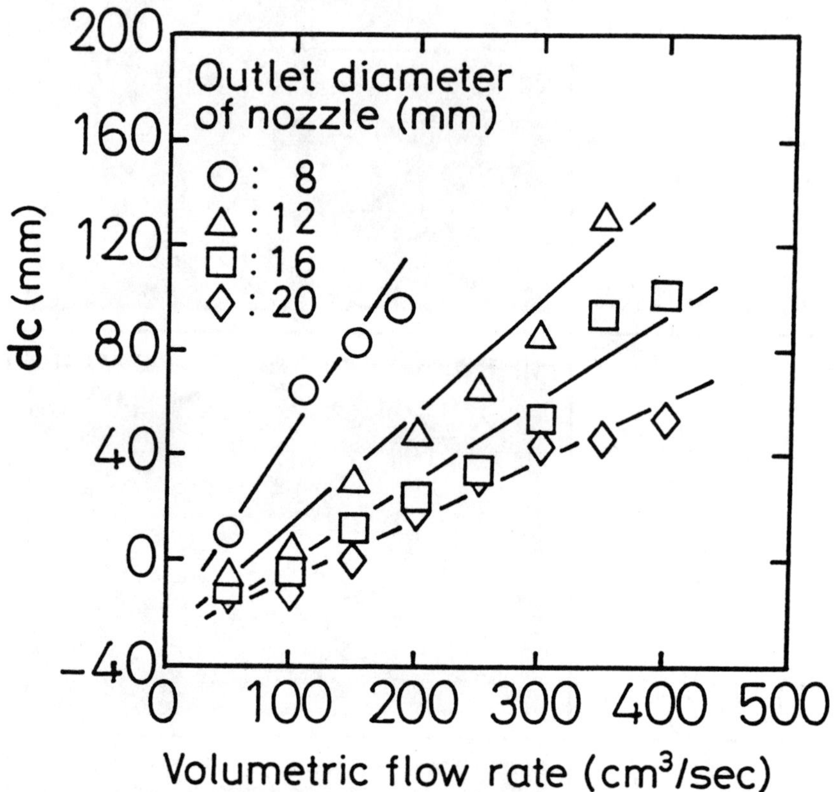

Figure 6 - Critical depth of submerged nozzle as a function of volumetric flow rates for various nozzle diameters

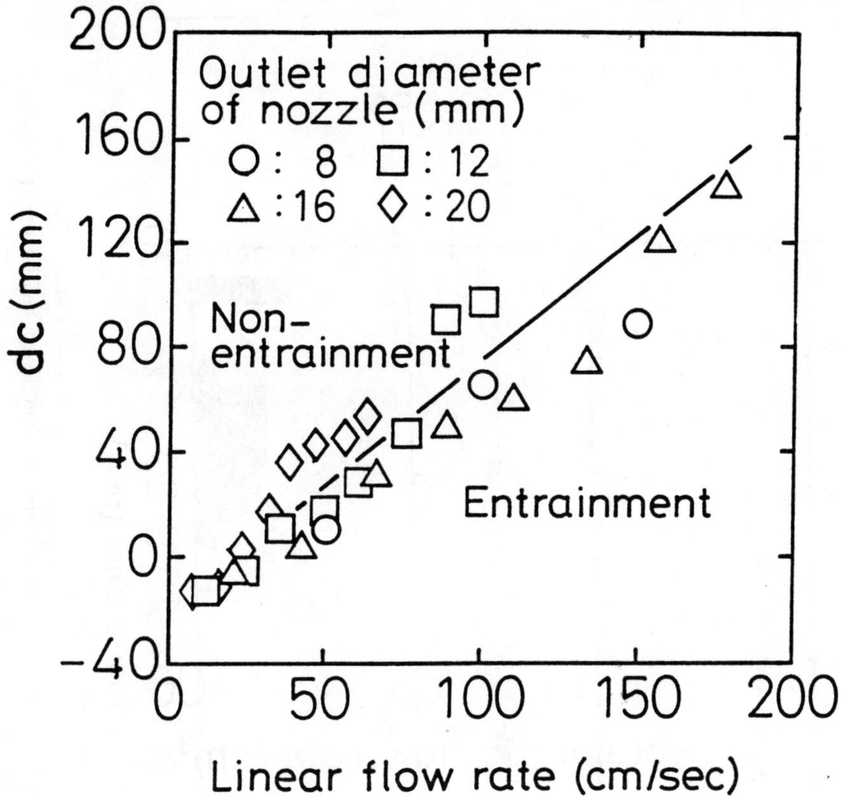

Figure 7 - Critical depth of submerged nozzle as a function of linear flow rates

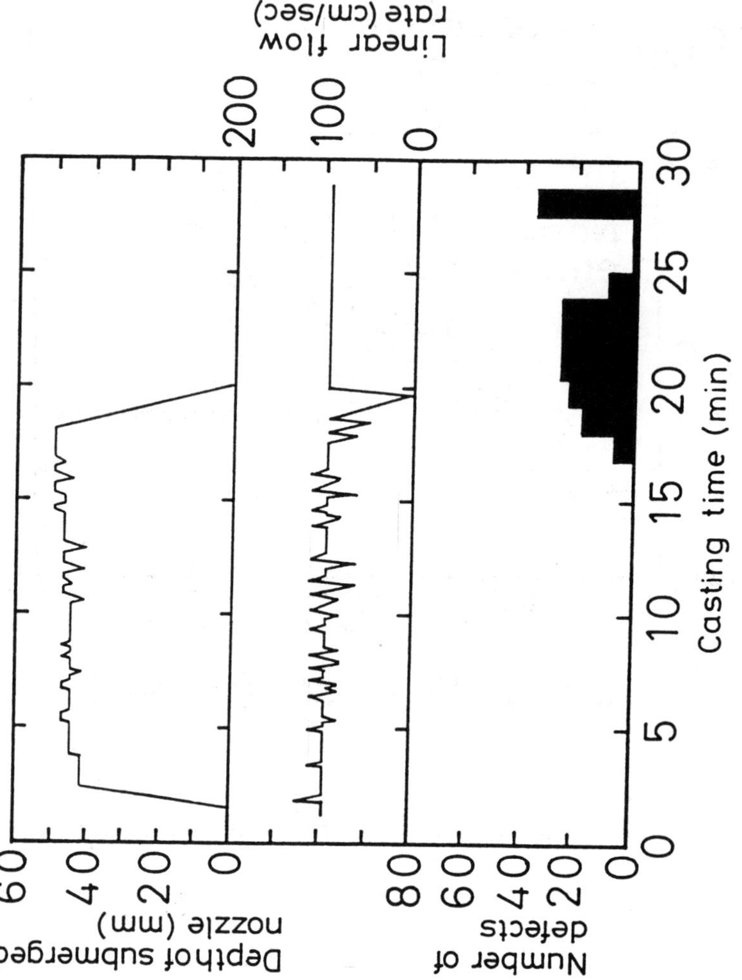

Figure 8 - Relationship between fluctuation of meniscus level and internal defects due to entrainment of molten flux

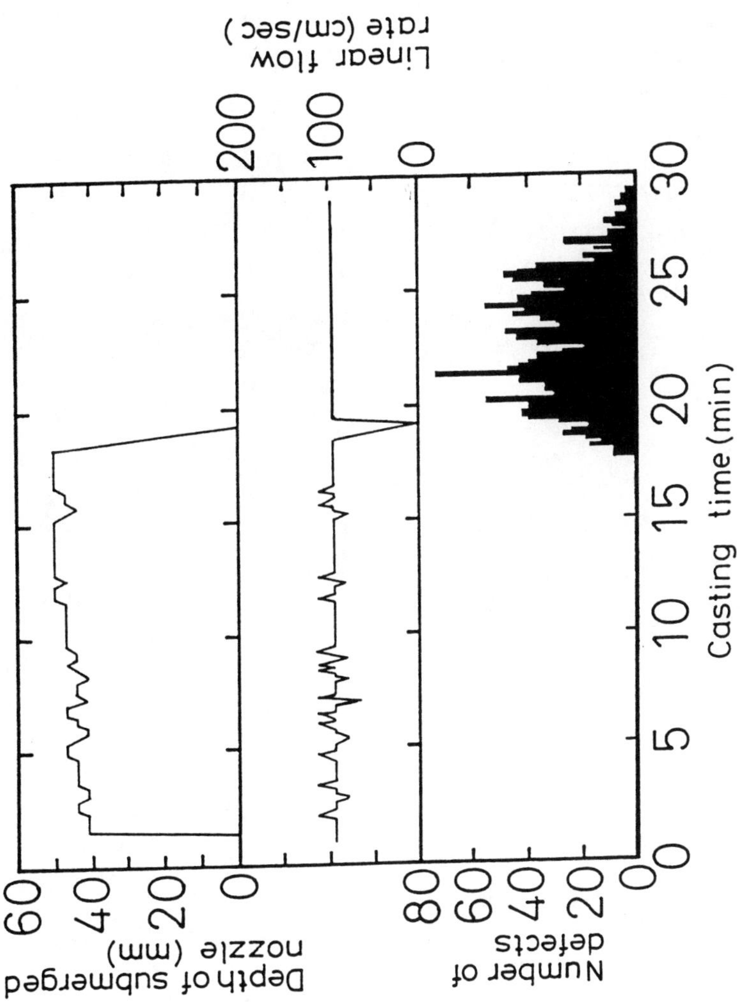

Figure 9 - Relationship between fluctuation of meniscus level and surface defects due to entrainment of molten flux

Authors' Index

Keyword Index